de Gruyter Textbook
Berger · Chaos and Chance

Arno Berger

Chaos and Chance

An Introduction to Stochastic Aspects
of Dynamics

Walter de Gruyter
Berlin · New York 2001

Author

Arno Berger
Institute of Mechanics
Vienna University of Technology
Wiedner Hauptstraße 8–10
1040 Vienna
Austria

Mathematics Subject Classification 2000: 37-01; 37A50, 37A30, 70K55
Keywords: Bifurcation, Attractor, Chaos, Ergodic theorems, Markov operator, Markov chain

⊚ Printed on acid-free paper which falls within the guidelines of the ANSI to ensure permanence and durability.

Library of Congress − Cataloging-in-Publication Data

Berger, Arno, 1968–
    Chaos and chance : an introduction to stochastic aspects of dyna-
mics / Arno Berger.
        p.   cm.
    Includes bibliographical references and index.
    ISBN 3-11-016990-8 (pbk.). − ISBN 3-11-016991-6 (hardcover)
    1. Differentiable dynamical systems.   2. Chaotic behavior in
systems.   I. Title.
    QA614.8 .B48   2001
    514'.352–dc21   2001042161

Die Deutsche Bibliothek − Cataloging-in-Publication Data

Berger, Arno:
Chaos and Chance : an introduction to stochastic aspects of dynamics / Arno Berger. − Berlin ; New York : de Gruyter, 2001
    (De Gruyter textbook)
    ISBN 3-11-016990-8 brosch.
    ISBN 3-11-016991-6 Gb.

Printed in Germany.
Cover design: Hansbernd Lindemann, Berlin.
Typesetting using the author's TEX files: I. Zimmermann, Freiburg.
Printing and Binding: WB-Druck GmbH & Co., Rieden/Allgäu.

Preface

Since the nineteen-seventies *chaos* has become a highly popular term which, possibly due to its inherent vagueness, provides a succinct notion for complexity and unpredictability of dynamical systems. Consequently, there is no lack of textbooks that are concerned with the stunning aspects of *chaotic dynamical systems*, and implications thereof in science, engineering and economics. On the other hand, the statistically oriented approach to dynamics has also flourished during the past few decades, as for example can be seen from a number of excellent textbooks on *ergodic theory* and its applications. Dating back to the first half of the last century, this approach in some sense introduces the notion of *chance* even to completely deterministic systems. By bringing the two viewpoints together, an illustrative impression of the both natural and fruitful interplay of chaos and chance, or, more formally, of the geometrical and the statistical approach to dynamics may be gained. So far, however, a pertinent introductory though mathematically reliable text has been hard to find. It is the aim of the book in hand to help close this notable gap.

In writing this book it has been my intention to draw the reader's attention to several interesting examples, thereby motivating more general (and abstract) notions as well as results in a setting as simple as possible. Illuminating dynamical systems abound, and it need not be difficult to get an intuitive feeling of the complexity inherent in them. As one class of examples among others we shall repeatedly and on different levels of sophistication deal with billiards. By its conceptual simplicity nothing could be more deterministic and hence predictable than a billiard, could it? Given a specific shape of a table we shall, however, observe that the future fate of a voyaging billiard ball may be completely unpredictable beyond a surprisingly small number of reflections. In this case, calculation-based predictions of the ball's further journey are doomed to be no more reliable than throwing a die. But how does *chance* emerge from a purely deterministic system, and how may it help us to better understand the latter? It will take us some time to conceptualize and thoroughly address these questions.

It is my firm belief that illustrative examples are indispensable for developing a general theory, not only as initial justification but also as lasting motivation. Accordingly, it has never been my intention (let alone ability) to present a comprehensive monograph on the topic. As mentioned earlier, a host of excellent advanced texts can be drawn on for more extensive and detailed study; Appendix B contains a short and somewhat biased list of books to this purpose. I believe, however, that the elementary insights gained here together with the mathematical tools developed will enable the reader to study these advanced texts more rewardingly and with greater relish.

This book comprises five chapters and two appendices containing background material and references to the literature. As Chapter One serves as an informal introduction and Chapter Five lets the reader take a look at more advanced topics, Chapters Two to Four should be considered the core part of this text. Chapter Two briefly introduces the topologically oriented approach to chaotic dynamics. Chapters Three and Four focus on the statistical description of dynamical systems in some more detail. Especially, their resemblances to and differences from special stochastic, i.e. chance-driven and hence explicitly random, processes are thoroughly discussed. As throughout, the emphasis is on specific examples rather than on general results. By and by the reader will thus come to think of chaos and chance as of two sides of the same coin.

This is primarily a mathematics text. In the first place it addresses advanced undergraduates and beginning graduates with a sound knowledge of calculus. Ideally, the reader should also be endowed with a basic knowledge of measure theory. Appendix A gathers the most relevant notions and results from measure theory that we shall rely on in the main text; anyone to whom the content of Appendix A seems familiar is certainly well prepared. Another desirable prerequisite concerns discrete dynamical systems: though essentially self-contained, Chapter Two proceeds at a good pace, which may be demanding for a complete novice. To better appreciate its content, a precedent exposition to an introductory book, e.g. parts of Devaney's highly readable text, may be helpful. Yet a lack of either of these desirable prerequisites should not completely discourage the aspirant. Especially students from applied sciences are often highly motivated to learn more about the mathematics behind the systems they encounter in their respective discipline. In fact, I feel confident that this text will be useful to these readers too, provided they either take for granted the presupposed mathematical facts or (even better) look them up in a textbook and thus enhance their knowledge. With a purpose in mind, investing in one's mathematical skills is certainly worthwhile at any level of proficiency!

In order to keep the text fluent, calculations and considerations which I consider elementary are often stated in a brief form or skipped altogether. On a first or cursory reading one may well proceed by just taking note of the facts presented, without pondering on each of them. The conscientious reader, however, will use paper and pencil in order to work out and carefully check the steps condensed into phrases like *a straightforward calculation confirms*, etc. This is particularly relevant to comprehending the proofs: although these are really complete proofs containing all the essential ideas, it may often be advisable to thoroughly reconsider the details in order to get a fully elaborated argument. Exercises have been added in moderate number to each chapter: while some of them are routine, others are challenges for the curious, providing possible points from where to launch further studies. Throughout the whole book, the reader may thus decide to what extent to delve into the matter. For all I know, it is only personal commitment and interest covering also the sometimes technical details that leads to a truly working knowledge of the field. Accordingly, I hope that the reader will gain some inspiration from thoroughly pondering on specific problems of dynamics. May you feel inspired to consult more advanced references and to penetrate more and more the fascinating and incredibly multifarious world of dynamical systems!

This book has grown out of courses I have repeatedly given at Vienna's University of Technology. In turning the original notes into a serious textbook I received help and advice from quite a number of people. To all of them I feel sincerely grateful. In particular, I was lucky to learn a lot about dynamics from Martin Blümlinger, Klaus Schmidt and Peter Szmolyan, who have let me benefit from their respective deep knowledge of the field. Peter also read parts of the manuscript and provided a host of corrections and improvements. For advice of both stylistic and general nature I am enormously indebted to Wuddy Grienauer as well as to Paul G. Seitz. The splendid illustration of *Snakes and Ladders* was kindly made available to me by Josie Porter and Tim Walters. Last but not least, I feel deep gratitude to Manfred Karbe for the most friendly and encouraging attention he patiently gave to this project.

Vienna, August 2001   *Arno Berger*

Contents

Chapter 1
Introduction

Even though defined in purely deterministic terms, dynamical systems from physics, biology, economics and other applied sciences may behave in quite a complicated and unpredictable manner, so that when looking at such systems we could easily get the impression of watching an experiment of *chance*. However, as there is no notion of chance in a completely deterministic setting, a seemingly erratic behaviour leads to several fundamental questions. How does uncertainty come about in this context and how does it affect our understanding of a system as a whole? What is the precise meaning of the word *chance* and, most important, how can we describe its emergence and implications? To pursue these questions and to find at least partial answers is the aim of this book. On our way we shall encounter numerous examples which will allow us to grasp the relevance of general notions and results. As this book primarily is about mathematics, several technicalities will have to be elaborated in some detail later in the text. In this introductory chapter, however, we are going to informally discuss a number of simple examples that will serve as starting-points and motivation for our further studies. The main point here is to observe how statistical and probabilistic aspects become important for the analysis of dynamical systems.

1.1 Long-time behaviour of mechanical systems

Very simple mechanical systems may evolve to a surprisingly complex behaviour if only they are observed for a sufficiently long period of time. We shall see and analyse such types of dynamics in a host of examples throughout this book. To get a first impression of the phenomena that may occur, let us start by looking at several *billiard systems*. A couple of fundamental notions may comfortably be introduced in the realm of these systems.

To explain what we mean by a billiard system let us consider a bounded open set Ω (the "table") in the plane which is connected and has a piecewise smooth boundary $\partial\Omega$. If we now put a small billiard ball somewhere on the table, make it move off with velocity $v \in \mathbb{R}^2$, $v \neq 0$ and assume that there is no loss of mechanical energy whatsoever (no friction, perfectly elastic reflection at the boundary etc.), then the resulting motion of the ball is easily described: inside Ω it moves with constant speed, whereas at the boundary it is reflected in such a way that *the angle of incidence equals the angle of reflection*. Since there is no loss of kinetic energy this motion goes on forever.

Let us first assume that the table Ω is rectangular with side-lengths a and b, respectively (see Figure 1.1). If we make the ball move off at time $t = 0$ with an angle α according to Figure 1.1, can we predict its position $x(t)$ for any $t > 0$? In fact, nothing could be simpler than this: observing that the projection of the ball's motion onto the x_1- and x_2-axis is given by a one-dimensional straight movement between walls with velocities $\pm v \cos \alpha$ and $\pm v \sin \alpha$, respectively, we could explicitly write down a formula for $x(t)$. It goes without saying that the angle α has to be chosen in such a way that the ball never hits a corner; most choices of α will satisfy this condition. There are, however, more challenging questions to ask: Will the trajectory eventually close, thus giving rise to a periodic motion? What will happen, if the trajectory fails to close? Can we – in the latter case – find a region of the table which is never hit by the ball?

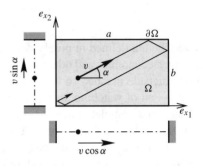

Figure 1.1. The rectangular table

To answer these questions we put our billiard system into a more tractable form by making the table twice as long and wide (cf. Figure 1.2). To benefit from this doubling procedure we consider straight motions with constant velocity on the doubled table subject to the following identification rule: for $0 \leq x_1 \leq 2a$ we consider the points $(x_1, 2b)$ and $(x_1, 0)$ as two guises of the same point; analogously we identify $(2a, x_2)$ and $(0, x_2)$ for all $0 \leq x_2 \leq 2b$. Concerning our straight motion this just says that we reenter from the bottom ($x_2 = 0$) if we have gone through the top ($x_2 = 2b$) and, analogously, reenter from the left if we have run out at the right. By means of

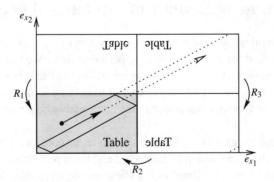

Figure 1.2. Each billiard trajectory uniquely corresponds to a straight line on the doubled table.

this identification and a repeated application of the reflections R_1, R_2, R_3 indicated in Figure 1.2 it is easy to see that each billiard trajectory within the rectangle Ω uniquely corresponds to a trajectory of the straight motion on the doubled table. Thereby we realize that the trajectory on the original table will close if and only if

$$p\, a \sin \alpha - q\, b \cos \alpha = 0 \quad \text{for some } (p, q) \in \mathbb{Z}^2, \ (p, q) \neq (0, 0).$$

Periodicity will therefore be observed if either $a/b \tan \alpha$ is a rational number or else $\cos \alpha = 0$. What happens if $a/b \tan \alpha$ is irrational? We cannot answer this question right now, but later we shall see that in this case every trajectory fills the rectangular table densely. In fact even more is true: we shall prove in Chapter Three that for $a/b \tan \alpha \in \mathbb{R} \backslash \mathbb{Q}$ the relation

$$\lim_{T \to \infty} \frac{1}{T} \int_0^T \mathbf{1}_Q\big(x(t)\big)\, dt = \frac{\text{area}(Q)}{ab} \tag{1.1}$$

holds for any rectangle $Q \subseteq [0, a] \times [0, b]$; here $\mathbf{1}_Q$ denotes the *indicator function* of Q (see Appendix A), so that the integrand equals one precisely if $x(t) \in Q$ and zero otherwise. Observe that (1.1) essentially is a statistical statement about the behaviour of the trajectory $\{x(t) : t \geq 0\}$. Indeed, since the quantity at the left is nothing else but the asymptotic relative frequency of the ball being in Q, relation (1.1) may be rephrased as follows: asymptotically, for a long time T of observation, the relative frequency ("probability") of the billiard ball finding itself in Q is given by the portion of the whole table that is covered by Q. The larger Q is, the more often the ball will be found there. This last result even holds for sets more general than rectangles. Yet is it really obvious that in the long run all the regions depicted in Figure 1.3 are visited with the same relative frequency?

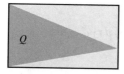 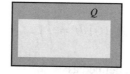 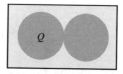

Figure 1.3. In the long run each shaded region Q is visited with the same relative frequency 0.5.

Let us now turn towards the *circular billiard*, that is the billiard on a disc, which probably has the simplest table with smooth boundary. From the left disc in Figure 1.4 it can be seen that the angle α enclosed by the billiard trajectory and the tangent to the boundary $\partial \Omega$ is the same at each point of reflection. As a consequence, the whole billiard trajectory is contained in the closed annulus

$$\overline{A}_\alpha := \{(x_1, x_2) : R^2 \cos^2 \alpha \leq x_1^2 + x_2^2 \leq R^2\},$$

and each segment of the trajectory is tangent to the circle with radius $R|\cos \alpha|$ concentric to $\partial \Omega$ (see the billiard tables in Figure 1.4). As for the rectangular billiard there are two types of behaviour a billiard trajectory may exhibit: if $\alpha/\pi \in \mathbb{Q}$ then the trajectory will close after a finite number of reflections, thus yielding a periodic motion. If on

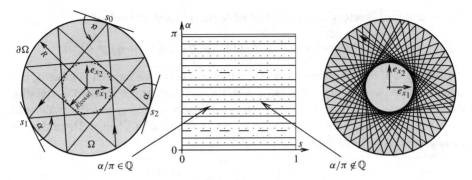

Figure 1.4. For the circular billiard the horizontal lines $[0, 1[\times\{\alpha\}$ are invariant under the associated billiard map T_{bill}.

the other hand α/π is irrational then the trajectory densely fills the annulus A_α; and we shall see that – in analogy to (1.1) –

$$\lim_{T\to\infty} \frac{1}{T} \int_0^T \mathbf{1}_Q(x(t))\, dt = \frac{1}{2\pi R \sin\alpha} \iint_Q \frac{dx_1\, dx_2}{\sqrt{x_1^2 + x_2^2 - R^2 \cos^2\alpha}} \qquad (1.2)$$

for every not-too-complicated set $Q \subseteq A_\alpha$ in this case.

Though elementary, a few observations are worth mentioning here. Firstly, (1.2) may be rewritten as

$$\lim_{T\to\infty} \frac{1}{T} \int_0^T \mathbf{1}_Q(x(t))\, dt = \iint_Q f_\alpha(x_1, x_2)\, dx_1\, dx_2$$

with the function $f_\alpha : A_\alpha \to \mathbb{R}$ defined as

$$f_\alpha(x_1, x_2) := \frac{1}{2\pi R \sin\alpha \sqrt{x_1^2 + x_2^2 - R^2 \cos^2\alpha}}.$$

Evidently $f_\alpha \geq 0$ and $\int_{A_\alpha} f_\alpha(x_1, x_2)\, dx_1 dx_2 = 1$; in probabilistic terms f_α therefore is a *density*. As we shall have occasion to observe again and again throughout this book, densities may be extremely useful for describing the long-term behaviour of dynamical systems.

A second observation is that a billiard system inside a table Ω whose boundary $\partial\Omega$ is a single smooth closed curve naturally induces a map on $\partial\Omega\times]0, \pi[$. To see this, parametrize $\partial\Omega$ by arc-length, take $(s, \alpha) \in \partial\Omega\times]0, \pi[$ and consider the billiard trajectory which emanates from the point with arc-length coordinate s and which encloses an angle α with the oriented local tangent (cf. Figure 1.4). It is natural then to assign to (s, α) the corresponding data of the next impact. Without loss of generality we may normalize $\partial\Omega$'s arc-length to one so that we end up with a *billiard map* T_{bill} which maps $[0, 1[\times]0, \pi[$ into itself. Since between any two reflection points the trajectory is just a straight line, the billiard map essentially contains all the information about the dynamics of the billiard system. In anticipation of our detailed studies in later chapters we now look at a few examples which indicate that the long-time behaviour of T_{bill} may differ considerably, depending on the specific shape of the table under consideration.

For the circular billiard it is straightforward to give an explicit formula for T_{bill}, namely

$$T_{\text{bill}} : (s, \alpha) \mapsto \left(s + \frac{\alpha}{\pi}(\text{mod } 1), \alpha\right)$$

for all $(s, \alpha) \in [0, 1[\times]0, \pi[$. It is a remarkable feature of this map that it does not alter the second coordinate and thus maps the straight line $[0, 1[\times\{\alpha_0\}$ onto itself for any $\alpha_0 \in]0, \pi[$ (see Figure 1.4). The restriction $T_{\text{bill}}|_{[0,1[\times\{\alpha_0\}}$ may therefore be considered a map on $[0, 1[$; it is in fact a *rotation* and thus the simplest example of a *circle map*, a class of maps we shall deal with in Sections 2.4 and 3.2.

Admittedly, the circular billiard is easy to survey, and the need for a statistically oriented analysis thereof may not be too pressing. A slight modification of the table may, however, suffice to yield much more complicated dynamics of the associated billiard map. Consider for example the table depicted in Figure 1.5 which differs from a disk only by a rectangle inserted between the two halves of the disk. As can be seen from Figure 1.5 this innocent surgery yields a billiard system quite different from the circular one; traditionally it is referred to as the *stadium billiard* for its resemblance to the shape of an athletic field. As is indicated by Figure 1.5, typical trajectories of the stadium billiard do not show much regularity, neither do the orbits of individual points under the associated billiard map T_{bill}. In fact, a statistical approach is required to demonstrate that there *is* a certain regularity in the long-time behaviour of this specific billiard.

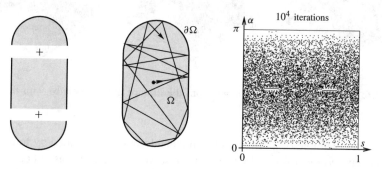

Figure 1.5. The *stadium billiard* and a typical orbit of T_{bill} (right)

The most basic statistical analysis certainly consists in drawing *histograms* of one or a few orbits of the billiard map T_{bill}. To this end, let us divide the space $[0, 1[\times]0, \pi[$ into a not-too-small number of squares S_i. Then we fix a point $x \in [0, 1[\times]0, \pi[$ and simply count how often $T_{\text{bill}}^k(x)$, i.e. the k-th iterate of x under T_{bill}, happens to fall into S_i for $0 \leq k < n$. More formally, we numerically evaluate the relative frequencies

$$a_i(x, n) := \frac{\#\{0 \leq k < n : T_{\text{bill}}^k(x) \in S_i\}}{n \ \text{area}(S_i)}$$

for various values of n. What help can the quantities a_i be for understanding the dynamics of T_{bill}? Most basically, $a_i \geq 0$ and

$$\sum_i a_i(x, n) \ \text{area}(S_i) \equiv 1,$$

so that we may interpret the family (a_i) as an approximation of a density. The larger a_i, the more often the iterates of x will visit S_i. Finally, we expect these quantities to tell us something about the long-time behaviour of T_{bill} for $n \to \infty$. How will $a_i(x, n)$ evolve as $n \to \infty$?

It turns out that the specific choice of x does not much affect the striking result displayed in Figure 1.6: in the long run each of the quantities a_i seems to converge! After we will have developed the statistical point of view in Chapter Three we shall not only be able to confirm this visual impression but also to explain the specific shape of the limit in Figure 1.6. Additionally, we shall see how these insights lead to a satisfactory description of the stochasticity inherent to the stadium billiard.

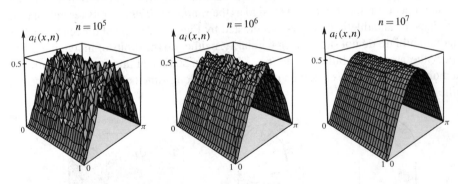

Figure 1.6. Empirical histograms drawn from the billiard map associated with the stadium billiard

As a final example of a billiard system consider a table bounded by four quarters of a circle curved *inward* (see Figure 1.7). Observe that this table is not convex. As a consequence, T_{bill} no longer is a continuous map, and it is not defined at the vertices of the table. Ignoring these technical difficulties for the time being we can perform a similar analysis as above, and we find that this billiard is by no means simpler than the stadium billiard (Figure 1.7). In fact, the mechanism which we can see at work here consists in an exponential instability of individual trajectories resulting in a *sensitive dependence on initial conditions* of the whole system. It is this latter effect which nowadays is prevalently considered as an essential ingredient of the notion of *chaos*, especially among physicists. Later we shall give a rigorous formalization, but for the specific billiard considered here we may informally discuss right away what sensitive dependence on initial conditions can mean for practical purposes.

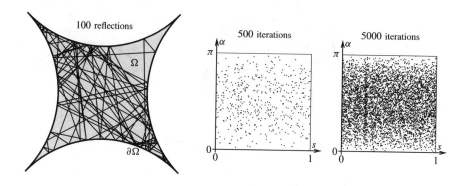

Figure 1.7. A trajectory for the dispersing billiard (left); the corresponding T_{bill}-orbit is likewise irregular.

Consider two billiard trajectories emanating from the same point in slightly different directions. Let δ denote the angle between these two directions; we consider δ as a perturbation parameter which is very small in absolute value. As a result of this small perturbation the two trajectories hit the boundary $\partial\Omega$ at different points which are $c\delta$ apart where c denotes a positive number depending on the local geometry of the boundary (see Figure 1.8). A heuristic argument shows that after reflection the directions of the trajectories will differ by an angle $(1 + 2c)\delta$. Since $|\delta|$ is very small this augmentation does not look critical. However, observe that the actual difference between the directions is multiplied by a factor larger than one at *each* reflection. As a dramatic consequence, small perturbations will grow more or less exponentially. In other words, even the smallest deviation δ will cause trajectories to significantly diverge after a frighteningly small number of reflections. Figure 1.8 provides a visualization of this effect with $\delta = 10^{-7}$: the two trajectories are close-by only for the first eleven reflections and then diverge completely.

Billiards like the one discussed here have been termed *dispersing*, a notion being self-explanatory by now. In the light of the above discussion we expect such billiards to behave chaotically in the long run. As we shall see in Chapter Three this is true in a precise sense, though working out the mathematical details is rather demanding.

Throughout this book we shall encounter and carefully analyse the effect of *sensitive dependence on initial conditions* for many systems. The practical impact of this effect has very clearly been seen already by the founders of modern dynamical systems theory.

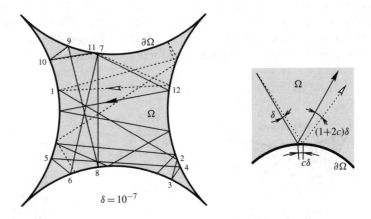

Figure 1.8. This billiard shows *sensitive dependence on initial conditions*.

In 1908 Poincaré wrote:

> *A very small cause, which escapes us, determines a considerable effect which we cannot ignore, and we then say that this effect is due to chance.*

As a consequence, our ability to predict the future evolution of individual trajectories (orbits) of such systems is tremendously limited. The best we can wish for in this situation is a meaningful *statistical* perspective of the dynamics, and we are going to develop this point of view from Chapter Three onward.

The last mechanical system we are going to introduce here has its roots in the work of Lagrange on celestial mechanics. Consider a system of $d + 1$ points labelled $0, 1, \ldots, d$ that perform a planar motion according to the following rule: the point 0 is fixed while for $k = 1, \ldots, d$ the point labelled k circles around the point labelled $k - 1$ with radius r_k and (absolute) angular velocity ω_k. One could think of a family of celestial bodies circulating around each other with constant angular velocities. What we are interested in here is the motion of the last point which may concisely be described by means of complex numbers according to

$$z(t) = a_1 e^{i\omega_1 t} + \ldots + a_d e^{i\omega_d t}$$

with $a_k \in \mathbb{C}$, $|a_k| = r_k$ for $k = 1, \ldots, d$. As might be imagined this motion can be quite complicated and non-uniform (especially for large d; cf. Figure 1.9). Writing $z(t) = r(t)e^{i\varphi(t)}$, Lagrange asked whether an *(asymptotic) average angular velocity* $\omega_\infty := \lim_{t \to \infty} \varphi(t)/t$ could be assigned to that system. Here the angle $\varphi(t)$ is assumed to be continuous unless $z(t) = 0$; in the latter case it may exhibit a jump of an absolute value of at most π. If it exists at all, the quantity ω_∞ will describe on average the long-time behaviour of the system of rotating points. Lagrange found that the special case of a dominant rod, i.e.

$$r_{k_0} > r_1 + \ldots + r_{k_0-1} + r_{k_0+1} + \ldots + r_d \quad \text{for some } k_0 \in \{1, \ldots, d\},$$

is easy to deal with: in this case $\lim_{t\to\infty} \varphi(t)/t$ does exist, and $\omega_\infty = \omega_{k_0}$. Using more advanced techniques, we shall see in Chapter Three that the average angular velocity

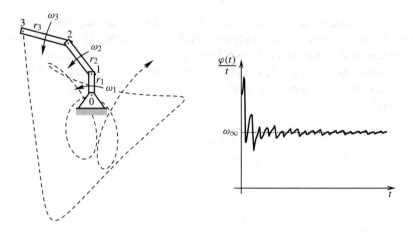

Figure 1.9. A typical trajectory for $d = 3$ (left) and the convergence towards ω_∞

exists under mild conditions. More precisely, one finds that

$$\omega_\infty = \sum_{k=1}^d p_k \omega_k \quad \text{with} \quad p_k = p_k(r_1, \ldots, r_d) \geq 0, \quad \sum_{k=1}^d p_k = 1;$$

therefore ω_∞ is a barycentric mean of the ω_k with coefficients depending only on the geometry. Although the motion of the last point may be very complicated in detail, we may nevertheless gain some non-trivial insight by means of statistical methods.

1.2   Iteration of maps

The most important mathematical objects through which an applied scientist typically encounters dynamical systems are *differential equations* and *maps*. Both types of mathematical models for real-world phenomena may exhibit complicated, apparently erratic behaviour. Throughout this book we shall, however, almost exclusively focus on maps. On the one hand, many problems in physics, biology, economics and other disciplines directly lead to discrete-time models, e.g. the size of a population quantified every year or the price of a share recorded every second. As we are going to illustrate below by an example, iteration procedures from mathematics may likewise give rise to discrete dynamical systems (cf. also Exercise 1.2). On the other hand, an analysis of differential equations often proceeds by extracting special maps from the continuous-time system. A very simple though typical example is the procedure outlined above by which the map T_{bill} is obtained from a continuous billiard system. Therefore, a basic knowledge about maps, their potential complexity as well as adequate statistical treatment may also be helpful for the thorough and more demanding study of differential equations.

Let us informally introduce a few phenomena which one may observe when innocently iterating a map. Assume for concreteness that we are given a polynomial p with real coefficients and degree higher than one. In order to find zeros of this polynomial, i.e. solutions of $p(x) = 0$, we are going to apply *Newton's method*. Recall from calculus that this method consists in choosing an initial point x_0 and then calculating inductively

$$x_{n+1} := x_n - \frac{p(x_n)}{p'(x_n)} \quad (n = 0, 1, \ldots);$$

$$(1.3)$$

here p' denotes the derivative of p. Introducing a map N_p according to

$$N_p(x) := x - \frac{p(x)}{p'(x)}$$

we may rewrite (1.3) as $x_{n+1} = N_p(x_n)$. From a dynamical systems point of view, Newton's method for the polynomial p results in iterating the associated map N_p. In calculus we learned that $p(\overline{x}) = 0$ if $x_n \to \overline{x}$ as $n \to \infty$. In practice one therefore chooses x_0 and keeps one's fingers crossed for the sequence $x_0, N_p(x_0), N_p^2(x_0), \ldots$ to converge. Let us illustrate how this method may succeed as well as fail even for quadratic polynomials.

If we specifically take for p the polynomial $q_-(x) := x^2 - 1$ then

$$N_{q_-}(x) = \frac{x^2 + 1}{2x} .$$

By an easy calculation (explicitly undertaken in Section 2.3) we find

$$\lim_{n \to \infty} x_n = \lim_{n \to \infty} N_{q_-}^n(x_0) = \begin{cases} 1 & \text{if } x_0 > 0, \\ -1 & \text{if } x_0 < 0. \end{cases}$$

Except for the origin (where N_{q_-} is not defined at all) every initial point wanders towards the nearest root of q_- under iteration of N_{q_-}. The Newton method therefore works perfectly well in this case.

Boosted by this success we now take $q_+(x) := x^2 + 1$. The fact that there are no real roots for this polynomial could already make us feel uncomfortable. Which "zeros" is the Newton method going to detect? In fact, we end up with some sort of disappointment: whichever initial point we choose, the sequence from (1.3) does not converge! Even worse, it typically looks completely erratic, as can be seen from Figure 1.10. Moreover, the phenomenon of sensitive dependence on initial conditions, which has been discussed earlier, is present in this system, too. No matter how close we choose two initial points x_0 and $\tilde{x}_0$, the corresponding iterates will inevitably diverge at some time in the not-too-far future (see Figure 1.10). In a nutshell, Newton's method does not work for the tame-looking polynomial $q_+(x) = x^2 + 1$!

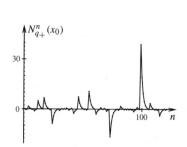
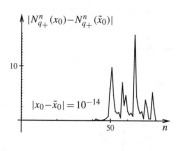

Figure 1.10. An orbit of the Newton map N_{q_+}, and a visualization of its sensitive dependence on initial conditions (right)

Before completely admitting defeat let us inspect this system from another perspective. Surely, the iteration scheme (1.3) behaves rather erratically, and realistically speaking we are not able to predict its future evolution for more than about fifty steps. (Actually the length of predictions one may trust depends on the computational power available; Section 3.4 will indicate how to quantify this prediction horizon.) So why not interpreting x_n as the outcome of an experiment of chance, i.e. as the actual value of a *random quantity* ξ_n? Assume for the sake of concreteness that ξ_n may be described in terms of a density f_n, which means that the probability of the event $\xi_n \leq a$ is given by

$$\mathbb{P}(\xi_n \leq a) = \int_{-\infty}^{a} f_n(x)\, dx$$

for all $a \in \mathbb{R}$ with $f_n \geq 0$ and $\int_{-\infty}^{\infty} f_n(x)\, dx = 1$. The important question now is: How does the density of $\xi_{n+1} := N_{q_+}(\xi_n)$, say f_{n+1}, look like? Observing that

$$\mathbb{P}(N_{q_+}(\xi_n) \leq a) = \mathbb{P}\left(\xi_n \leq a - \sqrt{a^2 + 1}\right) + \mathbb{P}\left(0 < \xi_n \leq a + \sqrt{a^2 + 1}\right)$$

for all $a \in \mathbb{R}$ we have

$$\int_{-\infty}^{a} f_{n+1}(x)\, dx = \int_{-\infty}^{a - \sqrt{a^2 + 1}} f_n(x)\, dx + \int_{0}^{a + \sqrt{a^2 + 1}} f_n(x)\, dx$$

and therefore obtain via differentiation the following explicit formula for f_{n+1}:

$$f_{n+1}(x) = \frac{f_n\left(x - \sqrt{x^2 + 1}\right)}{x^2 + 1 + x\sqrt{x^2 + 1}} + \frac{f_n\left(x + \sqrt{x^2 + 1}\right)}{x^2 + 1 - x\sqrt{x^2 + 1}}. \tag{1.4}$$

Right from the definition it is evident that f_{n+1} is also a density, i.e. $f_{n+1} \geq 0$ and $\int_{-\infty}^{\infty} f_{n+1}(x)\,dx = 1$. If we now fix an initial density f_0 and then investigate the sequence $f_0, f_1, f_2, \ldots$ we observe a surprisingly regular scenario which contrasts (or explains?) the randomness of individual orbits: the sequence $(f_n)_{n \in \mathbb{N}_0}$ always tends towards the same limit f^* rather quickly. As can be seen from Figure 1.11 this asymptotic behaviour does not depend on the specific choice of f_0.

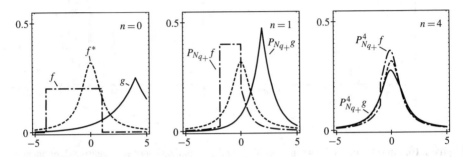

Figure 1.11. All densities are rapidly attracted towards f^* by $P_{N_{q+}}$.

In Chapter Four we shall describe the assignment $f_n \mapsto f_{n+1}$ by means of a linear operator $P_{N_{q+}}$, called the *Frobenius–Perron operator* associated with N_{q+}. Formally, we shall define

$$P_{N_{q+}} f(x) := \frac{f\left(x - \sqrt{x^2 + 1}\right)}{x^2 + 1 + x\sqrt{x^2 + 1}} + \frac{f\left(x + \sqrt{x^2 + 1}\right)}{x^2 + 1 - x\sqrt{x^2 + 1}} \tag{1.5}$$

by which we may rewrite (1.4) as $f_{n+1} = P_{N_{q+}} f_n$. As may be anticipated even on the basis of the present brief exposition, Frobenius–Perron operators turn out to be extremely useful for studying the iterations of maps within a statistical framework. For example, this can be seen if we perform a statistical analysis of a single orbit of N_{q+} by means of histograms – just as we did earlier for the stadium billiard. Thereby we unmistakably rediscover the shape of f^* (see Figure 1.12) for sufficiently long initial segments of that orbit. So f^* must be considered a key object in understanding the dynamics of the Newton map N_{q+}. In Chapter Four we shall identify f^* as the unique density that is invariant under $P_{N_{q+}}$, and we shall explicitly calculate it as

$$f^*(x) = \frac{1}{\pi(1 + x^2)}.$$

By inserting this expression into (1.5) the reader may wish to check that, as a matter of fact, $P_{N_{q+}} f^* = f^*$.

Among other things the above discussion shows that maps on the real line may generate quite different dynamical effects: one may encounter perfect regularity as well as complete randomness. The question arises whether there exist intermediate

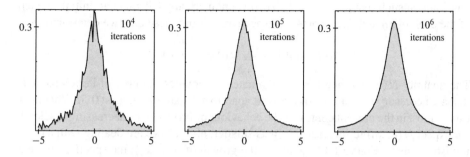

Figure 1.12. Histograms for the Newton map N_{q_+}

levels of complexity between these two extreme cases. In textbooks this question is –
traditionally and for good reasons – firstly addressed by studying the so-called *logistic
family*, that is the family of quadratic maps $F_\mu : x \mapsto \mu x(1 - x)$ with μ denoting a
positive real parameter. In Chapter Two we shall follow this tradition to some extent
because a variation of the parameter μ allows us to get an impression of how the
dynamics of apparently simple maps may become more and more complicated. Here
we merely indicate that this family does in fact incorporate both perfectly regular and
completely erratic dynamics.

As an example for regularity consider the case $\mu = 1$. It is easy to see that
$\lim_{n\to\infty} F_1^n(x_0) = 0$ precisely if $0 \le x_0 \le 1$; otherwise the iterates $F_1^n(x_0)$ wander
to the left towards infinity. On the other hand, let us have a look at the case $\mu = 4$. If we take x_0 from somewhere outside the unit interval $[0, 1]$, then $F_4^n(x_0)$ tends
to infinity as before. However, for typical orbits starting anywhere between zero
and one we find complete randomness, just as for the above example from Newton's
method. Not too surprisingly, a statistical analysis by means of histograms analogously
shows a regularizing effect in the long run (Figure 1.13). Finally, the map F_4 also
exhibits sensitive dependence on initial conditions which we may empirically quantify
as follows. First we fix a small positive number ε and choose an initial point x_0 from the

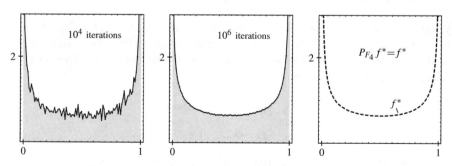

Figure 1.13. Histograms approach the invariant density f^* of P_{F_4}.

unit interval such that $\varepsilon < x_0 < 1$. Then we numerically calculate an initial segment of the orbit of x_0 and $\tilde{x}_0 := x_0 - \varepsilon$, respectively. From these data we determine

$$N(\varepsilon; x_0) := \min\{k : |F_4^k(x_0) - F_4^k(\tilde{x}_0)| \geq 0.5\}.$$

The quantity $N(\varepsilon; x_0)$ therefore denominates the smallest number of iterations of F_4 that are necessary to make x_0 and $\tilde{x}_0$ drift apart to a distance of at least 0.5. What interests us most in the present context is the behaviour of $N(\varepsilon; x_0)$ for decreasing ε. Clearly we expect $N(\varepsilon; x_0)$ to increase as ε gets smaller; after all we are dealing with a deterministic system. Figure 1.14 quantifies the growth of $N(\varepsilon; x_0)$; the specific choice of x_0 only marginally affects the observed results. A closer look at Figure 1.14 shows that the growth rate of $N(\varepsilon; x_0)$ is rather poor. In fact, as $\varepsilon \to 0$ it merely increases *logarithmically* – and not like ε^{-a} with some positive a as one might wish for. Therefore even for very small ε we observe rather moderate values of $N(\varepsilon; x_0)$; specifically $N(10^{-16}; 0.3) = 52$ in Figure 1.14 which means that two points with the tiny initial distance of 10^{-16} will totally separate within no more than 52 iterations of F_4. To put this in perspective recall that the precision of the usual double precision floating point arithmetic amounts to $2^{-53} \approx 1.110 \cdot 10^{-16}$. Our ability to reliably investigate the fate of individual points by means of digital computers is thus considerably limited.

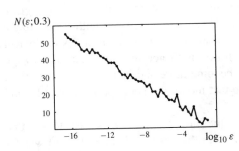

Figure 1.14. Quantifying F_4's sensitive dependence on initial conditions

Besides underpinning the importance of a statistical point of view this observation also raises the question whether numerical calculations with maps like F_4 are of any use at all. This question will be addressed in Section 2.7.

1.3   Elementary stochastic processes

By further pursuing our statistical approach to dynamics, in Chapter Four we shall encounter a class of elementary stochastic, i.e. chance-driven, processes. To exemplify how these processes may emerge from a treatment of dynamical systems let us consider a map T on the unit interval defined as

$$T(x) := \begin{cases} \frac{7}{9}(1 - 3x) & \text{if } 0 \leq x < \frac{1}{3}, \\ \frac{3}{2}(1 - x) & \text{if } \frac{1}{3} \leq x \leq 1, \end{cases} \tag{1.6}$$

the graph of which is depicted in Figure 1.15. In Section 4.3 this map will serve as a simple model of a driven pendulum. According to that interpretation the three disjoint intervals A_1, A_2, A_3 in Figure 1.15 correspond to impacts of the driving mechanism at low, medium and high speed, respectively. As we are going to see, typical orbits under T are highly erratic.

On the other hand, T is a *statistically stable* map; informally speaking, this means that its long-time behaviour may satisfyingly be described by an invariant density. (That is a density f^* fixed by the Frobenius–Perron operator P_T associated with T, see Figure 1.15 where again f^* is found via a histogram generated by a single orbit.) The speciality about T is that the invariant density f^* is constant on each of the intervals A_i. Recalling our description of the n-th iterate of T by means of a random quantity ξ_n we may express this fact as follows: conditional on being in a specific interval A_i, all outcomes in this interval are equally likely in the long run. In other words, conditional on $\xi_n \in A_i$ the random quantity ξ_n is equally distributed all over A_i.

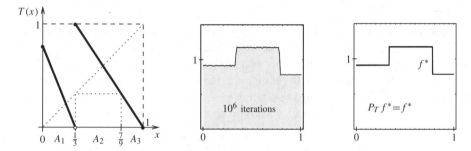

Figure 1.15. The map T as defined by (1.6), a histogram drawn from a single orbit (middle), and the invariant density f^* (right)

Assume now that someone watches the iteration of T, but rather than keeping an account of the precise numerical values of $T^n(x_0)$ he or she only notices in which of the intervals A_i these iterates may be found. More formally, the observer does not record the actual outcomes of ξ_n but merely observes the new random quantity

$$\eta_n := \rho(\xi_n)$$

where $\rho(x) := i$ precisely if $x \in A_i$. As a result of this rough observation the observer obtains a sequence of the digits 1, 2, 3 which looks completely random. In the long run, however, there is a certain regularity because the probability for the process to jump from any digit i to any other digit j is constant. Specifically, for large n we find the transition probability

$$p_{12} := \mathbb{P}(\eta_{n+1} = 2 | \eta_n = 1) = \mathbb{P}(T(\xi_n) \in A_2 | \xi_n \in A_1)$$

$$\approx \frac{\text{length}\,(T^{-1}(A_2) \cap A_1)}{\text{length}\, A_1} = \frac{4}{7}.$$

Furthermore $p_{11} = \frac{3}{7}$, $p_{13} = 0$, and the other transition probabilities may be calculated analogously. In the long run, the sequence of random quantities $(\eta_n)_{n \in \mathbb{N}_0}$ therefore behaves exactly like a stochastic process on the three states 1, 2, 3 with constant transition probabilities between them (cf. Figure 1.16).

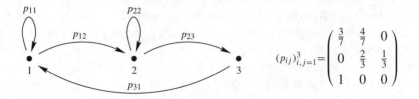

Figure 1.16. The probabilistic structure inherent to the map T from Figure 1.15

Stochastic processes like the one derived from the map T above are generally known as *Markov chains* and constitute the most elementary and well-understood class of all stochastic processes. A rich and rather complete theory has been developed for them during the first half of the twentieth century. Certainly we shall not delve into the details of the theory in the course of this book; however, in Section 4.4 we shall concentrate on a few characteristic properties of Markov chains which intimately relate them to dynamical systems.

Markov chains abound in applications. While referring to Chapter Four for more substantial support for this fact, the reader's imagination may right now be stimulated by the following simple example. The game *Snakes and Ladders* may be considered as a method to keep yourself awake during long train trips or boring lectures. You can play it alone or together with a number of travel companions or colleagues, respectively. Neither is there any interaction between players nor do you need anything like a strategy. Just throw a die and climb up whenever you reach the bottom end of a ladder. Stopping at the head of a snake forces you to slide down to its tail (cf. Figure 1.17). The details of each individual game are exclusively governed by chance. Some of the questions that arise in this context may be formulated as follows. Certainly, each game comes to an end with probability one, but how long will this take in the mean? Given $N \in \mathbb{N}$, what is the probability for the game to last precisely N steps? Having reached a certain field, how long will it take to reach the finish? These and a lot of other questions have to be tackled by purely probabilistic methods.

Although not obvious at first glance, some of these questions may naturally be related to dynamical systems. One such relation is established by studying the *long-time behaviour* of Markov chains. Specifically, if one plays a large number of games of *Snakes and Ladders* and afterwards draws histograms of the recorded data (e.g. the duration T of individual games), then one will observe a phenomenon of convergence which resembles the corresponding effect for histograms drawn from orbits of deterministic maps. Figure 1.18 shows a few results in this direction. In probabilistic terms, this convergence may be considered a Law of Large Numbers, but we shall see that there is a dynamical interpretation, too.

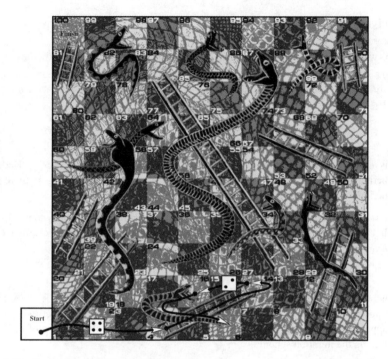

Figure 1.17. *Snakes and Ladders* (Illustration from the album Snakes and Ladders by SLAW. © 2001 Josie Porter & Tim Walters)

As suggested by the above exposition and made precise later, Markov chains provide a consistent way of thinking about the long-time behaviour of certain dynamical systems. Moreover we shall observe striking analogies between methods developed for deterministic and stochastic systems, respectively. In Chapter Four we shall thus be able to gain a unified view on phenomena which seem quite unconnected at first glance.

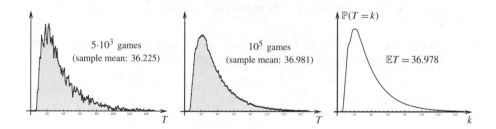

Figure 1.18. The long-time behaviour of *Snakes and Ladders*: empirical histograms for the random duration T, and its probability distribution (right)

Exercises

(1.1) As we have seen, orbits under the map N_{q_+} induced by Newton's method for the polynomial $q_+ : x \mapsto x^2 + 1$ may look rather irregular (recall for instance Figure 1.10). There are, however, many regular orbits too: given natural numbers l, m with $m \geq 2$ and $1 \leq l < 2^m - 1$ prove that each point

$$x_{l,m} := \cot \frac{\pi l}{2^m - 1} \in \mathbb{R}$$

is m-periodic, which means that it returns to itself after m iterations of N_{q_+}. Deduce that periodic points of N_{q_+} are *dense* in $\mathbb{R}$, i.e., given any $\varepsilon > 0$ and $y \in \mathbb{R}$ there exists a periodic point x such that $|x - y| < \varepsilon$.

(1.2) Iterative methods from applied mathematics may naturally bring forth dynamical systems. In the text we reverted to the example of Newton's method. As another example consider *Euler's method* of solving the initial value problem

$$\dot{x} = f(x), \quad x(0) = x_0 ; \tag{1.7}$$

here the dot ($\dot{}$) denotes differentiation with respect to time, and $f : \mathbb{R} \to \mathbb{R}$ is a continuous function. Fix now a step-size $h > 0$ and recall that Euler's method provides an approximate value x_n for $x(nh)$, that is the actual value of the solution of (1.7) at time nh, via the iteration scheme

$$x_{n+1} := x_n + hf(x_n) \quad (n \in \mathbb{N}_0) .$$

In the terminology of dynamical systems thus $x_n = T_h^n(x_0)$ with the map T_h defined according to

$$T_h : x \mapsto x + hf(x) .$$

Specifically consider now the case $f(x) = x(1 - x)$. If $x_0 > 0$ then the solution of (1.7) reads

$$x(t) = 1 - \frac{1 - x_0}{1 - x_0 + x_0 e^t} ,$$

hence $\lim_{t \to \infty} x(t) = 1$. Does the Euler method correctly reproduce this long-time behaviour, i.e., do we find $\lim_{n \to \infty} T_h^n(x_0) = 1$ for all $x_0 > 0$? Show that the answer is negative if h is chosen too large. Try to analyse the ultimate fate of $T_h^n(x_0)$ in the latter case.

(1.3) The following *shooting method* is probably the most simple-minded strategy to solve a *boundary value problem*. Consider the ordinary differential equation

$$\ddot{x} = 1 - \dot{x}^2$$

subject to the boundary conditions

$$x(0) = 0, \quad x(1) = 1 .$$

In order to solve this problem replace the boundary condition at 1 by a second condition at 0, i.e. $\dot{x}(0) = \xi_0 > 0$. Now solving the initial value problem

$$\ddot{x} = 1 - \dot{x}^2, \quad x(0) = 0, \quad \dot{x}(0) = \xi_n$$

and putting

$$\xi_{n+1} := \xi_n - \varepsilon(x(1) - 1) \quad (n \in \mathbb{N}_0) \tag{1.8}$$

naturally yields an iteration scheme; here $\varepsilon > 0$ denotes a fixed parameter. Does this scheme provide the solution to the original problem? More precisely, do we find $\lim_{n\to\infty}\xi_n = 1$ which is the correct value for $\dot{x}(0)$? You may wish to perform a few numerical experiments to find out whether the value of ε is of any importance for the dynamics of (1.8).

(1.4) The table below shows the powers 2^n for $n = 1, \dots, 40$. When looking at their leading digits one finds all numbers except 7 and 9. What about these two numbers? Will they ever appear as the leading digit of 2^n for some n?

n	2^n		n	2^n	
1	2	2	21	2097152	2
2	4	4	22	4194304	4
3	8	8	23	8388608	8
4	16	1	24	16777216	1
5	32	3	25	33554432	3
6	64	6	26	67108864	6
7	128	1	27	134217728	1
8	256	2	28	268435456	2
9	512	5	29	536870912	5
10	1024	1	30	1073741824	1
11	2048	2	31	2147483648	2
12	4096	4	32	4294967296	4
13	8192	8	33	8589934592	8
14	16384	1	34	17179869184	1
15	32768	3	35	34359738368	3
16	65536	6	36	68719476736	6
17	131072	1	37	137438953472	1
18	262144	2	38	274877906944	2
19	524288	5	39	549755813888	5
20	1048576	1	40	1099511627776	1

Figure 1.19. The initial digits of 2^n for $n = 1, \dots, 40$

Chapter 2
Basic aspects of discrete dynamical systems

Discrete dynamical systems are crucial for many fields of applied mathematics. Iterating a map $T : X \to X$ (or even a family of maps) turns out to be an appropriate model for a great variety of different phenomena. This universality of discrete systems is mainly due to the enormous generality tied into the notion of a *map T* on a *space X*. The provision of a reasonable structure therefore allows one to adapt the abstract situation to the concrete problem one is struggling with. During this text we shall repeatedly come to think about what we may consider *reasonable*. In the present chapter we adopt a rather geometric or topological point of view. Questions we are mainly concerned with will sound like: Where do individual points eventually go under the iteration of a map? What happens to them while they are going there? Do they share their ultimate fate with their neighbours? and the like. In investigating problems like these, a small but crucial portion of basic terminology from the theory of dynamical systems is developed. As throughout the whole text we shall focus on examples, thereby motivating abstract definitions and general results in the simplest possible setting.

2.1 Hyperbolicity and bifurcations

This section briefly introduces some basic notions and observations concerning the dynamics of maps on real intervals. In this elementary context the clarifying role of hyperbolicity not only is easily grasped but also hints at many of our investigations to follow. Guided by a specific example we shall then address the question how substantial changes of the dynamics may occur for systems depending upon parameters.

Let $T : X \to X$ be a C^2-map on the (possibly infinite) closed interval $X \subseteq \mathbb{R}$. We are interested in the behaviour of individual points $x \in X$ under iteration of T, and therefore denote by

$$O^+(x) := \{x, T(x), T^2(x), \dots\} = \{T^n(x) : n \in \mathbb{N}_0\}$$

the *(forward) orbit* of x. Here and in the sequel T^n will always stand for the n-fold composition of T with itself, $T^0 = \mathrm{id}_X$ by definition. If T is a diffeomorphism, we may likewise define the *backward orbit* $O^-(x) := \{T^{-n}(x) : n \in \mathbb{N}_0\}$, and also $O(x) := O^+(x) \cup O^-(x)$. In the one-dimensional setting chosen here, however, the dynamical behaviour of invertible maps is rather simple ([20]). For the time being we shall therefore concentrate on the study of forward orbits.

How can $O^+(x)$ look like? Undoubtedly, the simplest situation occurs if $O^+(x) = \{x\}$. In this case we call x a *fixed point* since $T^n(x) = x$ for every $n \in \mathbb{N}_0$. More generally, a point x fixed by T^p for some $p \in \mathbb{N}$ is called a *periodic point* of period p with $\min\{n \geq 1 : T^n(x) = x\}$ being the *prime* period of x. The sets of all fixed points and all points of period p in X are denoted by Fix T and $\mathrm{Per}_p\, T$, respectively; in addition we write $\mathrm{Per}\, T := \bigcup_{p \in \mathbb{N}} \mathrm{Per}_p\, T$. Finally, we call x *eventually periodic* if $T^n(x)$ is periodic for some $n \in \mathbb{N}$. Obviously $O^+(x)$ is finite if and only if x is eventually periodic. A point with finite forward orbit thus is easily analysed dynamically. Yet even if $O^+(x)$ is not finite there may be some regularity. If for example

$$\lim_{n \to \infty} T^{pn}(x) = x_0 \quad \text{for some } p \in \mathbb{N},\, x_0 \in X,$$

then the point x_0 is p-periodic by continuity, and x can therefore be called *forward asymptotic* to the periodic orbit $O^+(x_0)$. In order to become acquainted with this situation it is helpful to look at some examples.

A convenient tool to visualize the action of one-dimensional maps is *graphical analysis*. As can be seen from Figure 2.1, successive iterates of x may be found graphically using the graph of T: just start at the point (x, x), move vertically to $\big(x, T(x)\big)$ on the graph of T, then horizontally to $\big(T(x), T(x)\big)$ on the line $y = x$ and repeat this procedure. By means of such a graphical analysis one often gets a good impression of what happens dynamically. For example, Figure 2.1 suggests that fixed points may differ considerably with respect to the dynamics in their vicinity: the fixed point x_1 seems to attract all points nearby while the fixed point x_2 does not.

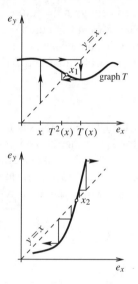

Figure 2.1. Graphical analysis

Definiton 2.1.   Let x_0 denote a fixed point of T. This fixed point is called *stable*, if within any open interval U which contains x_0 there exists an open interval V still containing x_0 such that $T^n(x) \in U$ for all $x \in V$ and $n \in \mathbb{N}_0$. Furthermore x_0 is said to be *attracting* if $\lim_{n \to \infty} T^n(x) = x_0$ for all $x \in V$. (A completely analogous definition may be given for periodic points by writing T^p instead of T.)

Notice that according to this definition there may be attracting fixed points which are not stable. Fixed points being both stable and attracting are sometimes termed *asymptotically stable*.

Contemplating on Figure 2.1 immediately gives the following result. Since the generalization to periodic points is evident, we may restrict ourselves to fixed points.

Lemma 2.2. *Let x_0 denote a fixed point of T with $|T'(x_0)| \neq 1$.*

(i) *If $|T'(x_0)| < 1$ then x_0 is stable and attracting;*

(ii) *If $|T'(x_0)| > 1$ then x_0 is unstable (i.e. not stable).*

Proof. Since T' is continuous there exists a bounded open interval V containing x_0 and a real number $\delta > 0$ such that $\left|\left|T'(x)\right| - 1\right| \geq \delta$ for all $x \in V$. If $|T'(x_0)| < 1$, the Mean Value Theorem yields $|T(x) - x_0| \leq (1-\delta)|x - x_0|$ for every $x \in V$ which implies (i). If $|T'(x_0)| > 1$ and $T(x) \in V$ we find that $|T(x) - x_0| \geq (1+\delta)|x - x_0|$, which shows that all points in $V \setminus \{x_0\}$ must eventually leave V. Consequently, x_0 cannot be stable. $\square$

Despite its simplicity this result allows us to completely understand the dynamics near a fixed (or periodic) point x_0 as long as $|T'(x_0)| \neq 1$. Note, however, that the alternative (ii) above does not necessarily prevent points near x_0 from coming back again sometime in the future.

The cases not captured yet therefore are those fixed points with $|T'(x_0)| = 1$. What can be said about such points? In order to answer this question we consider four maps $T_i : \mathbb{R} \to \mathbb{R}$ all satisfying $T_i(0) = 0$ and $|T_i'(0)| = 1$; more specifically, let

$$T_1(x) := -x, \quad T_2(x) := x + x^2, \quad T_3(x) := x + x^3, \quad T_4(x) := x - x^3. \quad (2.1)$$

Since $\mathrm{Per}_2\, T_1 = \mathbb{R}$, the stable fixed point at the origin is surrounded by points of period two for T_1, neither attracting nor repelling them. It is easy to see that $T_2^n(x) \to 0$ if $-1 \leq x \leq 0$; otherwise one finds $T_2^n(x) \to +\infty$. The fixed point 0 of T_2 is therefore at the same time attracting from the left and repelling to the right. Finally, the fixed point at the origin turns out to be repelling under T_3 and attracting under T_4 (cf. Figure 2.2).

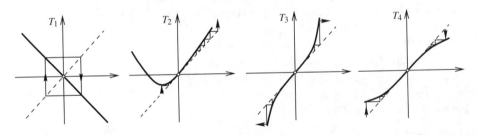

Figure 2.2. Four different non-hyperbolic fixed points for the maps T_i from (2.1)

Apparently, at a fixed point x_0 with $|T'(x_0)| = 1$ even the local dynamics ask for a refined analysis. This is in considerable contrast to the case $|T'(x_0)| \neq 1$ to which we assign a specific name.

Definiton 2.3. A periodic point x_0 of prime period p is called *hyperbolic*, if $|(T^p)'(x_0)| \neq 1$.

Summarizing we may say that hyperbolic fixed or periodic points are easy to analyse (at least locally) while this need not be true for their non-hyperbolic counterparts. As we come to see that the analysis of the latter will usually be quite difficult we could with good reasons ask whether non-hyperbolic points are of any practical importance. After all, such points will typically disappear or transform into hyperbolic ones if

T is replaced by a C^1-close map $\tilde{T}$, which means that both $|T(x) - \tilde{T}(x)|$ as well as $|T'(x) - \tilde{T}'(x)|$ are small for all x. Interpreting $\tilde{T}$ as a perturbation of T we conclude that non-hyperbolic fixed or periodic points will usually not survive C^1-small perturbations.

There are, however, at least two reasons for non-hyperbolic points to be important. On the one hand, an additional structure (e.g. symmetry) may generically prevent a system from exhibiting hyperbolic points. (This statement especially holds for higher-dimensional systems.) On the other hand, the appearance of non-hyperbolic points may be inevitable if one has to deal with maps depending on *parameters*, i.e., when analysing families of maps. As will become clear immediately, global changes of the dynamics within parameterized maps often involve a certain amount of non-hyperbolicity.

Let us consider the family of quadratic maps $Q_c : \mathbb{R} \to \mathbb{R}$ defined by $Q_c(x) := x^2 - c$ ($c \in \mathbb{R}$) which – in a different setting – has been the object of immense study (see [16, 20, 39, 40]; our aim, however, is not too ambitious here.) If $c < -\frac{1}{4}$, obviously $Q_c^n(x) \to +\infty$ for every x. Solving $Q_c(x) = x$ we find that a pair of fixed points $x_{1,2}(c) = \frac{1}{2}(1 \pm \sqrt{1 + 4c})$ is born from a single non-hyperbolic fixed point at $c = -\frac{1}{4}$. It is easy to see that for $c > -\frac{1}{4}$ the point x_1 is unstable while x_2 is stable as long as $c < \frac{3}{4}$. At $c = \frac{3}{4}$ the fixed point x_2 turns out to be non-hyperbolic again, and it will be unstable if $c > \frac{3}{4}$. The case $c = \frac{3}{4}$ therefore needs some further analysis. Figure 2.3 suggests looking at Q_c^2. Solving $Q_c^2(x) = x$ we

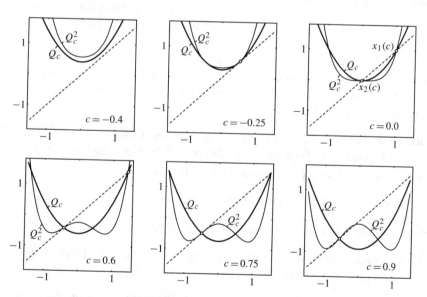

Figure 2.3. The morphogenesis of the family Q_c for $-0.4 \leq c \leq 0.9$

find two points $x_{3,4}(c) = \frac{1}{2}(\pm\sqrt{4c - 3} - 1)$ constituting an orbit of prime period two. Since $(Q_c^2)'(x_{3,4}) = 4(1 - c)$ these points are stable and hyperbolic if $\frac{3}{4} < c < \frac{5}{4}$. What happens if c is increased further? We will discuss this question later in a slightly different form. Figure 2.4 summarizes our findings, indicating by arrows the behaviour of the non-fixed points.

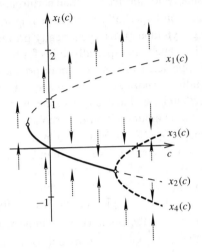

As pointed out earlier, sufficiently C^1-small perturbations will not remove a point's hyperbolicity. Qualitative local changes of a parameterized system's dynamical structure are therefore typically accompanied by the appearance of non-hyperbolic points. Such substantial changes are usually referred to as *bifurcations*. As can already be seen from our elementary calculations with Q_c, the study of bifurcations is crucial for the understanding of parameterized dynamical systems.

Subsequently we shall formalise the observations of Figures 2.3 and 2.4. To this end we shall exclusively consider families $(T_\lambda)_{\lambda \in \Lambda}$ of C^2-maps on X, where $\Lambda \subseteq \mathbb{R}$ denotes an open interval. The dependence on the parameter is assumed to be twice differentiable, too. In this specific setting we can give a fairly complete local analysis. (Without losing any generality we again restrict ourselves to the case of a fixed point.)

Figure 2.4. Elementary bifurcations in the quadratic family Q_c

The first and most basic observation drawn from Figure 2.3 may be expressed as follows: no dramatic changes can take place in the neighbourhood of a fixed point as long as the latter is hyperbolic. The fixed point itself may, at most, be shifted slightly if the parameter is perturbed.

Theorem 2.4. *Let $(T_\lambda)_{\lambda \in \Lambda}$ be a C^2-family of maps on X and let x_0 be a fixed point of T_{λ_0} with $T'_{\lambda_0}(x_0) \neq 1$. Then there exist intervals $J \subset \Lambda$ containing λ_0 and $U \subset X$ containing x_0 as well as a C^2-function $\xi : J \to U$ with*

(i) $\xi(\lambda_0) = x_0$ *and* $T_\lambda(\xi(\lambda)) = \xi(\lambda)$ *for all* $\lambda \in J$;

(ii) *for λ fixed, $\xi(\lambda)$ is the only fixed point of T_λ in U, and it is hyperbolic.*

Proof. All assertions immediately follow from the Implicit Function Theorem applied to $G(x, \lambda) := T_\lambda(x) - x$ since $G(x_0, \lambda_0) = 0$ and $\frac{\partial G}{\partial x}(x_0, \lambda_0) = T'_{\lambda_0}(x_0) - 1 \neq 0$. $\square$

Notice that Theorem 2.4 does not assume full hyperbolicity for the fixed point x_0 of T_{λ_0}: the case $T'_{\lambda_0}(x_0) = -1$ also is allowed.

For the family Q_c we furthermore observed different types of bifurcations arising from non-hyperbolic points. At $c = -\frac{1}{4}$ we had $Q'_c(x_{1,2}) = 1$, and a pair of fixed points was born. In order to guarantee this kind of behaviour in some generality a minor technical assumption has to be incorporated.

Theorem 2.5 (Saddle-node bifurcation). *Let $(T_\lambda)_{\lambda \in \Lambda}$ be as before and assume that x_0 is a fixed point for T_{λ_0} with $T'_{\lambda_0}(x_0) = 1$. If $T''_{\lambda_0}(x_0) \neq 0$ and $\frac{\partial T_\lambda}{\partial \lambda}|_{\lambda=\lambda_0}(x_0) \neq 0$, then there exists an interval U containing x_0 and a C^2-function $l : U \to \Lambda$ with*

(i) *$T_{l(x)}(x) = x$ and $T'_{l(x)}(x) \neq 1$ for all $x \in U \setminus \{x_0\}$;*

(ii) *$l(x_0) = \lambda_0$, $l'(x_0) = 0$, $l''(x_0) \neq 0$.*

(The family $(T_\lambda)_{\lambda \in \Lambda}$ undergoes a saddle-node bifurcation at λ_0.)

Proof. Again apply the Implicit Function Theorem to $G(x, \lambda) := T_\lambda(x) - x$, taking into account that $G(x_0, \lambda_0) = 0$ and $\frac{\partial G}{\partial \lambda}(x_0, \lambda_0) = \frac{\partial T_\lambda}{\partial \lambda}|_{\lambda=\lambda_0}(x_0) \neq 0$. A straightforward calculation yields $l''(x_0) = -T''_{\lambda_0}(x_0)/\frac{\partial T_\lambda}{\partial \lambda}|_{\lambda=\lambda_0}(x_0) \neq 0$. $\square$

Under the assumptions of Theorem 2.5 two branches of unstable ("saddles") and stable ("nodes") hyperbolic fixed points, respectively, originate from the fixed point x_0 if T_{λ_0} has slope one (see Figure 2.5). Note that all the assumptions of the theorem are met by the family Q_c at $c = -\frac{1}{4}$. We therefore observe a saddle-node bifurcation for this value of the parameter (compare Figures 2.4 and 2.5). At $c = \frac{3}{4}$ we watched

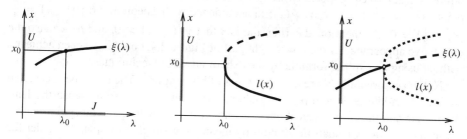

Figure 2.5. No bifurcation (left), saddle-node bifurcation (middle) and period-doubling bifurcation (corresponding to Theorems 2.4, 2.5 and 2.6, respectively)

a completely different scenario: a pair of two-periodic points emerging from a fixed point with slope -1. As before, we have to impose a condition of non-degeneracy when transforming this observation into a general result.

Theorem 2.6 (Period-doubling bifurcation). *Let again $(T_\lambda)_{\lambda \in \Lambda}$ be as before and assume that $T_\lambda(\xi(\lambda)) \equiv \xi(\lambda)$ with λ in some interval around λ_0 and the parameterization $\xi(.)$ being C^1. If for $x_0 := \xi(\lambda_0)$ the relations $T'_{\lambda_0}(x_0) = -1$ and $\frac{\partial}{\partial \lambda}(T^2_\lambda)'|_{\lambda=\lambda_0}(x_0) \neq 0$ hold then there exists an interval U containing x_0 and a C^1-function $l : U \to \Lambda$ with*

(i) *$T_{l(x)}(x) \neq x$ but $T^2_{l(x)}(x) = x$ for all $x \in U \setminus \{x_0\}$;*

(ii) *$l(x_0) = \lambda_0$, $l'(x_0) = 0$.*

(This type of local behaviour is called a period-doubling bifurcation.)

Proof. Unfortunately the Implicit Function Theorem does not directly apply to $G(x, \lambda) :=$ $T_\lambda^2(x) - x$ because $G(\xi(\lambda), \lambda)$ vanishes identically around λ_0. Therefore define locally at (x_0, λ_0)

$$H(x, \lambda) := \begin{cases} \dfrac{G(x, \lambda)}{x - \xi(\lambda)} & \text{if } x \neq \xi(\lambda), \\[2ex] \dfrac{\partial G}{\partial x}(x, \lambda) & \text{if } x = \xi(\lambda). \end{cases}$$

One can easily see that H is C^1 and $\frac{\partial H}{\partial x}(x_0, \lambda_0) = \frac{1}{2}\frac{\partial^2 G}{\partial x^2}(x_0, \lambda_0) = 0$. As $H(x_0, \lambda_0) = 0$ and $\frac{\partial H}{\partial \lambda}(x_0, \lambda_0) = \frac{\partial}{\partial \lambda}(T_\lambda^2)'|_{\lambda = \lambda_0}(x_0) \neq 0$, the Implicit Function Theorem yields a C^1-function $l : U \to \Lambda$ with $H(x, l(x)) \equiv 0$ and $l'(x_0) = 0$. If $x \neq \xi(l(x))$ we have $G(x, l(x)) = 0$ but $T_{l(x)}(x) \neq x$ because the (locally) unique fixed point of $T_{l(x)}$ is given by $\xi(l(x))$. □

Informally speaking, Theorem 2.6 describes how a two-periodic orbit may be born at a fixed point with slope minus one. It should be mentioned that with some additional differentiability for $(T_\lambda)_{\lambda \in \Lambda}$ more information on the *bifurcation curve* $\lambda = l(x)$, e.g. the sign of $l''(x_0)$, may be obtained.

For the quadratic family at $c_0 = \frac{3}{4}$ we have $x_2(c_0) = -\frac{1}{2}$, $Q'_{c_0}(x_2(c_0)) = -1$ as well as $\frac{\partial}{\partial c}(Q_c^2)'|_{c = c_0}(x_2(c_0)) = 2$ and thus definitely observe a period-doubling bifurcation at $c = c_0$. Especially, we can explicitly calculate the curve of two-periodic points as $c = x^2 + x + 1 =: \gamma(x)$; in accordance with Theorem 2.6 $\gamma(-\frac{1}{2}) = \frac{3}{4}$, $\gamma'(-\frac{1}{2}) = 0$. As mentioned earlier, the fact that $\gamma''(-\frac{1}{2}) > 0$, and hence also the qualitative appearance of the lower right part of Figure 2.4, could have been achieved without using an explicit formula by the help of advanced techniques.

Not too surprisingly, there are other types of bifurcations. The two types discussed so far, i.e. saddle-node and period-doubling, are, however, in some sense the least degenerate and thus most typical ones. Without pretending to make this point mathematically rigorous we shall illustrate it by means of a simple example. Consider the family $T_\lambda(x) := \lambda \sin x$ with $\lambda > 0$. The obvious fixed point $x_0 = 0$ is stable only as long as $\lambda < 1$ and becomes unstable for $\lambda > 1$, thereby splitting in *three* fixed points at $\lambda = 1$. Although $T_1'(0) = 1$ we have not encountered such a bifurcation yet. Since $T_1''(0) = 0$ and $\frac{\partial T_\lambda}{\partial \lambda}|_{\lambda = 1}(0) = 0$, the technical assumptions of Theorem 2.5 are not satisfied. The type of bifurcation we observe here is usually called *pitch-fork bifurcation*, the name being self-explanatory from Figure 2.6. Such a bifurcation frequently arises in systems with a certain amount of *symmetry*. For the family T_λ we obviously have the relation $T_\lambda(-x) = -T_\lambda(x)$. It is precisely this symmetry that makes the situation more degenerate. An arbitrarily (even C^∞-) small perturbation may drastically change the local structure near a pitch-fork bifurcation. To be more specific, consider the perturbed family

$$T_{\lambda, \varepsilon}(x) := \lambda \sin x + \varepsilon \quad \text{with } |\varepsilon| \ll 1. \tag{2.2}$$

As an effect of this perturbation the fixed points are shifted only slightly. Nevertheless the pitch-fork bifurcation has been replaced by an ordinary saddle-node bifurcation (for which all assumptions are fulfilled if $\varepsilon \neq 0$), while one branch of fixed points is not bifurcating at all (see Figure 2.6). On the other hand, a symmetric perturbation may in fact preserve the local structure. In contrast to these phenomena of non-robustness, saddle-node and period-doubling bifurcations cannot be destroyed by arbitrarily C^2-small perturbations.

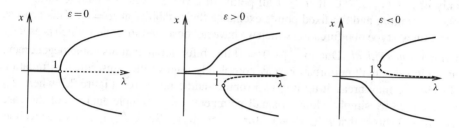

Figure 2.6. The bifurcation diagrams for (2.2) demonstrate that a pitch-fork bifurcation can be removed by an arbitrarily small perturbation.

Much of what has been discussed here in the one-dimensional context carries over to higher dimensions more or less directly. If for example $T : U \rightarrow U$ denotes a C^1-map of an open set $U \subseteq \mathbb{R}^d$ (or, more generally, of a smooth manifold), then hyperbolicity of a point x_0 of period p is naturally understood as $DT^p|_{x_0}$, the differential of T^p at x_0, having no eigenvalues of modulus one. As in the one-dimensional case hyperbolic points are dynamically well-behaved (at least locally), although now there is the possibility of contraction as well as expansion at the same time near such a point; in later sections we shall see examples referring to this. Again we need some non-hyperbolicity if we want the local structure to change qualitatively. The discussion of bifurcations in higher dimensions turns out to be more cumbersome; and it becomes increasingly intricate, if families of maps depending on more than one parameter are considered and *typical* bifurcations are to be classified. Questions of the like are dealt with by the highly developed discipline of *bifurcation theory* ([20, 27, 60]).

2.2 How may simple systems become complicated ?

Even very simple non-linear maps on $\mathbb{R}$ may exhibit very involved and multifarious dynamics. Whole books have been (and still are) devoted to this amazing fact ([16, 39]). Since we consider our present discussion mainly as a motivation for the statistical view on deterministic systems to be developed later, we will not delve into the depths of one-dimensional dynamics here. Instead, we will merely point out some aspects of a single family of maps. It turns out, however, that a number of important concepts may comfortably be introduced in the context of this specific example. It thereby also constitutes a good point from which to start when exploring the cited literature.

The family of one-dimensional maps on which we shall concentrate is the so-called *logistic family* defined by $F_\mu(x) := \mu x(1 - x)$. In case of negative μ, the dynamics of F_μ is equivalent to that of $F_{2-\mu}$ in a sense made precise later. We may therefore assume $\mu > 0$. For the two fixed points 0 and $x_\mu := 1 - \mu^{-1}$ we find $F'_\mu(0) = \mu$ and $F'_\mu(x_\mu) = 2 - \mu$. Consequently these points are, respectively, attracting and repelling hyperbolic if $0 < \mu < 1$. In this latter case $O^+(x)$ is bounded if and only if $x \in [x_\mu, \mu^{-1}]$. If $\mu \geq 1$ all points in $\mathbb{R}\backslash[0, 1]$ certainly tend towards $-\infty$ under iteration, and the fixed points exchange their stability at $\mu = 1$. This process constitutes a type of bifurcation which we have not observed so far and which is usually termed *transcritical*. Due to $\frac{\partial F_\mu}{\partial \mu}(0) = 0$ this bifurcation exhibits some degeneracy: an arbitrarily small perturbation may completely remove the transcritical bifurcation (locally) or may break it up into two robust saddle-nodes (cf. Figure 2.7 where F_μ is perturbed by simply adding a small parameter ε). A simple analysis of the case $1 < \mu < 3$ shows that $F^n_\mu(x) \to x_\mu$ for all $x \in]0, 1[$. Since $F_\mu(x_\mu) \equiv x_\mu$ as well as $F'_3(x_3) = -1$ and $\frac{\partial}{\partial \mu}(F^2_\mu)'(x_\mu) = 2(\mu - 2)(2\mu - 3)/\mu$, a period-doubling bifurcation takes place at $\mu = 3$. A straightforward calculation shows that the new orbit of period two is stable for $3 < \mu < 1 + \sqrt{6}$.

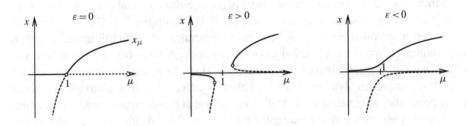

Figure 2.7. The transcritical bifurcation of F_μ at $\mu = 1$ may split in different ways.

What happens next? Obviously we cannot answer this question by means of brute force analysis: solving $F^n_\mu(x) = x$ necessitates finding zeros of a polynomial of degree 2^n. By means of advanced techniques like *renormalization* and *kneading theory* this difficulty may be surpassed, and a rather complete analysis of F_μ's dynamics for $3 \leq \mu \leq 4$ may be provided. Basically, it turns out that there exists a strictly increasing sequence $(\mu_n)_{n\in\mathbb{N}}$ of parameters at which a period-doubling bifurcation of $F^{2^{n-1}}_\mu$ occurs; specifically, $\mu_1 = 3$ and $\mu_2 = 1 + \sqrt{6}$. If $\mu_n < \mu < \mu_{n+1}$ there is precisely one attracting orbit of period 2^n for F_μ. This *period-doubling scenario* terminates at $\mu_\infty := \lim_{n\to\infty} \mu_n \approx 3.570$, and the dynamics of F_μ become more complicated for $\mu > \mu_\infty$.

Although we will not prove any of these facts here, we will give some empirical evidence for them. In order to do so we perform the following numerical experiment: we fix a value of μ, take an initial point in $]0, 1[$ and calculate a number (say a few thousands) of iterates. Neglecting the transient behaviour by disregarding the first (say five hundred) iterates, we then plot the remaining points in a (μ, x)-diagram. Finally we repeat this procedure with many different values of μ. The *orbit diagram* thus constructed will look the same for nearly every choice of the initial point. The result depicted in Figure 2.8 for different ranges of μ may therefore be considered a numerical description of the long-time behaviour of typical points.

The dynamical features of F_μ discussed above are clearly seen in Figure 2.8. On this scale, however, only a few period-doublings can be observed. This is mainly due to a famous result of Feigenbaum which states that, approximately, the difference $\mu_{n+1} - \mu_n$ between two successive bifurcation points shrinks exponentially with a rather small factor (in fact, $\mu_{n+1} - \mu_n \approx 2.069\,q^{-n}$ with $q = 4.669$, [61]). For $\mu > \mu_\infty$ the orbit diagram exhibits an extremely complicated structure. Due to the errors unavoidably occurring during numerical calculations with real numbers, we cannot draw mathematically rigorous conclusions immediately from pictures like Figure 2.8. The latter nevertheless strongly suggests that there are many phenomena to observe and questions to answer. One may forever wonder about the amazing complexity of the logistic family's orbit diagram! (In fact, many substantial features are shared by a wide class of one-dimensional maps [16, 20, 39].)

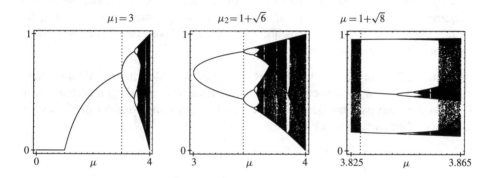

Figure 2.8. A few details from the orbit diagram of the logistic family

In the sequel we shall merely concentrate upon one single aspect of the orbit diagram. Figure 2.8 suggests the existence of a three-periodic orbit at $\mu \approx 3.835$. Taking into account that there are no other points visible in the orbit diagram for this specific parameter value, we expect this orbit to be attracting. This conjecture is heavily supported by looking at the graph of F_μ^3 (see Figure 2.9). An elementary but tedious calculation confirms that F_μ^3 undergoes a saddle-node bifurcation at $\mu = 1 + \sqrt{8} \approx 3.828$ ([58]). Consequently there is an attracting three-periodic orbit for parameters slightly larger than $1 + \sqrt{8}$. The reason for highlightening the appearance of period three stems from the following simple result.

Lemma 2.7. *Let $T : X \to X$ be continuous. If there exists a point with prime period three, then, for every $p \in \mathbb{N}$, there is a point with prime period p.*

Proof. Let $O^+(x_1) = \{x_1, x_2, x_3\}$ and assume $x_1 < x_2 < x_3$ without loss of generality. Set $I := [x_1, x_2]$ and $J := [x_2, x_3]$. Since $T(J) \supseteq J$ as well as $T(I) \supseteq J$ and $T^2(I) \supseteq I$ there

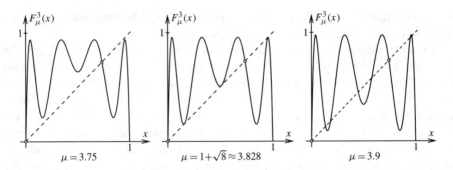

$\mu = 3.75$ $\mu = 1 + \sqrt{8} \approx 3.828$ $\mu = 3.9$

Figure 2.9. At $\mu = 1 + \sqrt{8}$ the map F_μ^3 undergoes a saddle-node bifurcation.

must be a fixed point of T in J and a point of prime period two in I, by virtue of the Intermediate Value Theorem. Let $p \geq 4$. Again by means of the Intermediate Value Theorem we can find a sequence of intervals $I_1, \ldots, I_{p-1}$ contained in J and $I_p \subseteq I$ such that $T(I_i) = I_{i+1}$ for $i = 1, \ldots, p-1$ and $T(I_p) = J$. Consequently there is a fixed point for T^p in I_1. It is easy to see that p is in fact the prime period of this point. $\square$

Lemma 2.7 clearly shows that there is something special about period three. Nevertheless it is only a special case of a more general (and amazing) theorem due to Sharkovsky which we shall quote without proof. To this end consider the following ordering of the natural numbers:

$$1 \lhd 2 \lhd 2^2 \lhd \ldots \lhd 2^r \lhd \ldots \lhd 2^s(2t-1) \lhd \ldots \lhd 2^s \cdot 5 \lhd 2^s \cdot 3 \lhd \ldots$$

$$\ldots \lhd 2^{s-1}(2t-1) \lhd \ldots \lhd 2^{s-1} \cdot 5 \lhd 2^{s-1} \cdot 3 \lhd \ldots$$

$$\ldots \lhd 2(2t-1) \lhd \ldots \lhd 2 \cdot 5 \lhd 2 \cdot 3 \lhd \ldots \lhd 2t-1 \lhd \ldots \lhd 5 \lhd 3 .$$

This is in fact a total ordering of $\mathbb{N}$ with 3 being the largest element. For example the numbers 1,...,12 line up as

$$1 \lhd 2 \lhd 4 \lhd 8 \lhd 12 \lhd 10 \lhd 6 \lhd 11 \lhd 9 \lhd 7 \lhd 5 \lhd 3 .$$

Theorem 2.8 (Sharkovsky). *Let $T : X \to X$ be continuous. Suppose T has a periodic point with prime period p. Then for each $q \lhd p$ the map T also has a periodic point of prime period q.*

There are elementary proofs of this theorem relying only on the two facts already used above: the Intermediate Value Theorem and the intimate relation between the usual ordering and topology of $\mathbb{R}$. Working out the details, however, necessitates some rather tricky arguments of combinatorial nature. We therefore refer to the literature ([20, 31]) for a proof, restricting ourselves to a few remarks here.

First of all, Sharkovsky's theorem is remarkable for the strange hierarchy it imposes on possible periods of one-dimensional maps. Since the structure of $\mathbb{R}$ and the continuity of T are crucial for the proof, one expects that none of the assumptions may be omitted. One can see from simple examples that this indeed is true. Finally, Sharkovsky's theorem is sharp: if $p \lhd q$ and $p \neq q$ there are maps having periodic points with prime period p but not having any point of prime period q (see Exercise 2.1).

Having thus pointed out the importance of period three, we are concerned with a serious question. At parameter values around $\mu = 3.835$ we observed an attracting three-periodic orbit for F_μ. The orbit diagram does not show any other periodic points for these values of μ although – by our above discussion – there should be quite a lot of them. So, where are all these points, and why cannot we see them in Figure 2.8? Clearly, the periodic points from the period-doubling scenario are still present. But the powers of two only constitute the lower end of the Sharkovsky ordering. What about the other periods like $5, 6, 7, 9 \ldots$? We have to develop some technical tools in order to satisfyingly answer these questions.

2.3   Facing deterministic chaos

Many of the following considerations are motivated by the preceding discussion of the logistic family, and we shall eventually return to this object of study. However, since most of the following pertains to wider fields of dynamical systems theory, we shall prefer a slightly more general framework. Accordingly, for the rest of this chapter, X will generally denote an arbitrary metric space rather than just an interval. If it matters, the specific form of the metric d on X will always be clear from the context.

2.3.1   Symbolic dynamical systems

First we introduce an important class of dynamical systems that might seem highly artificial at the beginning. Later we shall see that *symbolic dynamical systems* not only provide a useful tool for analysing complicated systems but also allow to emphasize some unifying features of deterministic and stochastic dynamics. Fix a natural number $l \geq 2$ and let I denote the finite set $\{1, \ldots, l\}$ endowed with the discrete topology. The set of sequences in I,

$$\Sigma_l := I^{\mathbb{N}_0} := \left\{ (x_k)_{k \in \mathbb{N}_0} : x_k \in I \text{ for all } k \in \mathbb{N}_0 \right\}$$

is given a metric structure by defining a distance between two sequences according to

$$d\left((x_k)_{k \in \mathbb{N}_0}, (y_k)_{k \in \mathbb{N}_0}\right) := \begin{cases} 0 & \text{if } (x_k) = (y_k), \\ 2^{-\min\{k \in \mathbb{N}_0 : x_k \neq y_k\}} & \text{if } (x_k) \neq (y_k). \end{cases} \tag{2.3}$$

There are of course many other (non-equivalent) metrics inducing the product topology on Σ_l and making Σ_l a compact space. The metric chosen here is especially simple: two sequences are apart at most 2^{-n} precisely if their first n entries agree; the further two sequences agree, the nearer they are. For us, the most important map acting on Σ_l is the *(left) shift* $\sigma : \Sigma_l \to \Sigma_l$ defined by

$$\sigma\big((x_k)_{k\in\mathbb{N}_0}\big) := (x_{k+1})_{k\in\mathbb{N}_0} = (x_1, x_2, \dots).$$

This map simply shifts all sequences to the left by one step, thereby cancelling the entry x_0. Since $d\big(\sigma(x), \sigma(y)\big) \leq 2d(x, y)$ the shift map clearly is continuous. For notational simplicity we have refrained from ornamenting σ by a subscript l. Since there is a natural embedding $\Sigma_l \hookrightarrow \Sigma_m$ for $l \leq m$, writing σ at all times will cause no confusion.

The dynamics of σ might look somehow artificial at first glance. Later we shall revise this opinion. At the moment, however, we are going to define a host of interesting – though even more artificial – subsystems of (Σ_l, σ). For this purpose let A be an $l \times l$ matrix having as entries only zeros and ones. A subset $\Sigma_{l,A}$ of Σ_l is defined by

$$\Sigma_{l,A} := \big\{(x_k) \in \Sigma_l : a_{x_k x_{k+1}} = 1 \text{ for all } k \in \mathbb{N}_0\big\}.$$

One may express $a_{ij} = 1$ as "state j may follow state i". The entries of A therefore determine whether an element of Σ_l belongs to $\Sigma_{l,A}$. Obviously $\Sigma_{l,A} = \Sigma_l$ if $a_{ij} = 1$ for all $(i, j) \in I^2$. On the other hand $\Sigma_{l,A}$ may also be empty. This is the case if and only if the matrix is nilpotent (i.e. $A^n = 0$ for some $n \in \mathbb{N}$). In the sequel we shall more restrictively assume that A does not contain any row consisting of zeros only – states corresponding to such rows may obviously be removed.

Lemma 2.9. $\Sigma_{l,A}$ *is a closed subset of* Σ_l, *and* $\sigma(\Sigma_{l,A}) \subseteq \Sigma_{l,A}$.

Proof. The complement of $\Sigma_{l,A}$ in Σ_l is open since non-admissible sequences remain so when being perturbed "far at the right". It is obvious that $(x_{k+1})_{k\in\mathbb{N}_0}$ belongs to $\Sigma_{l,A}$ if $(x_k)_{k\in\mathbb{N}_0}$ does. $\square$

As a consequence, we may separately investigate the dynamics of the restricted shift $\sigma_A := \sigma|_{\Sigma_{l,A}}$ on $\Sigma_{l,A}$. The system $(\Sigma_{l,A}, \sigma_A)$ is usually called a *subshift of finite type* or a *topological Markov chain*. It is, however, not until Chapter Four that we shall appreciate the latter notion. More generally, for any closed subset Σ of Σ_l that satisfies $\sigma(\Sigma) \subseteq \Sigma$ the system $(\Sigma, \sigma|_\Sigma)$ is called a *symbolic dynamical system*. Subshifts of finite type therefore constitute the simplest (and probably most important) class of symbolic dynamical systems.

Historically symbolic dynamics has been invented to provide a tool for roughly describing and thereby understanding more complicated phenomena. Originating from the work of Morse and Hedlund this discipline has gone through a rapid development. Today it is an independent field of mathematics giving theoretical basis to such applied disciplines as cryptology and data-compression ([36]). Our use of symbolic dynamics will be close to the original idea which may be sketched as follows.

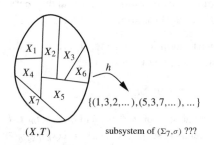

$\{(1,3,2,\dots),(5,3,7,\dots),\dots\}$

(X,T) subsystem of (Σ_7,σ) ???

Figure 2.10. Subshifts may arise from roughly recording orbits.

Suppose a (complicated) continuous map T on a space X to be given. In order to obtain a first insight into the dynamics we divide X into a finite number of closed pieces X_i and investigate the history of individual points relative to the decomposition $(X_i)_{i \in I}$. To each point $x \in X$ we can assign a sequence $h(x)$ in Σ_I, the n-th entry of which tells us in which of the pieces $T^n(x)$ is to be found. Therefore $h(x)$ gives a protocol (an itinerary) of the entire journey of x under iteration of T, and $\sigma\big(h(x)\big) = h\big(T(x)\big)$ (cf. Figure 2.10). Unfortunately, in practice there may (and generally will) arise several problems from this procedure: often h will fail to be continuous or even sufficiently well defined if for example some points admit a representation by more than one itinerary. Even if these difficulties do not occur, $h(X)$ need not constitute a subshift of finite type. However, in many situations this blunt approach turns out to be surprisingly successful. Our further investigations will give strong support to this statement.

What can be said about the dynamics of $(\Sigma_{I,A}, \sigma_A)$? Obviously, a point x in $\Sigma_{I,A}$ is p-periodic precisely if $x_{k+p} = x_k$ for all $k \in \mathbb{N}_0$, i.e., the entries x_k are p-periodic in which case we write $x = (x_0, \dots, x_{p-1})$. The number of fixed points in $\Sigma_{I,A}$ is easily seen to equal the trace of A, or $\#\operatorname{Fix}\sigma_A = \sum_{i \in I} a_{ii} = \operatorname{tr} A$. This fact generalizes.

Lemma 2.10. $\#\operatorname{Per}_p \sigma_A = \operatorname{tr} A^p$ for all $p \in \mathbb{N}$.

Proof. It is obvious that the p-periodic sequence $(i_0, i_1, \dots, i_{p-1}, i_0, \dots)$ is an element of $\Sigma_{I,A}$ if and only if $a_{i_0 i_1} \cdot \dots \cdot a_{i_{p-1} i_0} = 1$. The total number of p-periodic points beginning with i_0 therefore equals $(A^p)_{i_0 i_0}$. $\square$

The cardinality of $\operatorname{Per}_p \sigma_A$ therefore grows exponentially with p if A has at least one eigenvalue with modulus greater than one. Many periodic points may then be found in $\Sigma_{I,A}$. Take as an example

$$A := \begin{pmatrix} 1 & 1 & 1 \\ 0 & 1 & 1 \\ 0 & 1 & 1 \end{pmatrix}.$$

The points in $\Sigma_{3,A}$ are easily seen to be either $(1,1,\dots)$ or to consist of a finite number of 1's followed by an arbitrary sequence of 2's and 3's. Therefore $\#\operatorname{Per}_p \sigma_A = 1 + 2^p$ indeed grows exponentially. Nevertheless the non-periodic points beginning with a finite block of 1's do not have any periodic points in their vicinity. This is clearly due to the fact that 2 and 3 may follow 1 but not vice versa. In order to find periodic points *everywhere* in $\Sigma_{I,A}$ we must rule out this possibility.

Definiton 2.11. An $l \times l$ matrix B, the elements of which are non-negative real numbers, is *irreducible* if for any pair (i, j) of indices $(B^N)_{ij} > 0$ for some $N = N(i, j) \in \mathbb{N}$.

The matrix A from above obviously is not irreducible. In general, we have the following simple fact, the converse of which, however, need not be true.

Theorem 2.12. *If A is irreducible then* Per σ_A *is dense in* $\Sigma_{l,A}$.

Proof. We take the opportunity to introduce a useful notation. For $m \in \mathbb{N}_0$ and $i_0, \dots, i_m \in I$ we define the *cylinder* $[i_0, \dots, i_m]$ to be the (possibly empty) set $[i_0, \dots, i_m] := \{(x_k)_{k \in \mathbb{N}_0} \in \Sigma_{l,A} : x_0 = i_0, \dots, x_m = i_m\}$. Cylinders are open sets and form a basis of $\Sigma_{l,A}$'s topology. In order to prove the theorem it therefore suffices to show that every non-empty cylinder intersects Per σ_A. Since A is irreducible, there exists a periodic point of period $m + N(i_m, i_0)$ in $[i_0, \dots, i_m]$ and therefore Per $\sigma_A \cap [i_0, \dots, i_m] \neq \emptyset$. $\square$

Another important question is whether one can split the system $(\Sigma_{l,A}, \sigma_A)$ into smaller pieces that may be studied separately. In that case our analysis of the whole system would simply consist of analysing the smaller systems and putting together the results. There are several ways of imposing indecomposability conditions on dynamical systems. We shall come across a few of them in this book. For the time being we adopt a topological version.

Definiton 2.13. Let T be a continuous map on a metric space X. We call T *topologically transitive* if for every pair U, V of non-empty open sets $T^{-n}(U) \cap V \neq \emptyset$ for some $n \in \mathbb{N}$.

This definition clearly formalizes indecomposability: if T is not topologically transitive then there are open sets U_0, V_0 such that no point in V_0 will ever be mapped to U_0. In general it may be difficult to decide whether a given map is topologically transitive or not. When looking at the subshift $(\Sigma_{l,A}, \sigma_A)$ the answer is simple.

Theorem 2.14. $(\Sigma_{l,A}, \sigma_A)$ *is topologically transitive if and only if A is irreducible.*

Proof. Again it suffices to study cylinders. Let A be irreducible and consider two cylinders $[i_0, \dots, i_k]$ and $[j_0, \dots, j_m]$. Straightforwardly one has $\sigma_A^{-(N(j_m, i_0)+m)}([i_0, \dots, i_k]) \cap [j_0, \dots, j_m] \neq \emptyset$. Conversely, if σ_A is topologically transitive, then for any $(i, j) \in I^2$ we can find a number $N(i, j)$ such that the intersection $\sigma_A^{-N(i,j)}([j]) \cap [i]$ is not empty, and hence $(A^{N(i,j)})_{ij} > 0$. $\square$

We finish our analysis of subshifts by pointing out another property which seems trivial in this context. Assume that A is irreducible and that there exists an aperiodic point x in $\Sigma_{l,A}$. Given $\varepsilon > 0$ there is a periodic point y with $d(x, y) < \varepsilon$. But since $x \neq y$, the iterates of these two points will at some time have the distance one no matter how small ε has been chosen. In the long run, nearby points diverge completely under the action of σ_A (at least for a while). It is this effect which is commonly considered essential for a dynamical system to be chaotic. Our formalization reads as follows.

Definiton 2.15. A continuous map T on a metric space X exhibits *sensitive dependence on initial conditions*, if there exists a real number $\delta > 0$ with the following property: for every $x \in X$ and $\varepsilon > 0$ we can find a point y with $d(x, y) < \varepsilon$ but $d\big(T^N(x), T^N(y)\big) \geq \delta$ for some $N \in \mathbb{N}$.

Although this definition looks rather technical, it formalizes our experience with the system $(\Sigma_{l,A}, \sigma_A)$. For the latter we may take $\delta = 1$, which is in fact the diameter of the whole space $\Sigma_{l,A}$; different points eventually diverge as far as possible. Observe that for these particular systems the minimal number of iterates necessary to disperse two distinct points as far apart as possible is easily calculated: indeed, if $x \neq y$ then

$$d\big(\sigma^N(x), \sigma^N(y)\big) = 1 \quad \text{with } N := -\log_2 d(x, y).$$

Setting $d(x, y) = \varepsilon$ thus yields $N = -\log_2 \varepsilon$, a quantity that grows rather slowly with decreasing ε. When struggling with a real-world problem, a result like this clearly is of great practical relevance because it concisely describes how fast small perturbations are magnified. Although we generally expect N to grow with decreasing ε, incorporating a specific growth-rate into the above definition would have been too restrictive. Nevertheless, we shall repeatedly see that the rather slow growth of N, which at first glance seems to be a special feature of $(\Sigma_{l,A}, \sigma_A)$, substantially carries over to more complicated (and realistic) systems. Due to this fact our ability to predict the future fate of individual points for such systems turns out to be tremendously limited.

For us, subshifts of finite type will mainly serve as tools for modelling other systems. We therefore have to define in which way a dynamical system can be understood as a model of another one. It comes as no surprise that there are many sensible notions formalizing the equivalence of dynamical systems in one sense or the other (see for instance the discussion in [30, 31]). The notion which we adopt is restrictive but simple.

Definiton 2.16. Let T, S be continuous maps on metric spaces X, Y, respectively. Assume that there is a continuous map $h : X \to Y$ satisfying $h\big(T(x)\big) = S\big(h(x)\big)$ for all $x \in X$, i.e. the diagram

$$
\begin{array}{ccc}
X & \xrightarrow{\;T\;} & X \\
\downarrow{\scriptstyle h} & & \downarrow{\scriptstyle h} \\
Y & \xrightarrow{\;S\;} & Y
\end{array}
$$

commutes. If h is a homeomorphism, (X, T) and (Y, S) are said to be *topologically conjugate*. If h is merely onto, (X, T) is said to be *semi-conjugate* onto (Y, S). In the latter case we call (Y, S) a *factor* of (X, T).

Obviously, semi-conjugacy constitutes a much weaker relation between dynamical systems than conjugacy does. The latter not only requires the spaces X and Y to be homeomorphic, but also intimately connects basic dynamical features in these spaces. For example, there is a one-to-one correspondence of orbits, $h\big(O^+(x)\big) = O^+\big(h(x)\big)$

for every $x \in X$, and all periods are preserved. Informally, the homeomorphism h may be interpreted as a mere introduction of new coordinates in the space X. Accordingly, $S = h \circ T \circ h^{-1}$ should be considered as the new form of T under these new coordinates. Topologically conjugate systems therefore dynamically look the same in a very strict sense. Deciding whether two systems are conjugate or not may consequently be a rather difficult task. It is mainly this fact that motivates the weaker notion of semi-conjugacy. While conjugacy defines an equivalence among dynamical systems, being semi-conjugate merely establishes a reflexive and transitive relation. Existence of a semi-conjugacy generally does not imply a one-to-one correspondence of orbits nor a preservation of periods. The system (Y, S) may therefore be regarded a simplified model of (X, T). Finding such a model that is neither too crude nor too complicated usually is a main point in this context.

Since we intend to use symbolic dynamics as a source of models, the following question has to be settled: Up to what extent do the dynamical properties of the model reflect the properties of the original system? A partial answer is contained in the following

Proposition 2.17. *Let (X, T) be semi-conjugate onto (Y, S). If T is topologically transitive then so is S. If Per T is dense in X, then Per S is dense in Y, too.*

Observe that sensitive dependence on initial conditions – the third dynamical feature discussed above – is generally not preserved, not even by conjugacies. Since we feel that, nevertheless, sensitive dependence on initial conditions is practically important, we shall circumvent this disturbing fact by means of the following theorem. Notice that X consists of just one single periodic orbit if T is topologically transitive and X is finite.

Theorem 2.18. *Let T be a topologically transitive continuous map on an infinite metric space X. Assume further that Per T is dense in X. Then T exhibits sensitive dependence on initial conditions.*

Proof. In a first step we shall show that there exists a universal number $\delta > 0$ with the following property: for every $x \in X$ there is a periodic point a_x with $d(x, O^+(a_x)) \geq 4\delta$. Since X is infinite there are two periodic points a_1, a_2 not belonging to the same orbit. Take $\delta := \frac{1}{8} \inf_{m,n \in \mathbb{N}_0} d(T^m(a_1), T^n(a_2))$. By the triangle inequality at least one of the points a_1, a_2 may be chosen as a_x, and therefore δ has the desired property.

The second step now consists in showing that this very δ may be taken to establish sensitive dependence on initial conditions. Take $x \in X$ and, without loss of generality, $0 < \varepsilon < \delta$. Since periodic points are dense, there is a periodic point b with period p and $d(x, b) < \varepsilon$. On the other hand there is a periodic point a_x with $d(x, O^+(a_x)) \geq 4\delta$. Furthermore T is continuous, so $d(y, a_x) < \mu$ implies $d(T^i(y), T^i(a_x)) \leq \delta$ for sufficiently small $\mu < \delta$ and all $i \in \{0, \ldots, p\}$. Finally, due to topological transitivity there is a point y and a natural number M with $d(y, x) < \varepsilon$ and $d(T^M(y), a_x) < \mu$. Setting $N := \lceil \frac{M}{p} \rceil p$ and putting pieces together, we obtain

$$4\delta \leq d(x, T^{N-M}(a_x)) \leq d(x, T^N(b)) + d(T^N(b), T^N(y)) + d(T^N(y), T^{N-M}(a_x))$$
$$\leq \delta + d(T^N(b), T^N(y)) + d(T^{N-M}(T^M(y)), T^{N-M}(a_x)).$$

Therefore $2\delta < d\left(T^N(b), T^N(y)\right) \le d\left(T^N(b), T^N(x)\right) + d\left(T^N(x), T^N(y)\right)$ but on the other hand $d(x, b) < \varepsilon$ as well as $d(x, y) < \varepsilon$. $\square$

Neglecting trivial cases, topological transitivity and dense periodic orbits together imply quite complicated dynamics. It is easy to see that the conclusion of the above theorem may fail if any of the assumptions is omitted.

2.3.2   The emergence of chaos

We have now at our disposal all concepts necessary for rigorously analysing the logistic map F_μ at $\mu \approx 3.84$. To be more specific, we fix $\mu = 3.839$ for the following considerations and drop the subscript μ whenever possible; especially we write F instead of $F_\mu = F_{3.839}$. Since $F^n(x) \to -\infty$ if $x \in \mathbb{R}\setminus[0, 1]$, the more interesting things are sure to happen in the unit interval. Looking in detail at the graph of F^3 (Figure 2.11) we find a stable as well as an unstable three-periodic orbit $\{a_1, a_2, a_3\}$ and $\{b_1, b_2, b_3\}$, respectively. A numerical calculation yields

$$a_1 = 0.150, \quad a_2 = 0.489, \quad a_3 = 0.959, \quad (F^3)'(a_i) = -0.788;$$
$$b_1 = 0.169, \quad b_2 = 0.539, \quad b_3 = 0.954, \quad (F^3)'(b_i) = 2.668;$$

though we shall not explicitly need these numbers for the argument to follow. Three further points $\hat{b}_i$ are defined as the nearest to b_i with $F^3(\hat{b}_i) = b_i$ (cf. Figure 2.11). With these points we label four compact intervals as

$$I_0 := [0, \hat{b}_1], \quad I_1 := [b_1, \hat{b}_2],$$
$$I_2 := [b_2, \hat{b}_3], \quad I_3 := [\hat{b}_3, 1]; \tag{2.4}$$

in addition we set $A := \bigcup_{i=0,\ldots,3} I_i$ and $B := [0, 1]\setminus A$. The following relations between forward images of these sets may be checked without too much effort:

$$F(I_0) \cap A = I_0 \cup I_1, \quad F(I_1) = I_2,$$
$$F(I_2) \cap A = I_1 \cup I_2, \quad F(I_3) = I_0.$$

On the other hand $F(B) \subseteq B$, and points in B are attracted towards the orbit $\{a_1, a_2, a_3\}$ under iteration of F. Therefore B provides an example of what is usually called a *trapping region*: a point entering this set will never leave it again. Clearly, B does not contain any periodic point different from the a_i. Within the interval I_0 we find the obvious fixed point

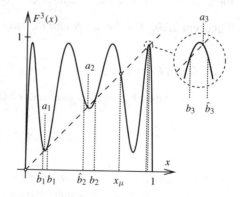

Figure 2.11. The endpoints of the intervals I_i as defined by (2.4)

at the origin. If $x \in I_0 \setminus \{0\}$, then for some n either $F^n(x) \in B$ or $F^n(x) \in I_1 \cup I_2$. In either case x will never return to I_0. Since a similar argument applies to I_3, we are led to conclude that any periodic point different from the obvious ones is necessarily located in the set

$$\Lambda := \left\{ x \in [0, 1] : O^+(x) \subseteq I_1 \cup I_2 \right\} = \bigcap_{k \in \mathbb{N}_0} F^{-k}(I_1 \cup I_2). \qquad (2.5)$$

As Λ is constructed by means of a nested sequence of non-empty compact sets it is certainly not empty. Our definition of Λ, however, does not make transparent any details of the set's structure. When looking at Figure 2.12 one feels reminiscent of the construction of the classical Cantor middle-thirds set (see also Figure 5.7): at stage n a number of small subintervals is removed consisting of those points that irreversibly leave $I_1 \cup I_2$ under the n-th iteration of F. We thus expect Λ to be a rather dusty set.

Definiton 2.19. A compact set $C \subseteq \mathbb{R}$ is called a *Cantor set* if it is perfect (i.e., every point in C is an accumulation point of C) and totally disconnected (i.e. containing no interval).

As Figure 2.12 suggests, Λ will probably be a Cantor set. Before rigorously proving this, let us have a quick look at the dynamics on Λ. Since $F(\Lambda) \subseteq \Lambda$, we can study F on Λ without taking notice of the ambient space. To each $x \in \Lambda$ we may assign a sequence $(y_k)_{k \in \mathbb{N}_0} \in \Sigma_2$ according to

$$y_k = i \quad \text{if and only if } F^k(x) \in I_i \ (i = 1, 2),$$

thereby defining a map $h : \Lambda \to \Sigma_2$ via $h(x) := (y_k)_{k \in \mathbb{N}_0}$. There is an obvious restriction on the sequences generated this way: an entry 1 must always be followed by an entry 2 because $F(I_1 \cap \Lambda) \subseteq I_2 \cap \Lambda$. Since we feel that there are no other restrictions, we may hope that symbolic dynamics will give us some insight. In fact, the structure of Λ and the dynamics of F on that set can be unravelled completely.

Theorem 2.20. *For the set Λ and the map $h : \Lambda \to \Sigma_2$ defined above the following holds:*

(i) Λ *is a Cantor set;*

(ii) h *defines a topological conjugacy between $(\Lambda, F|_\Lambda)$ and $(\Sigma_{2,A}, \sigma_A)$ where*

$$A := \begin{pmatrix} 0 & 1 \\ 1 & 1 \end{pmatrix}.$$

Proof. We start with a preliminary observation. The set of points x in $I_1 \cup I_2$ where $|(F^3)'(x)| \le 1$ consists of three intervals. It is straightforward to show that each of these intervals is mapped into B by F^3. Consequently $|(F^3)'(x)| \ge \alpha > 1$ for all $x \in \Lambda$. Bearing this in mind we see that for any $n \in \mathbb{N}$ the intersection $\bigcap_{k=0}^n F^{-k}(I_1 \cup I_2)$ consists of a finite number of intervals with diameter eventually decreasing to zero. Since the endpoints of these intervals clearly belong to Λ, every point in the latter set is an accumulation point. We already know that Λ is non-empty and compact. By virtue of the Mean Value Theorem it cannot contain any interval, which implies (i).

Now take $x, y \in \Lambda$; if $h(x) = h(y)$ we must have $[x, y] \subseteq \Lambda$ because $F^n(x)$ and $F^n(y)$ always lie on the same side of $\frac{1}{2}$. Therefore $x = y$, and h is one-to-one. Proving the continuity of h is straightforward: given $\varepsilon > 0$, take $n \in \mathbb{N}$ with $2^{-n} < \varepsilon$ and let $\delta(\varepsilon) > 0$ be half the minimal mutual distance between the intervals building the set $\bigcap_{k=0}^{n} F^{-k}(I_1 \cup I_2)$; if $|x-y| < \delta(\varepsilon)$ then the first n digits of $h(x)$ and $h(y)$ agree, implying $d(h(x), h(y)) \leq 2^{-n} < \varepsilon$. To show that h is onto, let a sequence $(y_k)_{k \in \mathbb{N}_0} \in \Sigma_{2,A}$ be given. The intervals $J_n := I_{y_0 y_1 \ldots y_n} := \bigcap_{k=0}^{n} F^{-k}(I_{y_k})$ form a nested sequence of non-empty compact sets (see Figure 2.12). From this we obtain $\bigcap_{n=0}^{\infty} J_n \neq \emptyset$. Any x from this intersection satisfies $h(x) = (y_k)_{k \in \mathbb{N}_0}$. Due

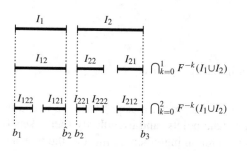

$$\bigcap_{k=0}^{1} F^{-k}(I_1 \cup I_2)$$

$$\bigcap_{k=0}^{2} F^{-k}(I_1 \cup I_2)$$

Figure 2.12. The very first steps in the construction of the set Λ

to compactness h is in fact a homeomorphism. Since $h(x) = (y_k)_{k \in \mathbb{N}}$ is equivalent to $x \in \bigcap_{k=0}^{\infty} F^{-k}(I_{y_k})$, we finally find $F(x) \in \bigcap_{k=1}^{\infty} F^{-(k-1)}(I_{y_k}) = h^{-1} \circ \sigma_A \circ h(\{x\})$, showing that $h \circ F = \sigma_A \circ h$. $\square$

We have thus arrived at a complete understanding of F_μ at $\mu = 3.839$: any point in $]0,1[$ either is eventually mapped into the Cantor set Λ or is attracted by the unique stable three-periodic orbit. On Λ the dynamical behaviour of F_μ is equivalent to a subshift of finite type in a very strong sense. Recall that we originally searched for all the periodic points guaranteed by Sharkovsky's theorem. By now we know that – except for 0 and $\{a_1, a_2, a_3\}$ – they must be looked for in the Cantor set Λ. Can we find them there? By means of our symbolic model $(\Sigma_{2,A}, \sigma_A)$ it is easy to give an affirmative answer to this question. Obviously $\text{Per}_p \sigma_A \neq \emptyset$ for every $p \in \mathbb{N}$. Moreover, there are periodic points with *prime* period p for every natural p. It is instructive to have a closer look at the cases $p = 1, 2, 3$. We find

$$\# \text{Fix} \, \sigma_A = 1 \, ; \qquad \text{Fix} \, \sigma_A = \{(\overline{2})\} = \{h(x_\mu)\} \, ,$$

$$\# \text{Per}_2 \, \sigma_A = 3 \, ; \qquad \text{Per}_2 \, \sigma_A \backslash \text{Fix} \, \sigma_A = \{(\overline{1,2}), (\overline{2,1})\} \, ,$$

$$\# \text{Per}_3 \, \sigma_A = 4 \, ; \qquad \text{Per}_3 \, \sigma_A \backslash \text{Fix} \, \sigma_A = \{(\overline{1,2,2}), (\overline{2,2,1}), (\overline{2,1,2})\}$$
$$= \{h(b_1), h(b_2), h(b_3)\} \, .$$

The matrix A clearly is irreducible implying that σ_A is topologically transitive and $\text{Per} \, \sigma_A$ is dense in $\Sigma_{2,A}$. Since Λ is infinite we conclude that F exhibits sensitive dependence on initial conditions. We therefore are aware of a dynamical complexity for the simple logistic map that we would hardly have dreamed of in the beginning. It may be suspected by now that we could find even more complexity when switching to other values of the parameter μ. Although this surmise is true to a large extent we shall not pursue that theme further here but only take a short look at the case $\mu \geq 4$. Before doing so we formalize the complicated dynamics we have observed for σ_A and $F|_\Lambda$.

Definiton 2.21. A continuous map T on a metric space X is said to be *chaotic* if it has the following properties:

(i) T is topologically transitive;

(ii) Per T is dense in X;

(iii) T exhibits sensitive dependence on initial conditions.

There are, of course, other ways of assigning a precise meaning to the term *chaos*. Coexistence of regularity (here: many periodic points) and irregularity (here: sensitive dependence on initial conditions) is a common ingredient to most of the relevant notions. Definition 2.21 originally is due to [20]. It has been chosen mainly for its simplicity. In addition, its topological flavour makes it suitable for the approach taken in this chapter. Observe that a map acting on a finite space is certainly not chaotic. Due to our earlier results (Theorem 2.18) property (iii) in the above definition is always implied by (i) and (ii) and could thus have been omitted. By incorporating (iii) nevertheless, we want to highlight the importance of sensitive dependence on initial conditions for our understanding of chaos.

As indicated before, other formalizations of the term *chaos* have been conceived. Although we shall exclusively use Definition 2.21 throughout this book let us mention another, quite popular version. In [11] a continuous map T on a metric space X is termed chaotic if $T^m(X_0) \subseteq X_0$ for some closed subset $X_0 \subseteq X$ and $m \in \mathbb{N}$, and the system $(X_0, T^m|_{X_0})$ semi-conjugates onto (Σ_2, σ). This comes closest to the original denomination coined in the realm of one-dimensional maps during the nineteen-seventies. It is mainly for the clear geometric picture it evokes in higher dimensions and general metric spaces that we prefer Definition 2.21. Regarding the examples discussed so far both notions are equivalent.

Let us now complete our analysis of the logistic map by considering the cases $\mu \geq 4$. Postponing the special case $\mu = 4$ for a little while we shall for the time being assume $\mu > 4$. Contrary to the situations studied before there is now the possibility for points in the unit interval to leave the latter and wander to infinity. Those points staying in $[0,1]$ for at least one step are located in the disjoint union of two intervals $F_\mu^{-1}([0, 1]) =: I_1 \cup I_2$ (cf. Figure 2.13). If we again set $\Lambda_\mu := \{x \in [0, 1] : O^+(x) \subseteq I_1 \cup I_2\}$, the further analysis proceeds in much the same way as for $\mu = 3.839$. There are no restrictions whatsoever on the sequences obtained by coding the journeys of the points in Λ_μ. We therefore expect the dynamics of F_μ on Λ_μ to be topologically conjugate to the full shift (Σ_2, σ). That this is indeed the case can be proved in an analogous way as before – if one makes use of the following fact which seems to be evident but is quite difficult to establish ([39]). Given $\mu > 4$ there exists a natural number $N(\mu)$ such that $|(F_\mu^{N(\mu)})'(x)| > 1$ for all $x \in \Lambda_\mu$. (Some elementary calculus shows that one can take $N = 2$ if $\mu \geq 4.035$, and $N = 1$ if $\mu > 2 + \sqrt{5} \approx 4.236$.) Taking this result for granted, Λ_μ is easily shown to be a Cantor set, and F_μ clearly acts chaotically on Λ_μ. We have thus found quite a lot of chaotic members of the logistic family.

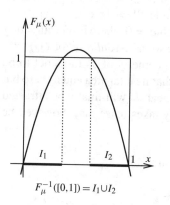

$$F_\mu^{-1}([0,1]) = I_1 \cup I_2$$

Figure 2.13. Points may leave the unit interval if $\mu > 4$.

The question arises as to what extent our findings are of any practical significance: after all, the Cantor sets we found have total length zero when considered as subsets of $\mathbb{R}$. One could argue that the probability of a randomly chosen point to be in such a Cantor set equals zero. Consequently we will never observe the chaotic behaviour in reality. Numerical experiments seem to be in favour of this argument: no matter which point in the open unit interval is given to the computer, the iterates always either tend to infinity (if $\mu > 4$) or are attracted by the three-periodic orbit $\{a_1, a_2, a_3\}$ (if $\mu = 3.839$). In fact, without having recourse to special tricks we will not see any of the periodic orbits located within the Cantor set for a simple reason: all of them are unstable! Although the chaotic motion we found is thus not visible directly, it may nevertheless cause effects that are visible. For example, points near the chaotic set may considerably differ in their long-time behaviour. Due to continuity and sensitive dependence on initial conditions, such points will probably diverge to completely different parts of the space. A rather tame example of this phenomenon is depicted in Figure 2.14. Observe that we have $|F_\mu^n(x_1) - F_\mu^n(x_2)| \not\to 0$, although both points x_1 and x_2 are finally attracted by the same periodic orbit and their initial distance is rather small.

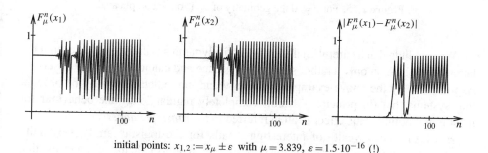

initial points: $x_{1,2} := x_\mu \pm \varepsilon$ with $\mu = 3.839$, $\varepsilon = 1.5 \cdot 10^{-16}$ (!)

Figure 2.14. Asymptotically, the orbits of $x_{1,2}$ are both three-periodic; nevertheless $|F_\mu^n(x_1) - F_\mu^n(x_2)| \not\to 0$ as $n \to \infty$.

If more than one long-time scenario exists, the situation can be even worse: the ultimate fate of points near the chaotic set turns out to be completely unpredictable.

Again we give just a simple illustration of this effect which may in fact be observed for many physical systems. Consider the family Q_c dealt with earlier, but allow the argument as well as the parameter c to be complex numbers. If $c = 0$, the resulting dynamics in $\mathbb{C}$ is easily analysed: we find $Q_0^n(z) \to 0$ if $|z| < 1$ and $Q_0^n(z) \to \infty$ if $|z| > 1$. Furthermore, Q_0 is chaotic on the unit circle $S^1 = \{z \in \mathbb{C} : |z| = 1\}$. This situation substantially persists for sufficiently small $c \neq 0$: there is one attracting fixed point z_c near the origin and a compact set J_c (the so-called *Julia set* of Q_c) with $Q_c(J_c) = J_c$ on which Q_c is chaotic. If $c \neq 0$, the geometry of J_c turns out to be rather complicated. Again J_c is a *thin* set in the sense that its total area equals zero but it is definitely not a Cantor set for small $|c|$. A point near J_c will finally be attracted by z_c or infinity – which of these possibilities actually takes place may, however, be undecidable in practice (see Figure 2.15).

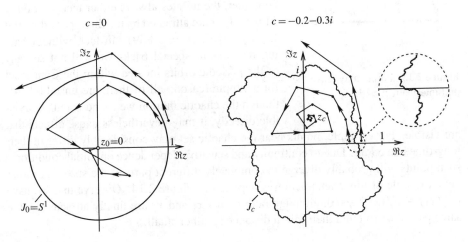

Figure 2.15. For $c \neq 0$ the geometry of J_c is quite complicated.

We conclude that a careful analysis of chaotic dynamics is indispensable even if the latter takes place on only a rather small part of space and cannot be directly observed. As indicated by the above examples, erraticity and unpredictability may be brought into systems that (in principle) exhibit completely regular long-time behaviour for most initial points. This effect is often referred to as *transient chaos*.

Having found a wealth of interesting details for the logistic family, we finally consider the case $\mu = 4$. The analysis is simple: a short calculation confirms that $h : S^1 \to [0, 1]$ with $h(z) := (\Im z)^2$ satisfies $h \circ Q_0 = F_4 \circ h$. We have thus found a semi-conjugacy of a chaotic system onto $([0, 1], F_4)$. Observe that contrary to the cases studied so far, F_4 is chaotic on the whole unit interval. Already in the Introduction we got an impression of F_4's extreme sensitivity on initial conditions (recall Figure 1.14).

Another, more ingenious way of demonstrating the chaoticity of F_4 would have been as follows. Define a homeomorphism of the unit interval by $h(x) := \frac{1}{2}(1 - \cos \pi x) =$

$\left(\sin(\pi x/2)\right)^2$ and observe that

$$T(x) := h^{-1} \circ F_4 \circ h(x) = \begin{cases} 2x & \text{if } 0 \leq x \leq \frac{1}{2}, \\ 2 - 2x & \text{if } \frac{1}{2} < x \leq 1; \end{cases} \quad (2.6)$$

the map T is – by definition – topologically conjugate to F_4 and easily seen to be chaotic on $[0,1]$.

2.3.3 Newton's method for polynomials: a case study

As an application of the terminology and techniques discussed so far let us now have a look at a few dynamical systems arising from Newton's method. More specifically, assume we are given a real polynomial p of degree higher than one,

$$p(x) = a_l x^l + a_{l-1} x^{l-1} + \ldots + a_1 x + a_0 \quad \text{with } a_0, a_1, \ldots, a_l \in \mathbb{R}, a_l \neq 0, \quad (2.7)$$

where $l \geq 2$. Recall that Newton's method for finding roots of p, i.e. solutions of $p(x) = 0$, consists in choosing an initial point $x_0 \in \mathbb{R}$ and then hoping for the recursion

$$x_{n+1} := x_n - \frac{p(x_n)}{p'(x_n)} \quad (n \in \mathbb{N}_0) \quad (2.8)$$

to converge. From a dynamical systems point of view Newton's method thus consists in iterating the associated map

$$N_p(x) := x - \frac{p(x)}{p'(x)}. \quad (2.9)$$

It will become clear in a moment that the limit of (2.8) yields a root of p indeed – provided this limit exists at all. As we shall find ourselves compelled to note, the dynamics of (2.9) may be very complicated.

In principle we would like to regard N_p as a map on the real line but evidently there is a slight problem: N_p is a rational map (i.e. a quotient of two polynomials) and may thus have poles. This difficulty arises at *critical points* of p, that is, at points $\bar{x}$ where $p'(\bar{x}) = 0$. If such a critical point is also a root of p, we know from elementary calculus that p may be given the form $p(x) = (x - \bar{x})^m q(x)$ for some $m \geq 2$ and an appropriate polynomial q with $q(\bar{x}) \neq 0$. In this case N_p may be continuously extended to $\bar{x}$ by $N_p(\bar{x}) := \bar{x}$. Critical points which are *not* zeros of p clearly cannot be treated in this way. As will be outlined below this difficulty, however, is not a serious one and may elegantly be circumvented. We will simply study the dynamics of N_p on the unit circle (where we shall write N_{p,S^1} instead) rather than on the real line.

Recall that we denote by S^1 the set of complex numbers of modulus one. Define now a map Φ from the punctured unit circle $S^1 \backslash \{1\} \subseteq \mathbb{C}$ to the real numbers according to

$$\Phi : \begin{cases} S^1 \backslash \{1\} & \to & \mathbb{R}, \\ z & \mapsto & \dfrac{\Im z}{1 - \Re z}. \end{cases}$$

In geometric terms, Φ describes the *stereographic projection* from $S^1\backslash\{1\}$ onto $\mathbb{R}$ (see Figure 2.16). Φ is a smooth invertible map wherever it is defined, its inverse being

$$\Phi^{-1}: \begin{cases} \mathbb{R} & \to & S^1\backslash\{1\}, \\ x & \mapsto & -e^{-2i\arctan x} . \end{cases}$$

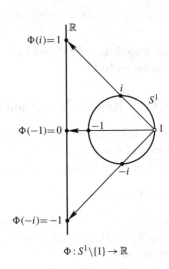

Figure 2.16. The projection map Φ

By means of Φ and its inverse we can reasonably convert the Newton map N_p into a map N_{p,S^1} defined on S^1. To this purpose observe that by the definition (2.9) we have $|N_p(x)| \to +\infty$ for $x \to \pm\infty$. Clearly, we intend to define $N_{p,S^1}(z) := \Phi^{-1} \circ N_p \circ \Phi(z)$ whenever the latter expression makes sense. For N_{p,S^1} to be continuous we therefore set $N_{p,S^1}(1) := 1$. Furthermore, for any $z_0 \neq 1$ which is mapped to a critical point by Φ, i.e. $p'(\Phi(z_0)) = 0$, we define

$$N_{p,S^1}(z_0) := \begin{cases} 1 & \text{if } p(\Phi(z_0)) \neq 0, \\ z_0 & \text{if } p(\Phi(z_0)) = 0. \end{cases}$$

We have thus defined N_{p,S^1} on all of S^1; it is easily seen that this map is in fact smooth. Moreover, concerning our original map N_p we notice that

$$N_p(x) = \Phi \circ N_{p,S^1} \circ \Phi^{-1}(x)$$

holds on $\mathbb{R}$ with the exception of at most $l - 1$ points. If $\Phi(z_0)$ is a zero of p with multiplicity m, then a straightforward calculation yields

$$N'_{p,S^1}(z_0) = \frac{m - 1}{m} < 1,$$

so that z_0 is an attracting fixed point for N_{p,S^1}. Analogously we find

$$N'_{p,S^1}(1) = \frac{l}{l - 1} > 1,$$

showing that 1 is a repelling fixed point. Evidently, there are no other fixed points of N_{p,S^1} whatsoever. Before turning towards examples we summarize our observations.

Proposition 2.22. *Let p denote a real polynomial of degree higher than one. The Newton map N_p associated with p uniquely induces a smooth map N_{p,S^1} on the unit circle such that $N_p(x) = \Phi \circ N_{p,S^1} \circ \Phi^{-1}(x)$ holds on $\mathbb{R}$ except for a finite number of points. Apart from the repelling fixed point at 1 every fixed point of N_{p,S^1} has the form $\Phi^{-1}(x_0)$ where x_0 denotes a root of p; these latter fixed points are attracting.*

Let us now look at some examples partly sketched in the Introduction. At first we shall deal with *quadratic* polynomials. Let p denote a polynomial of degree two. It is easily shown (see Exercise 2.4) that N_{p,S^1} is topologically conjugate to N_{q,S^1} where q is one of the three polynomials

$$q_+(x) := x^2 + 1, \quad q_\bullet(x) := x^2, \quad q_-(x) = x^2 - 1.$$

The quadratic case thus boils down to three special problems. In order to conveniently study the dynamics of N_{q,S^1} by means of graphical analysis we shall treat the latter map as a map $\tilde{N}_q$ on $[0, 1[$. More formally, $\tilde{N}_q(\xi)$ is understood as the unique number in $[0, 1[$ satisfying $N_{q,S^1}(e^{2\pi i\xi}) = e^{2\pi i\tilde{N}_q(\xi)}$. The reader should not be confused by the variety of Newton-like maps associated with the polynomial q: the original map N_q is defined on $\mathbb{R}$ but has the unpleasant property that there might be poles where the further iterations are not defined. On the other hand N_{q,S^1} is a smooth map on S^1, the dynamics of which incorporates the dynamics of N_q in a sense made precise earlier. As N_{q,S^1} is hard to directly visualize we instead use $\tilde{N}_q$ for graphical analysis.

For the polynomial $q_\bullet$ a short calculation yields

$$\tilde{N}_{q_\bullet}(\xi) = \frac{1}{2} - \frac{1}{\pi} \arctan\left(\frac{1}{2}\cot \pi\xi\right) \quad \text{if } 0 < \xi < 1$$

and clearly $\tilde{N}_{q_\bullet}(0) = 0$. Analogously we find for q_-

$$\tilde{N}_{q_-}(\xi) = \begin{cases} \frac{1}{\pi}\arctan\left(\sin(2\pi\xi)\right) & \text{if } 0 \leq \xi \leq \frac{1}{2}, \\ 1 + \frac{1}{\pi}\arctan\left(\sin(2\pi\xi)\right) & \text{if } \frac{1}{2} < \xi < 1. \end{cases}$$

The graphs of these functions are depicted in Figure 2.17. (Admittedly, the graph of $\tilde{N}_{q_-}$ does not look smooth. The discontinuity is, however, solely due to the definition of $\tilde{N}_q$; as we know, N_{q,S^1} is a smooth map.) From Figure 2.17 the dynamics of $N_{q_\bullet,S^1}$ and N_{q_-,S^1} is rather clear: by $N_{q_\bullet,S^1}$ all points in $S^1\backslash\{1\}$ are attracted towards -1. As

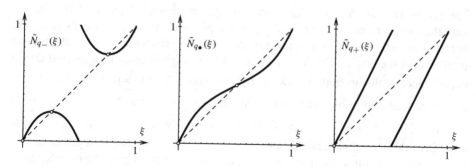

Figure 2.17. The maps induced by Newton's method for q_-, $q_\bullet$ and q_+, respectively

far as N_{q_-,S^1} is concerned we deduce that i attracts all points in the upper half of S^1 whereas $-i$ attracts all points in the lower half. Correspondingly, the original Newton map N_{q_-} on $\mathbb{R}$ attracts all points in $]0, +\infty[$ to the root $+1$ of q_- whereas the points in $]-\infty, 0[$ wander towards the root -1.

If we now perform a similar analysis for N_{q_+,S^1} we end up with a little surprise: we find $N_{q_+,S^1}(z) = z^2 = Q_0(z)$ so that N_{q_+,S^1} is chaotic on S^1 (see Figure 2.17 and also Figure 2.15). If we remove from S^1 all points which eventually are mapped onto 1 by N_{q_+,S^1} then we can rigorously observe chaos on the real line, too. More formally, defining

$$X := \Phi\big(S^1\backslash\{e^{2^{1-m}\pi i l} : l, m \in \mathbb{N}_0, 0 \le l < 2^m\}\big)$$
$$= \mathbb{R}\backslash\{\cot \pi l 2^{-m} : l, m \in \mathbb{N}, 1 \le l < 2^m\}$$

we have $N_{q_+}(X) = X$, and N_{q_+} is chaotic on X. Clearly, this explains why we could not find any regularity for the iterates of the Newton map associated to q_+ in the Introduction. Heuristically, as q_+ has no real roots which could be detected by N_{q_+}, typical points do not find a place to settle down under the iteration scheme (2.8): as a consequence, they wander around in a chaotic manner.

Let us now turn towards Newton's method for *cubic* polynomials. As will soon become clear, the overall picture is much more complicated here than for the quadratic case. For example, the whole period-doubling scenario discussed in the realm of the logistic map will show up again. Let p denote a cubic polynomial. Again, it is not difficult to show (Exercise 2.4) that N_{p,S^1} is topologically conjugate to N_{c,S^1} where c either is one of the three special polynomials

$$c_+(x) := x^3 + x, \quad c_\bullet(x) := x^3, \quad c_-(x) := x^3 - x, \tag{2.10}$$

or else is a member of the one-parameter family

$$c_\gamma(x) := x^3 + \gamma x + 1 \quad (\gamma \in \mathbb{R}). \tag{2.11}$$

The graphs of $\tilde{N}_{c_+}, \tilde{N}_{c_\bullet}$ and $\tilde{N}_{c_-}$, respectively, are shown in Figure 2.18. In this case the dynamics of N_{c_+,S^1} and $N_{c_\bullet,S^1}$ is simple: except for 1 all points in S^1 are attracted towards -1. Correspondingly, N_{c_+} and $N_{c_\bullet}$ attract all of $\mathbb{R}$ to the unique (real) root at the origin. For N_{c_-,S^1} the situation is slightly more complicated due to the existence of the two unstable two-periodic points $\Phi^{-1}\big(\pm\frac{1}{\sqrt{5}}\big)$: with the exception of a countable family of points which lie within the two arcs $\Phi^{-1}\big([-\frac{1}{\sqrt{3}}, -\frac{1}{\sqrt{5}}]\big)$ and $\Phi^{-1}\big([\frac{1}{\sqrt{5}}, \frac{1}{\sqrt{3}}]\big)$ all points in $S^1\backslash\{1\}$ are ultimately attracted by N_{c_-,S^1} towards one of the three asymptotically stable fixed points $\pm i, -1$; these fixed points in turn correspond to the three real roots $\pm 1, 0$ of c_-. When starting near the unstable two-periodic orbit it may, however, be difficult to decide which of the fixed points the iteration will head for in the long run (cf. Exercise 2.4).

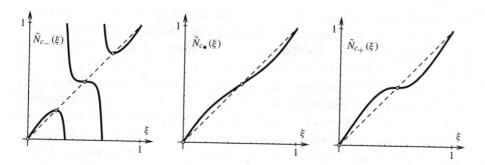

Figure 2.18. The maps induced by the cubic polynomials c_-, $c_\bullet$ and c_+, respectively

Contrary to the rather tame behaviour of the first three polynomials in (2.10) the dynamical morphogenesis of N_{c_γ, S^1} from (2.11) is much more involved. Since things are already complicated if c_γ has only one real root we shall restrict ourselves to that case: an elementary calculation confirms that c_γ has just one real root if $\gamma > -\frac{3}{2}\sqrt[3]{2} \approx -1.890$. If $\gamma \geq 0$ the dynamics of N_{c_γ, S^1} is readily analysed: all points in $S^1 \backslash \{1\}$ are attracted by the unique fixed point corresponding to the real root of c_γ. (Figure 2.19 shows the graphs of some $\tilde{N}_{c_\gamma}$.) If $\gamma < 0$ more interesting effects may be observed.

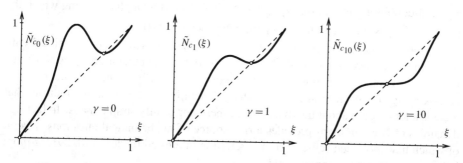

Figure 2.19. Depending on the parameter γ the maps $\tilde{N}_{c_\gamma}$ induced by the cubic polynomials c_γ may exhibit regular (above) and chaotic behaviour.

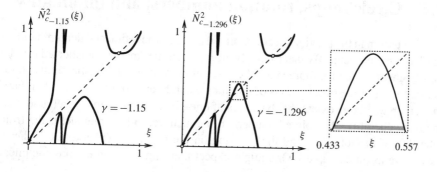

For example, when looking at the graphs of the second iterates at $\gamma = -1.150$ and $\gamma = -1.296$ we find strong evidence that there is a saddle-node bifurcation for $\tilde{N}^2_{c_\gamma}$ between these two values of γ (Figure 2.19). Moreover, it is easily verified that for $\gamma = -\sqrt[3]{2} \approx -1.260$ the point -1 is two-periodic and $\left(N^2_{c_\gamma,S^1}\right)'(-1) = 0$. Going back to the original Newton map N_{c_γ} on $\mathbb{R}$ we conclude that there exists a whole interval I around the origin such that the points in I are attracted towards the periodic orbit $\left\{0, \frac{1}{\sqrt[3]{2}}\right\}$. When taking a point from I as an initial value, Newton's method (2.8) will therefore not deliver any reasonable "zero" of c_γ. But things could get even worse. A closer look at Figure 2.19 suggests that there is an interval $J \subseteq [0, 1[$ which is mapped onto itself by $\tilde{N}^2_{c_\gamma}$ in much the same way as the unit interval is by F_4. A magnification (Figure 2.19) strongly supports this impression with J being approximately $[0.440, 0.552]$. Although we shall not rigorously work out the details it should be quite obvious by now that N_{c_γ,S^1} will act chaotically on parts of S^1 for $\gamma \approx -1.296$. Again, when going back to N_{c_γ} we thus find an interval around the origin where N_{c_γ} is chaotic. Moreover, on varying γ slightly above -1.296 we expect a procedure similar to the period-doubling scenario for F_μ to take place. A refined analysis confirms that this expectation is indeed correct ([29]). We remark in closing that the overall picture of N_{c_γ,S^1} will be much more complicated than the corresponding one for F_μ, even if $\gamma > -\frac{3}{2}\sqrt[3]{2}$. For example a stable periodic orbit of period three may be found for N_{c_γ,S^1} near -1 if $\gamma \approx -0.766$. On decreasing γ one locally observes a period-doubling scenario for $N^3_{c_\gamma,S^1}$ which is followed by a phase during which the real root of c_γ is the only attracting point whatsoever ([29]).

However sketchy our treatment of the dynamics of Newton maps associated with polynomials of degree three, it may have convinced the reader that in applying Newton's method (2.8) one has to choose the starting point x_0 carefully to end up at the root one is looking for (or even at any root at all). Generally our observations indicate that there is a lot to learn about the dynamics of these apparently simple maps. In fact, the dynamics of Newton maps provides a rich source of mathematical questions, several of which have not been settled yet. We refer the interested reader to [8, 29] for more information on this fascinating field.

2.4   Circle maps, rotation numbers, and minimality

Our discussion in the previous sections aimed at the introduction and denomination of a few basic phenomena. By means of advanced techniques more sophisticated problems may be investigated for more general classes of maps than the logistic family. It turns out that most of the observations of qualitative nature are not restricted to that family. In fact, up to a large extent the dynamics of one-dimensional maps is well understood in great generality. As pointed out earlier, the main reason for this fact is the fruitful interplay of the topology and the ordering on the real line. (Sharkovsky's theorem may serve as an example.) One might expect that things become more complicated and involved when we turn our attention towards higher dimensions. In the remaining sections of this chapter we shall discuss several examples showing that this indeed is the case: even when having recourse to more powerful mathematical tools the systems that can be completely understood become more and more artificial. Without any claim of comprehensiveness the subsequent sections indicate that it may be advantageous to modify our point of view on dynamical systems not least for practical reasons, as we shall do in Chapter Three.

In this section we deal with special maps on the unit circle S^1 and applications thereof. Let T denote a homeomorphism of S^1. Due to the specific structure of S^1 the map T may exhibit aspects of recurrence not shared by homeomorphisms of $\mathbb{R}$. In order to analyse the dynamics of T, the notion of a lift proves useful. A continuous map $L_T : \mathbb{R} \to \mathbb{R}$ is said to be a *lift* of T, if it makes the diagram

$$
\begin{array}{ccc}
\mathbb{R} & \xrightarrow{\;L_T\;} & \mathbb{R} \\
\pi \downarrow & & \downarrow \pi \\
S^1 & \xrightarrow{\;T\;} & S^1
\end{array}
\qquad \text{with} \qquad \pi : \begin{cases} \mathbb{R} \to S^1 \\ x \mapsto e^{2\pi i x} \end{cases}
$$

commute. Equivalently, $T(e^{2\pi i x}) = e^{2\pi i L_T(x)}$ for all $x \in \mathbb{R}$, and (S^1, T) is a factor of $(\mathbb{R}, L_T)$. Obviously $L_T(x)$ is unique up to an additive integer. We call T *orientation-preserving* if one (and hence every) lift is increasing, and *orientation-reversing* otherwise. Although a lift will be dynamically simple, it can be used for measuring the average rotation under T.

Lemma 2.23. *Let L_T be a lift of the homeomorphism T of S^1. For $x \in \mathbb{R}$ the limit*

$$
\rho(L_T) := \lim_{n \to \infty} \frac{L_T^n(x)}{n}
$$

exists and does not depend on x.

Proof. Since L_T^n is a lift of T^n we have $|L_T^n(x) - L_T^n(y)| \leq 1$ if $|x - y| < 1$. Consequently, if it exists at all, $\rho(L_T)$ does not depend on the choice of x. Suppose there is a p-periodic point z_0 for T and take $x_0 \in \mathbb{R}$ with $\pi(x_0) = z_0$. Obviously $L_T^p(x_0) = x_0 + q$ for some $q \in \mathbb{Z}$. From this we deduce that $\lim_{n \to \infty} \frac{L_T^{pn}(x_0)}{pn} = \frac{q}{p}$ if T is orientation-preserving. In the orientation-reversing case it is easily seen that T has two fixed points and thus $\rho(L_T) = 0$. If $p > 1$, notice that $L_T^k - \mathrm{id}_{\mathbb{R}}$ has period one for any $k \in N$ and set $c_p := \max_{k=1}^{p-1} \max_{x \in \mathbb{R}} |L_T^k(x) - x| < \infty$. This immediately yields

$$
\left| \frac{L_T^n(x_0)}{n} - \frac{L_T^{p\lfloor \frac{n}{p} \rfloor}(x_0)}{p\lfloor \frac{n}{p} \rfloor} \right| \leq \frac{c_p}{n} + \frac{p}{n} \left| \frac{L_T^{p\lfloor \frac{n}{p} \rfloor}(x_0)}{p\lfloor \frac{n}{p} \rfloor} \right| ;
$$

therefore $\lim_{n \to \infty} \frac{L_T^n(x_0)}{n} = \frac{q}{p}$. If there are no periodic points for T, we can find a sequence $(k_n)_{n \in \mathbb{N}}$ satisfying $k_n < L_T^n(x) - x < k_n + 1$ for all $x \in \mathbb{R}$. Choosing x to be

$0, L_T^n(0), \ldots, L_T^{(m-1)n}(0)$, respectively, and adding up all the resulting inequalities, we obtain $\frac{k_n}{n} < \frac{L_T^{mn}(0)}{mn} < \frac{k_n+1}{n}$. Taking into account the original inequality, we see that

$$\left| \frac{L_T^{mn}(0)}{mn} - \frac{L_T^n(0)}{n} \right| < \frac{1}{n}$$

and, after interchanging m and n,

$$\left| \frac{L_T^m(0)}{m} - \frac{L_T^n(0)}{n} \right| < \frac{1}{m} + \frac{1}{n} .$$

Consequently, $\left(\frac{L_T^n(0)}{n} \right)_{n \in \mathbb{N}}$ is a Cauchy sequence and hence converges. □

Let L_T and $\tilde{L}_T$ be two lifts of T. Since $\tilde{L}_T(x) - L_T(x) = k$ for some integer k, we have $\tilde{L}_T^n(x) = L_T^n(x) + nk$ if T is orientation-preserving, and $\tilde{L}_T^n(x) = L_T^n(x) + \frac{1}{2}(1 - (-1)^n)k$ if T is orientation-reversing. In any case we observe that $\rho(\tilde{L}_T) - \rho(L_T) \in \mathbb{Z}$.

Definiton 2.24. Let T be a homeomorphism of S^1 and L_T a lift of T. The number

$$\rho(T) := \rho(L_T) - \lfloor \rho(L_T) \rfloor \in [0, 1[$$

only depends on T. It is called the *rotation number* of T.

In accordance with Lemma 2.23, $\rho(T)$ may be interpreted as a quantity describing the average net rotation of S^1 under T. This becomes more apparent if one introduces the continuous real-valued function $f : S^1 \to \mathbb{R}$ according to

$$f(e^{2\pi i x}) := L_T(x) - x .$$

Intuitively, f measures – up to an additive integer constant – the normalized angle between $z := \pi(x)$ and $T(z)$. We find that

$$\rho(L_T) = \lim_{n \to \infty} \frac{1}{n} \sum_{k=0}^{n-1} f(T^k(z)) \tag{2.12}$$

which supports the above interpretation of $\rho(L_T)$, and also of $\rho(T)$, as averages. Expressions similar to (2.12) will extensively be discussed in later chapters.

The simplest examples of orientation-preserving homeomorphisms of S^1 are provided by rotations. The map

$$R_\vartheta : \left\{ \begin{array}{ccc} S^1 & \to & S^1 \\ z & \mapsto & e^{2\pi i \vartheta} z \end{array} \right.$$

simply rotates S^1 counter-clockwise by an angle $2\pi\vartheta$. Since $L_{R_\vartheta}(x) = x + \vartheta$ is a lift of R_ϑ we find $\rho(R_\vartheta) = \vartheta - \lfloor \vartheta \rfloor$, confirming the aforementioned interpretation. The dynamics of R_ϑ differ considerably for rational and irrational ϑ, respectively. If $\vartheta = p/q$, then all points are q-periodic, whereas for $\vartheta \notin \mathbb{Q}$ there are no periodic points at all. In the latter case, some sort of indecomposability even stronger than topological transitivity can be observed.

Theorem 2.25. *Let R_ϑ be a rotation of S^1 with $\vartheta \notin \mathbb{Q}$. Then $\overline{O^+(z)} = S^1$ for every $z \in S^1$, i.e., every orbit is dense.*

Proof. Let $\varepsilon > 0$ be given. Due to ϑ's irrationality, $O^+(1)$ is infinite and therefore accumulates somewhere in S^1. We may thus find natural numbers $m < n$ such that the arc-length between $R_\vartheta^m(1)$ and $R_\vartheta^n(1)$ is less then ε. Since R_ϑ^{n-m} preserves distances on S^1 the orbit $O^+(1)$ is ε-dense, and hence dense as ε has been arbitrary. The observation that the orbit of an arbitrary point is just a rotated version of $O^+(1)$ completes the proof. $\square$

A system where every orbit is dense is usually called *minimal*. Although it may exhibit complicated dynamics, clearly such a system never is chaotic in the sense of Definition 2.21. Summarizing our considerations, we may say that (S^1, R_ϑ) is minimal if and only if $\vartheta \notin \mathbb{Q}$.

As an application of Theorem 2.25 we will answer a question which we first encountered in the Introduction (cf. Exercise 1.4): What are the digits that the powers of two can begin with? The following fact is crucial for answering this question: the natural number k constitutes the beginning of the decimal representation of 2^n if and only if $k \cdot 10^l \leq 2^n < (k+1) \cdot 10^l$, or equivalently $l + \log_{10} k \leq n \log_{10} 2 < l + \log_{10}(k+1)$ for some $l \in \mathbb{N}_0$. Therefore 2^n begins with k precisely if

$$R_{\log_{10} 2}^n(1) \in \pi\left([\log_{10} k, \log_{10}(k+1)[\right) =: A_k. \tag{2.13}$$

Since A_k is an arc of positive length and $\log_{10} 2$ is irrational (even transcendental over $\mathbb{Q}$), we conclude that every natural number will sometime appear as an initial segment of the decimal representation of 2^n. When looking at the tables in Figures 1.19 and 2.20 one is led to suspect that nevertheless some numbers are more likely to occur than others. For example, it is not until $n = 53$ that $k = 9$ appears for the first time while

n	2^n		n	2^n	
41	2199023255552	2	51	2251799813685248	2
42	4398046511104	4	52	4503599627370496	4
43	8796093022208	8	53	9007199254740992	9
44	17592186044416	1	54	18014398509481984	1
45	35184372088832	3	55	36028797018963968	3
46	70368744177664	7	56	72057594037927936	7
47	140737488355328	1	57	144115188075855872	1
48	281474976710656	2	58	288230376151711744	2
49	562949953421312	5	59	576460752303423488	5
50	1125899906842624	1	60	1152921504606846976	1

Figure 2.20. The initial digits of 2^n (continuation of Figure 1.19)

$k = 8$ has already been observed five times till then. We shall settle this problem in Section 3.3.

Many observations concerning the rotation number carry over from rotations (where they are more or less trivial) to the general situation. Since the rotation number of an orientation-reversing homeomorphism necessarily equals zero we shall concentrate on orientation-preserving homeomorphisms from now on.

Theorem 2.26. *Let T be an orientation-preserving homeomorphism of S^1. Then:*

(i) $\rho(T^m) = m\rho(T) \pmod 1$ *for* $m \in \mathbb{N}_0$;

(ii) $\rho(T) \in \mathbb{Q}$ *if and only if* $\operatorname{Per} T \neq \emptyset$;

(iii) *If S is topologically conjugate to T, then $\rho(S) = \rho(T)$ or $\rho(S) = 1 - \rho(T)$, depending on whether the conjugacy is orientation-preserving or -reversing, respectively;*

(iv) *Given $\varepsilon > 0$ there exists a $\delta(\varepsilon) > 0$ such that for any orientation-preserving homeomorphism S with $\sup_{z \in S^1} |S(z) - T(z)| < \delta(\varepsilon)$ the rotation numbers $\rho(S)$ and $\rho(T)$ differ by less then ε modulo integers, in other words, the assignment $T \mapsto \rho(T)$ is C^0-continuous.*

Proof. Property (i) is obvious. We have already shown that a periodic point forces $\rho(T)$ to be rational. Assume in turn that $\rho(T) \in \mathbb{Q}$. Without loss of generality we may take $\rho(T) = 0$, and we are going to show that $\operatorname{Fix} T \neq \emptyset$. Let L_T be a lift of T with $L_T(0) \in [0, 1]$. If L_T has a fixed point we are done; otherwise we may assume $L_T(x) > x$ for all $x \in \mathbb{R}$. There are two possibilities: either $1 < L_T^m(0) \leq 2$ for *some* $m \in \mathbb{N}$ or $L_T^m(0) \in [0, 1]$ for *all* m. In the first case we deduce $\frac{1}{m} < \frac{L_T^{mn}(0)}{mn} \leq \frac{2}{m}$, contradicting $\rho(T) = 0$. In the second case $L_T^n(0)$ converges monotonically to a point in $]0,1]$ which obviously is a fixed point for L_T. In order to prove (iii) let h be a topological conjugacy and L_h a lift of h. It is easy to see that $L_h \circ L_T \circ L_h^{-1}$ is a lift of $S = h \circ T \circ h^{-1}$. We will assume $0 \leq L_h(0) < 1$ and first take h to be orientation-preserving. We find $|L_h^{\pm 1}(x) - x| \leq 2$ for all $x \in \mathbb{R}$. Using the fact that $|L_T^n(x) - L_T^n(y)| \leq 2$ if $|x - y| \leq 2$ we obtain

$$|(L_h \circ L_T \circ L_h^{-1})^n(x) - L_T^n(x)|$$
$$\leq |L_h \circ L_T^n \circ L_h^{-1}(x) - L_T^n \circ L_h^{-1}(x)| + |L_T^n \circ L_h^{-1}(x) - L_T^n(x)| \leq 4,$$

which yields $\rho(S) = \rho(T)$. If h is orientation-reversing we have the corresponding estimate $|L_h^{\pm 1}(x) + x| \leq 2$ and analogously get $\rho(S) + \rho(T) = 0 \pmod 1$. Concerning the verification of (iv), let $\varepsilon > 0$ be given and take $\frac{2}{n} < \varepsilon$. Clearly, we have $k - 1 < L_T^n(0) < k + 1$ for some $k \in \mathbb{Z}$. Choosing $\delta > 0$ sufficiently small, every homeomorphism with $\sup_{z \in S^1} |S(z) - T(z)| < \delta$ has a lift L_S which also satisfies $k - 1 < L_S^n(0) < k + 1$. As in the proof of Lemma 2.23 we deduce

$$\left| \frac{L_T^{mn}(0)}{mn} - \frac{L_S^{mn}(0)}{mn} \right| < \frac{2}{n} < \varepsilon$$

for every m, and thus $|\rho(T) - \rho(S)| < \varepsilon$. $\square$

The rotation number $\rho(T)$ provides an example of an invariant for the conjugacy class of a circle homeomorphism and may therefore be used as a tool for classifying such maps. The question arises whether rotation numbers are *complete* invariants, i.e. whether non-conjugate maps have different rotation numbers. It is easily seen that this is *not* the case for rational rotation numbers. When concentrating on the irrational case, the situation again looks completely different.

Theorem 2.27. *Let T be an orientation-preserving homeomorphism of S^1 with an irrational rotation number $\rho(T)$. Then T is semi-conjugate onto $R_{\rho(T)}$. If T is topologically transitive, then T and $R_{\rho(T)}$ are conjugate.*

Proof. Let L_T be a lift of T, and define a subset of $\mathbb{R}$ by $M := \{m + L_T^n(0) \ : \ m, n \in \mathbb{Z}\}$ as well as a map $H_0 : M \to \mathbb{R}$ by $H_0(m + L_T^n(0)) := m + n\rho(L_T)$. In a first step we are going to show that H_0 is strictly increasing. To this end observe that the validity of $m_1 + L_T^{n_1}(x) < m_2 + L_T^{n_2}(x)$ for *some* $x \in \mathbb{R}$ implies that this inequality holds for *all* $x \in \mathbb{R}$. (Otherwise T would have periodic points.) Therefore $m_1 + L_T^{n_1}(0) < m_2 + L_T^{n_2}(0)$ implies $L_T^{n_1 - n_2}(x) - x < m_2 - m_1$ for all $x \in \mathbb{R}$ if $n_1 > n_2$. (The case $n_1 < n_2$ is dealt with analogously.) Consequently $L_T^{n_1 - n_2}(0) < m_2 - m_1$ and by induction also $L_T^{k(n_1 - n_2)}(0) < k(m_2 - m_1)$. Therefore $\rho(L_T) < \frac{m_2 - m_1}{n_1 - n_2}$ or, equivalently, $H_0(m_1 + L_T^{n_1}(0)) < H_0(m_2 + L_T^{n_2}(0))$; here strict inequalities hold by virtue of the irrationality of $\rho(L_T)$. Similarly, if $x, y \in M$ with $x > y$ we see that $H_0(x) > H_0(y)$. The relation $H_0 \circ L_T(x) = H_0(x) + \rho(L_T)$ obviously holds on M. Since $H_0(M)$ is dense in $\mathbb{R}$, the map H_0 is seen to have a continuous extension $H_1 : \overline{M} \to \mathbb{R}$ still satisfying $H_1 \circ L_T(x) = H_1(x) + \rho(L_T)$. Furthermore, H_1 can uniquely be extended to a continuous, monotone map $H_2 : \mathbb{R} \to \mathbb{R}$ which is onto. Since the relation $H_2 \circ L_T = H_2 + \rho(L_T)$ extends from $\overline{M}$ to $\mathbb{R}$, we also have $H_2(x + 1) = H_2(x) + 1$ for all $x \in \mathbb{R}$. Consequently, there is a continuous, monotone map $h : S^1 \to S^1$ induced by H_2, i.e. $h \circ \pi = \pi \circ H_2$; furthermore, h is onto. From $\pi \circ H_2 \circ L_T = R_{\rho(T)} \circ \pi \circ H_2$ we deduce $h \circ T = R_{\rho(T)} \circ h$. If T is topologically transitive, we may assume $\overline{O^+(1)} = S^1$ without loss of generality. In this case the set M defined above is dense in $\mathbb{R}$, and thus h is a circle homeomorphism. $\square$

The theorem just proved provides a basis for the partial classification of circle maps. It is known that every C^2-diffeomorphism of S^1 with irrational rotation number is topologically transitive and hence conjugate to a rotation (Denjoy's theorem). However, there do exist examples of irrational C^1-diffeomorphisms which are *not* conjugate to the corresponding rotation. Differentiability thus plays a decisive role in the classification of circle maps. Even without giving any details (which may be found e.g. in [31]), the power of the concept of rotation numbers should have become clear by now. Although it was mainly the additional rich structure of S^1 which led to the success of the concept, one must not forget that the latter was originally introduced by looking at circle maps in a *statistical* manner via (2.12): assign to each point its *average* net rotation per step.

2.5   Glimpses of billiards

Despite their conceptual simplicity billiard systems may exhibit a very complicated dynamical behaviour, as we have already seen from examples in the Introduction. In the present section we are going to develop a description of convex planar billiards which will prove useful in later chapters. For the time being let Ω denote a bounded convex set in the plane, the boundary $\partial\Omega$ of which is a closed C^2-curve (see Figure 2.21). We think of Ω as a billiard table where we wish to study the motion of a small billiard ball. In the absence of any frictional effects the ball moves along straight lines with constant velocity until it hits $\partial\Omega$; there it is reflected according to the law that *the angle of incidence equals the angle of reflection.* This motion naturally induces a map on $\partial\Omega\times]0,\pi[$, usually referred to as the *billiard map*, in the following way: the pair $(x,\alpha)\in\partial\Omega\times]0,\pi[$ uniquely determines a segment of a billiard trajectory which emanates from x and encloses an angle α with the positively oriented tangent to $\partial\Omega$ in x. We can thus unambiguously assign to (x,α) the corresponding pair (x',α') of the billiard ball's next reflection (see Figure 2.21). For notational convenience we normalize the arclength of $\partial\Omega$ to one and thus obtain the billiard map in the form

$$
T_{\text{bill}} : \begin{cases} [0,1[\times]0,\pi[& \to & [0,1[\times]0,\pi[, \\ (s,\alpha) & \mapsto & (s',\alpha'); \end{cases}
$$

here s denotes the normalized arclength coordinate along the positively oriented boundary. For specific tables explicit formulas for T_{bill} can be provided. We shall, however, do so only exceptionally because dealing with these formulas may be rather cumbersome, and significant dynamical insights may be gained without that level of concreteness. For us it is more important to realize that T_{bill} is an invertible C^1-map. In fact, the derivative of T_{bill} can be expressed neatly by means of a few geometric quantities. Denoting by κ and κ' the curvature of $\partial\Omega$ at s and s', respectively, and by l the length of the trajectory segment joining s and s', a straightforward calculation (Exercise 2.14) yields

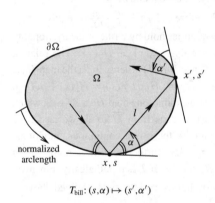

Figure 2.21. The billiard map

$$
DT_{\text{bill}} = \frac{\partial(s',\alpha')}{\partial(s,\alpha)} = \begin{pmatrix} \frac{l\kappa-\sin\alpha}{\sin\alpha'} & \frac{l}{|\partial\Omega|\sin\alpha'} \\ |\partial\Omega|\frac{l\kappa\kappa'-\kappa\sin\alpha'-\kappa'\sin\alpha}{\sin\alpha'} & \frac{l\kappa'-\sin\alpha'}{\sin\alpha'} \end{pmatrix} ; \qquad (2.14)
$$

here $|\partial\Omega|$ denotes the total length of $\partial\Omega$. Formula (2.14) will prove useful on several occasions.

We shall now turn towards specific examples. Besides the classical and beautiful details presented here, dynamical issues concerning billiards will repeatedly appear throughout this book. Perhaps the simplest convex table with smooth boundary is a disc with boundary $\mathcal{C}$. According to Figure 2.22 T_{bill} is explicitly given as

$$T_{\text{bill}} : (s, \alpha) \mapsto \left(s + \frac{\alpha}{\pi} \, (\text{mod } 1), \, \alpha\right).$$

Two elementary observations are worth mentioning. First, given a specific billiard trajectory starting at (s_0, α_0) each of its segments is tangent to the same inner circle with radius $|\cos \alpha_0|$ times the radius of $\mathcal{C}$. Consequently, there is a circular region in the middle of the table which is never crossed by the trajectory unless $\alpha_0 = \frac{\pi}{2}$. A curve which either is touched by all segments of a billiard trajectory or not by any is called a *caustic*. As far as the circular billiard is concerned each circle $\mathcal{C}^*$ concentric to $\mathcal{C}$ provides a caustic for a whole family of billiard trajectories.

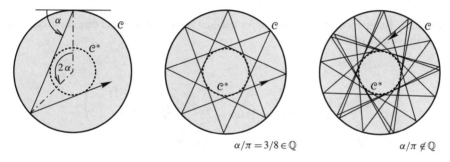

$$\alpha/\pi = 3/8 \in \mathbb{Q} \qquad \qquad \alpha/\pi \notin \mathbb{Q}$$

Figure 2.22. Each circle concentric to $\mathcal{C}$ is a caustic for the circular billiard.

A second observation is that $[0, 1[\times]0, \pi[$ decomposes into *invariant curves*: indeed, $T_{\text{bill}}([0, 1[\times \{\alpha\}) = [0, 1[\times \{\alpha\}$. In mechanical terms this observation reflects the fact that besides kinetic energy also the angular momentum of the billiard ball is conserved. As a consequence we may study T_{bill} separately on each invariant curve $[0, 1[\times \{\alpha\}$. But $T_{\text{bill}}|_{[0,1[\times \{\alpha\}}$ is just a rotation $R_{\alpha/\pi}$ dealt with in the last section. We therefore can easily classify all possible trajectories of the circular billiard. If α_0/π is rational, say $\alpha_0 = p\pi/q$, then the trajectory starting at (s_0, α_0) with arbitrary s_0 will close at the q-th reflection, the resulting motion thus being periodic. If on the other hand α_0/π is not a rational number then the trajectory densely fills the annulus between $\mathcal{C}^*$ and $\mathcal{C}$.

Things become slightly more complicated for the *elliptic* billiard. Nevertheless, it turns out that the two crucial observations from above carry over to that system in some sense. Let the ellipse $\mathcal{E}$ be defined by $(x/a)^2 + (y/b)^2 = 1$ with $a > b$, and count the normalized arclength s from the left vertex in counterclockwise direction. Before working out the details let us notice some significant differences between the circular and the elliptic billiard. For example, periodic points no longer come in one-parameter families: there are only two orbits of period two for T_{bill}, which correspond to billiard trajectories along the long and short principal axes, respectively (Figure 2.23). From formula (2.14) we deduce that for these periodic orbits

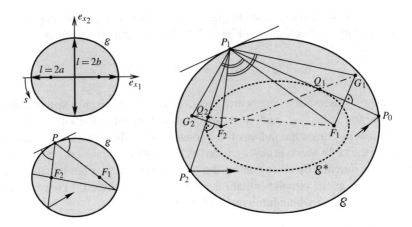

Figure 2.23. Two-periodic trajectories (upper left) and geometric details of the caustic construction for the elliptic billiard

$$DT_{\text{bill}}^2 = \begin{pmatrix} 2\kappa^2 l^2 - 4\kappa l + 1 & 2\frac{l}{|\varepsilon|}(\kappa l - 1) \\ 2|\varepsilon|\kappa(\kappa l - 1)(\kappa l - 2) & 2\kappa^2 l^2 - 4\kappa l + 1 \end{pmatrix}.$$

The eigenvalues of DT_{bill}^2 at these points are the roots of

$$\lambda^2 - 2(2\kappa^2 l^2 - 4\kappa l + 1)\lambda + 1 = 0.$$

Distinct real solutions of this equation are obtained if and only if $\kappa l(\kappa l - 2) > 0$; in this case, one eigenvalue has modulus greater than one. For the trajectory along the long axis we have $\kappa l = 2(a/b)^2 > 2$; therefore $\{(0, \frac{\pi}{2}), (\frac{1}{2}, \frac{\pi}{2})\}$ constitutes an unstable hyperbolic two-periodic orbit for T_{bill}. Along the short axis we find $\kappa l = 2(b/a)^2 < 2$ and thus obtain two complex conjugate eigenvalues for DT_{bill}^2 along $\{(\frac{1}{4}, \frac{\pi}{2}), (\frac{3}{4}, \frac{\pi}{2})\}$. We shall see that this orbit is in fact stable.

One important feature of the circular billiard discussed above concerned the existence of caustics. Indeed, there are caustics for the elliptic billiard too, and we are going to describe them. To this end we fix one specific billiard trajectory and label three consecutive points of reflection by P_0, P_1, $P_2 \in \mathcal{E}$. Furthermore we assume that none of the segments $P_0 P_1$, $P_1 P_2$ intersect the segment $F_1 F_2$ which joins the two foci. Let G_1 (G_2 respectively) denote the image of F_1 (F_2 respectively) under an orthogonal reflection with respect to the line $P_0 P_1$ ($P_1 P_2$ respectively; cf. Figure 2.23). Now recall the basic geometric fact that for any point $P \in \mathcal{E}$ the lines $P F_1$ and $P F_2$ enclose the same angle with the tangent to $\mathcal{E}$ at P. Taking into account this fact as well as the billiard property we find that the triangles $F_1 P_1 G_2$ and $G_1 P_1 F_2$ are congruent. If we denote by Q_1 the intersection of $F_2 G_1$ with $P_0 P_1$, and analogously by Q_2 the intersection of $F_1 G_2$ with $P_1 P_2$ we end up with the relation

$$d(F_2, Q_2) + d(Q_2, F_1) = d(F_2, Q_1) + d(Q_1, F_1).$$

(Here as throughout this section, d stands for the usual Euclidean distance in the plane.) The points Q_1 and Q_2 therefore lie on the same ellipse $\mathcal{E}^*$; the foci of $\mathcal{E}^*$ are the same as the foci of $\mathcal{E}$, and the segment $P_0 P_1$ is tangent to $\mathcal{E}^*$ in Q_1, as is the segment $P_1 P_2$ in Q_2. In other words, the ellipse $\mathcal{E}^*$ is a caustic for our specific billiard trajectory.

Before drawing further conclusions from this insight let us briefly mention the cases excluded in the above argument. If $P_0 P_1$ intersects $F_1 F_2$ then so does $P_1 P_2$. An argument similar to the one just outlined shows that in this case each segment of the billiard trajectory touches a *hyperbola* $\mathcal{H}^*$ confocal to $\mathcal{E}$. (The segment might have to be elongated as the point of tangency may lie outside $\mathcal{E}$; see Exercise 2.15.) If $P_0 P_1$ runs trough F_1 then $P_1 P_2$ passes through F_2, and each further segment of the trajectory will pass through one of the foci. This is evident from Figure 2.23 and reminds us why these points are called *foci*. Summarizing, we have found two families of caustics, ellipses and hyperbolas confocal with the original ellipse $\mathcal{E}$. For the sake of brevity we shall henceforth refer to these as to the elliptic and hyperbolic cases, respectively. Additionally, there is the special case of trajectories passing through the foci.

Let us now deal with some implications of the existence of caustics. First, we clearly may find sets in $[0, 1[\times]0, \pi[$ which are invariant with respect to T_{bill}. For example, take an elliptic caustic $\mathcal{E}^*$ and define $M(\mathcal{E}^*)$ to be the set of all points (s, α) such that the corresponding billiard trajectory emanating from that point is tangent to $\mathcal{E}^*$; then clearly $T_{\text{bill}}\big(M(\mathcal{E}^*)\big) = M(\mathcal{E}^*)$. Recall that the lines $\alpha = $ const. are invariant for the circular billiard. The elliptic billiard is slightly more complicated. However, as before the sets $M(\mathcal{E}^*)$ and $M(\mathcal{H}^*)$ are (unions of) smooth curves (see Figure 2.24).

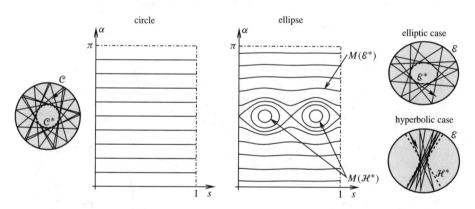

Figure 2.24. The space $[0, 1[\times]0, \pi[$ is partitioned into T_{bill}-invariant curves for both the circular (left) and the elliptic billiard.

For the circular billiard we have been able to completely analyse T_{bill} by means of rotations. When looking at Figure 2.24 one could get the impression that this should be possible for the elliptic billiard, too. In the sequel we shall describe how to relate the elliptic billiard to circle rotations. Admittedly, the final argument relies on the statistical view on dynamics and has thus to await the next chapter. However, the importance and applicability of circle rotations will already be highlighted by the subsequent elegant geometric argument which follows [34]. For the sake of simplicity we exclusively deal with the elliptic case here and only mention in passing the differences for the hyperbolic case.

Let us again fix a billiard trajectory inside $\mathcal{E}$ which has the confocal ellipse $\mathcal{E}^*$ as a caustic. As in Figure 2.23 we label by P_0, P_1, P_2 three consecutive points of reflection and by Q_1, Q_2 the points of tangency with $\mathcal{E}^*$ in between (see Figure 2.25). The lengths of the long axes of $\mathcal{E}$ and $\mathcal{E}^*$ are denoted by $2a$ and $2a^*$, respectively. Drawing two circles $\mathcal{K}$ and $\mathcal{K}^*$ with center F_2 and radii $2a$ and $2a^*$, respectively, we can find a homeomorphism Ψ of the plane which acts on $\mathcal{E}$ and $\mathcal{E}^*$ as depicted in Figure 2.25: for $Q \in \mathcal{E}^*$ we define $\Psi(Q)$ to be the unique point of intersection of $\mathcal{K}^*$ with the

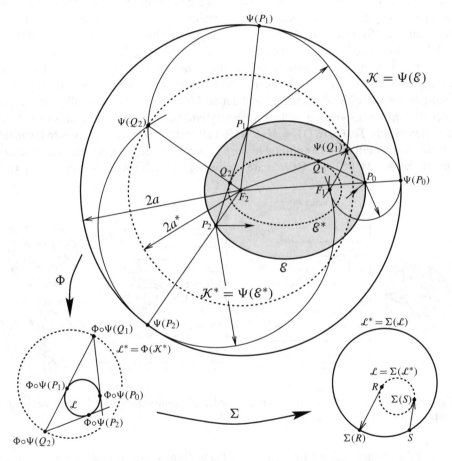

Figure 2.25. Geometric analysis of the elliptic billiard

half-line through Q emanating from the focus F_2. Analogously, for $P \in \mathcal{E}$ the image $\Psi(P)$ is defined as the intersection of $\mathcal{K}$ with the half-line joining F_2 and P. Clearly, for Ψ a smooth map may be chosen but we shall neither use this fact nor any explicit formula for Ψ. We now observe that $d(\Psi(Q_i), Q_i) = d(F_1, Q_i)$ for $i = 1, 2$. By virtue of the elementary geometric fact of equal angles already used earlier we deduce that $d(\Psi(Q_i), P_1) = d(F_1, P_1)$. Furthermore, due to the definition of $\mathcal{K}$ and $\mathcal{E}$

$$2a = d(\Psi(P_1), F_2) = d(\Psi(P_1), P_1) + d(P_1, F_2) = d(F_1, P_1) + d(P_1, F_2)$$

holds, hence also $d(\Psi(P_1), P_1) = d(F_1, P_1)$. The four points $\Psi(Q_1), \Psi(Q_2), \Psi(P_1)$ and F_1 therefore belong to the same circle centered at P_1. Moreover, this circle touches $\mathcal{K}$ in $\Psi(P_1)$ because F_2 (the center of $\mathcal{K}$), P_1 and the common point $\Psi(P_1)$ fall onto one straight line. Clearly, the present observations remain valid for any segment of the billiard trajectory: for $n \in \mathbb{N}$ the corresponding points $\Psi(Q_n), \Psi(Q_{n+1}), \Psi(P_n)$ and F_1 lie on a circle which has P_n as its center and touches $\mathcal{K}$ in $\Psi(P_n)$.

Though elementary the next step in our analysis is ingenious: we perform an *inversion* Φ with respect to a circle centered at F_1. More formally, for any point X in the plane different from F_1 the image $\Phi(X)$ is defined as the unique point on the half-line from F_1 through X which satisfies $d(\Phi(X), F_1) \cdot d(X, F_1) = \delta$. Here δ denotes an arbitrary positive constant; for reasons to become clear immediately we use the specific value

$$\delta := 2a^* - \frac{d(F_1, F_2)^2}{2a^*} > 0. \qquad (2.15)$$

The reader who has not seen the geometric procedure of inversion before may wish to play about with the map $z \mapsto z/|z|^2$ which provides a prototypical example in the complex plane. Especially, it can be seen that the inverted image of a circle is either a straight line or again a circle depending on whether the original circle did contain the origin (i.e. the center of inversion) or not. It is mainly this geometric fact which we shall rely on.

How does an application of Φ transform the geometric picture? On the one hand, the images $\mathcal{L} := \Phi(\mathcal{K})$ and $\mathcal{L}^* := \Phi(\mathcal{K}^*)$ are still circles, though no longer concentric, with the smaller $\mathcal{L}$ corresponding to the larger $\mathcal{K}$. On the other hand, the circle joining $\Psi(Q_1), \Psi(Q_2), \Psi(P_1)$ and F_1 turns into a straight line which is tangent to $\mathcal{L}$ at $\Phi \circ \Psi(P_1)$. The overall situation is depicted in Figure 2.25. Generally, any two consecutive points $\Phi \circ \Psi(Q_n), \Phi \circ \Psi(Q_{n+1})$ are joined by a straight line which is tangent to $\mathcal{L}$ at $\Phi \circ \Psi(P_n)$. Clearly, these straight lines are no longer trajectory segments of any reasonable billiard; they just provide a simple method of finding consecutive points of reflection (at $\mathcal{E}$) and tangency (at $\mathcal{E}^*$) via the correspondence $\Phi \circ \Psi$.

The final step we perform is rather a cosmetic one. Intuitively we would prefer the outer circle $\mathcal{L}^*$ (rather than the inner $\mathcal{L}$) to play the role of the original boundary $\mathcal{E}$. This can easily be achieved: define a smooth map Σ which interchanges these two circles as depicted in Figure 2.25. Each point $R \in \mathcal{L}$ is mapped to the intersection of $\mathcal{L}^*$ with the counter clockwise tangent to $\mathcal{L}$ at R. On the other hand, each point in $\mathcal{L}^*$ is sent to the point of tangency with $\mathcal{L}$. Finally, a simple calculation confirms that the radius of $\mathcal{L}^*$ equals one; the mysterious choice of δ in (2.15) has been motivated by this objective.

We are now in a position to put together all our findings. As above, let $M(\mathcal{E}^*)$ denote the set of all points in $[0, 1[\times]0, \pi[$ giving rise to billiard trajectories with caustic $\mathcal{E}^*$. We observed earlier that $M(\mathcal{E}^*)$ is invariant under T_{bill} so that we may separately study $T_{\text{bill}}|_{M(\mathcal{E}^*)}$ which we denote by T for the sake of brevity. Let (s_{P_n}, α_{P_n}) correspond to the trajectory segment emanating from $P_n \in \mathcal{E}$ and touching $\mathcal{E}^*$ in Q_{n+1}. Defining a homeomorphism J of $\mathcal{L}^*$ according to Figure 2.26 we have

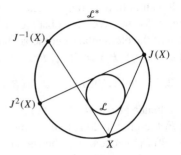

Figure 2.26. The circle map J

$$\Sigma \circ \Phi \circ \Psi\big(T(P_n)\big) = \Sigma \circ \Phi \circ \Psi(P_{n+1}) = J \circ \Phi \circ \Psi(Q_{n+1}) = J\big(\Sigma \circ \Phi \circ \Psi(P_n)\big).$$

Consequently, T is conjugate to the circle homeomorphism J via $\Sigma \circ \Phi \circ \Psi$. Admittedly, writing down an explicit formula for J may be cumbersome. Our observations, however, may be summarized as follows.

Proposition 2.28. *Let $\mathcal{E}^*$ denote an elliptic caustic for the billiard inside $\mathcal{E}$. Then $(M(\mathcal{E}^*), T_{bill}|_{M(\mathcal{E}^*)})$ is topologically conjugate to (S^1, J) where the orientation-preserving circle homeomorphism J is defined according to Figure* 2.26.

In Section 3.2 we shall prove that J is topologically conjugate to a rotation of S^1, and we shall calculate its rotation number. As a corollary of Proposition 2.28 we shall then arrive at a complete understanding of the billiard inside the ellipse $\mathcal{E}$ in the elliptic case. Given $\mathcal{E}$ and the confocal ellipse $\mathcal{E}^*$, all trajectories having $\mathcal{E}^*$ as a caustic either close after the same number of reflections or densely fill the ring-like region between $\mathcal{E}^*$ and $\mathcal{E}$.

In passing let us briefly touch upon the hyperbolic case. Given $\mathcal{E}$ and the hyperbolic caustic $\mathcal{H}^*$, Figure 2.24 suggests that one has to check $T_{\text{bill}}^2|_{M(\mathcal{H}^*)}$ rather than $T_{\text{bill}}|_{M(\mathcal{H}^*)}$ for its potential relations to circle maps. In fact, it is possible to relate $T_{\text{bill}}^2|_{M(\mathcal{H}^*)}$ to the map J by following the lines of the above analysis; the details, however, turn out to be more elaborate ([34]).

We close this section on billiards with a classical result due to Birkhoff concerning the existence of periodic points for T_{bill}. Recall that we found lots of periodic trajectories for the circular billiard while the situation for the ellipse was not as clear. The following general theorem asserts that there always are many periodic points. The proof also explains why billiard systems in [31] are first encountered amongst *variational aspects of dynamics*.

Theorem 2.29. *For any bounded convex table $\Omega \subseteq \mathbb{R}^2$ with C^2-boundary and any $p \in \mathbb{N}\setminus\{1\}$ there is a p-periodic point for the associated billiard map.*

Proof. Take $p \geq 2$ and let Y_p denote the product space $Y_p := \prod_{i=1}^{p} \partial\Omega$ together with the metric $d_p(x, y) := \sum_{i=1}^{p} d(x_i, y_i)$ for $x = (x_i)_{i=1}^{p}$, $y = (y_i)_{i=1}^{p} \in Y_p$. Clearly (Y_p, d_p) is a compact metric space. The continuous function

$$\varphi_p : \begin{cases} Y_p & \to & \mathbb{R} \\ x & \mapsto & d(x_1, x_2) + \ldots + d(x_{p-1}, x_p) + d(x_p, x_1) \end{cases}$$

therefore attains a maximum φ^*, at $x^* = (x_1^*, \ldots, x_p^*)$ say. We are going to show that the $x_i^* \in \partial\Omega$ are in fact reflection points of a billiard trajectory. Consider the ellipse $\mathcal{E}_i$ with foci x_{i-1}^*, x_{i+1}^* and the length of the long axis equal to $d(x_{i-1}^*, x_i^*) + d(x_i^*, x_{i+1}^*)$; here formally $x_0^* := x_p^*$ and $x_{p+1}^* := x_1^*$. Clearly $x_i^* \in \mathcal{E}_i$. If $\mathcal{E}_i$ and $\partial\Omega$ intersected at x_i^* with non-zero angle then we could slightly vary x_i^* along $\partial\Omega$ to produce a value of φ larger than φ^*. Therefore $\mathcal{E}_i$ is tangent to $\partial\Omega$ in x_i^*, and the angle of incidence equals the angle of reflection there. Since i has been arbitrary, we have found a billiard trajectory of period p. (Periodic points with *prime* period p have to be looked for by more sophisticated tools [31].) $\square$

On convex tables with C^2-boundary we can thus find infinitely many periodic points for T_{bill}. For the circular and elliptic billiards we explicitly found these points, but the above result also holds e.g. for the stadium sketched in the Introduction (Figure 1.5). There are, however, no *stable* periodic trajectories for the stadium billiard (cf. Exercise 2.10), and a thorough analysis of the stadium is *much* harder than for the ellipse. In fact, it is not without assuming a statistical point of view that the dynamics of the stadium or the dispersing billiard depicted in Figure 1.7 can satisfyingly be understood. In later chapters we shall be able to say more about these systems.

2.6 Horseshoes, attractors, and natural extensions

Two by now classical systems due to Smale demonstrate how simple geometric procedures can result in complicated dynamical behaviour. In this section we shall deal with the *horseshoe* as well as with the *solenoid* for both their archetypical simplicity and wealth of potential generalizations. Building on our discussion of the logistic family, the analysis of the so-called *horseshoe map* turns out to be quite easy. Consider a map $T = T_2 \circ T_1$ which acts on the stadium-shaped set $X := D_1 \cup E \cup D_2 \subseteq \mathbb{R}^2$ by first expanding it horizontally and compressing it vertically (T_1) and then folding it back into X (via T_2; see Figure 2.27). This geometric description suffices for completely analysing the dynamics, and we will not give any formula for T. Since the left half-disk D_1 is contracted into itself, there is a unique fixed point $x^* \in D_1$. Obviously, $T^n(x) \to x^*$ for every $x \in D_1 \cup D_2$. The points not tending to x^* therefore are precisely those in the set

$$\Lambda_+ := \bigcap_{n \in \mathbb{N}_0} T^{-n}(E) = \{x : T^n(x) \in E \text{ for all } n \geq 0\} \tag{2.16}$$

which is easily seen to be the product of a horizontal Cantor set and a vertical interval

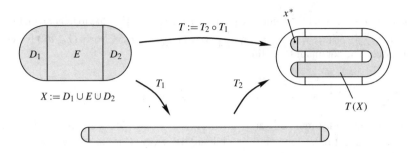

Figure 2.27. Smale's classical horseshoe

(cf. Figure 2.28). Since T is one-to-one and $T(\Lambda_+) \subseteq \Lambda_+$, it is in fact a homeomorphism of the compact set

$$\Lambda := \bigcap_{n \in \mathbb{N}_0} T^n(\Lambda_+) = \Lambda_+ \cap \Lambda_- ,$$

with $\Lambda_- := \bigcap_{n \in \mathbb{N}_0} T^n(E)$ being the product of a horizontal interval and a vertical Cantor set.

Assuming nothing more than uniform bounds on the expansion and contraction rates of T (as we did implicitly via our geometric definition), it is not at all difficult

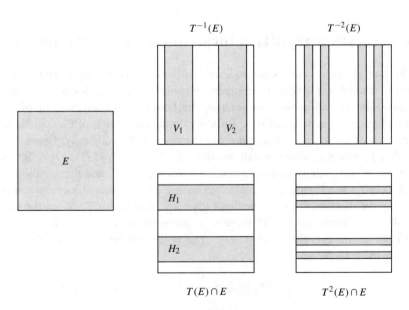

Figure 2.28. The first steps towards the sets Λ_+ and Λ_-

to relate the dynamics of T on Λ to symbolic dynamics. To this end, let $\Sigma_{l,\mathbb{Z}}$ denote the space of all *bi-infinite* sequences in $I = \{1, \ldots, l\}$ and endow this space with the metric

$$d\big((x_k)_{k\in\mathbb{Z}}, (y_k)_{k\in\mathbb{Z}}\big) := \begin{cases} 0 & \text{if } (x_k) = (y_k), \\ 2^{-\min\{|k| \,:\, x_k \neq y_k\}} & \text{if } (x_k) \neq (y_k). \end{cases}$$

Most of our observations concerning shift spaces and subshifts carry over to this context without any modification. As the most noticeable difference, we observe that the shift-map is now a homeomorphism, its inverse being the shift to the right. Having at our disposal this technical device, we can assign to $x \in \Lambda$ a sequence $(y_k)_{k\in\mathbb{Z}} \in \Sigma_{2,\mathbb{Z}}$ according to

$$y_k = i \quad \text{if and only if } T^k(x) \in V_i \quad (i = 1, 2);$$

here $V_{1,2}$ denote the two vertical strips in Figure 2.28. It is straightforward to show that $h : \Lambda \to \Sigma_{2,\mathbb{Z}}$ with $h(x) := (y_k)_{k\in\mathbb{Z}}$ defines a topological conjugacy between $(\Lambda, T|_\Lambda)$ and $(\Sigma_{2,\mathbb{Z}}, \sigma)$. The dynamics of the map T in the stadium X can therefore be summarized quite easily: all points in $X \setminus \Lambda_+$ are attracted by the stable fixed point x^*, whereas the points in Λ_+ are forward asymptotic to the Cantor set Λ. On the latter, T acts chaotically. Since the area of Λ_+ equals zero, we again have found an example of a transient chaotic system. Observe that the complicated dynamics is due to a combination of two effects: simultaneous expansion and contraction combined with a non-linear bending. Historically horseshoe-like behaviour was first (though implicitly) observed for flows arising in celestial mechanics by Poincaré at the end of the nineteenth century. It was not until the late nineteen-sixties that it became a paradigmatic scenario due to its striking geometric simplicity ([56]). Finally, it should be pointed out that an analogous analysis applies to many other maps of horseshoe-type. For example the two systems given geometrically in Figure 2.29 can easily be analysed by using subshifts of finite type (cf. Exercise 2.13).

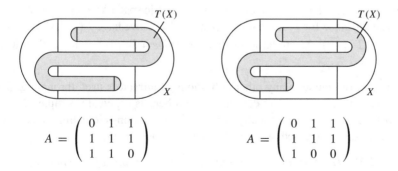

Figure 2.29. Two horseshoes giving rise to subshifts of finite type (see Exercise 2.13)

So far we have nearly exclusively seen transient chaotic behaviour in real-world (i.e. not symbolic) dynamical systems. In order to exhibit an example of chaos attracting nearby points, we shall discuss another example due to Smale, the so-called *solenoid*. Let X be the solid two-torus $S^1 \times D^2 \subseteq \mathbb{R}^3$ with D^2 denoting a disc of radius one. We label the points of X by pairs (z, w) of complex numbers satisfying $|z| = 1$ and $|w| \leq 1$, respectively, and define a smooth map $T : X \to X$ by

$$T(z, w) := \left(z^2, \frac{z}{2} + \frac{w}{10}\right).$$

Geometrically, T stretches the torus in tangential (z-)direction and heavily contracts it in radial (w-)direction. The image $T(X)$ therefore is a solid torus which wraps twice around itself (cf. Figure 2.30 where the radial contraction has been reduced for the sake of visual presentation). Since T is one-to-one and $T(X) \subseteq X$, we may define a non-empty compact set Γ, the *solenoid*, by $\Gamma := \bigcap_{n \in \mathbb{N}_0} T^n(X)$. Obviously, Γ is invariant, i.e. $T(\Gamma) = \Gamma$, and all points in X are attracted towards Γ.

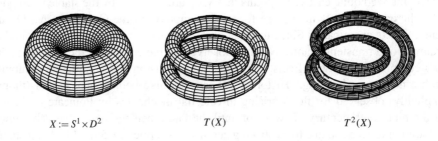

$X := S^1 \times D^2$ $\qquad\qquad\qquad$ $T(X)$ $\qquad\qquad\qquad$ $T^2(X)$

Figure 2.30. The construction of the solenoid

Definiton 2.30. Let T be a continuous map on a metric space X. A compact set $A \subseteq X$ is called an *attractor* for T, if for some open set $U \supseteq A$ we have $T(\overline{U}) \subseteq U$ as well as $A = \bigcap_{n \in \mathbb{N}_0} T^n(\overline{U})$. An attractor is called *transitive* if $T|_A$ is topologically transitive.

There are, of course, many other definitions of attractors used throughout the literature (see [20, 27, 31]). The definition given here is especially simple. Note that non-transitive attractors will usually consist of several non-interacting pieces. As far as the solenoid is concerned, we have a rather complete result.

Theorem 2.31. *The solenoid Γ is a transitive attractor, and $T|_\Gamma$ is chaotic.*

Proof. Since $\Gamma \cap (\{z\} \times D^2) \neq \emptyset$ for every $z \in S^1$, we only have to show that Per T is dense in Γ and $T|_\Gamma$ is topologically transitive. For notational convenience we define tubes $C_{\delta,m} \subseteq X$ (see Figure 2.31) by

$$C_{\delta,m}(z_0, w_0) := \{(z, w) : |z - z_0| < \delta, \ |w - w_0| < 10^{-m}\}$$

with $0 < \delta \ll 1$ and $m \in \mathbb{N}$. In order to prove denseness of $\operatorname{Per} T$, take $(z_0, w_0) \in \Gamma$ and U an open set containing (z_0, w_0). For $U = \Gamma \cap U_0$ with U_0 being an open set in X, there is some tube $C_{\delta,m}(z_0, w_0) \subseteq U_0$. Since $T^l(X)$ wraps around exactly 2^l times, $T^l(C_{\delta,m}(z_0, w_0))$ will completely pierce $C_{\delta,m}(z_0, w_0)$ if $\delta 2^l > \pi 2^{m+2}$. Therefore $T^l(\{z^*\} \times D^2) \subseteq \{z^*\} \times D^2$ for some z^* satisfying $|z_0 - z^*| < \delta$, and $C_{\delta,m}(z_0, w_0)$ contains a point of period l.

Topological transitivity is proved along the same lines: $T^l(C_{\delta_1,m_1})$ will intersect any C_{δ_2,m_2} for sufficiently large l. $\square$

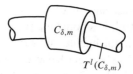

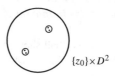

Figure 2.31. The tube $C_{\delta,m}$ (schematically) and a cross-section in the disc $\{z_0\} \times D^2$ (below)

Contrary to the systems before, we have now found a chaotic attractor. The complicated dynamics on Γ may be considered the long-time behaviour of *all* points in X. According to Definition 2.30 the most elementary attractors are provided by asymptotically stable periodic orbits. Attractors like Γ that are neither finite nor simply a closed curve often are classified under the notion of a *strange attractor*. However, the structure of Γ is not strange at all: locally we find a product of an interval and a Cantor set. In the following we shall give a more transparent description of the dynamics of T on Γ by means of a construction used throughout the theory of dynamical systems.

Let T denote a continuous map on a metric space X and define another metric space according to

$$\hat{X} := \{(x_k)_{k \in \mathbb{N}_0} : x_k = T(x_{k+1}) \text{ for all } k \in \mathbb{N}_0\},$$

$$\hat{d}\big((x_k)_{k \in \mathbb{N}_0}, (y_k)_{k \in \mathbb{N}_0}\big) := \sum_{k=0}^{\infty} 2^{-k} \frac{d(x_k, y_k)}{1 + d(x_k, y_k)}.$$

Observe that $\hat{X}$ again is a space of sequences, although the construction here is somewhat different from Σ_l. Two continuous maps naturally defined on $\hat{X}$ are given by

$$\hat{T}\big((x_k)_{k \in \mathbb{N}_0}\big) := (T(x_0), x_0, x_1, \dots) \quad \text{and} \quad \sigma\big((x_k)_{k \in \mathbb{N}_0}\big) := (x_1, x_2, x_3, \dots).$$

It is easy to see that $\hat{T}$ and σ are mutually inverse and hence are both homeomorphisms. Since T has been assumed to be merely continuous, the relation between T and $\hat{T}$ needs some clarification. To this end observe that $\hat{X}$ may be empty (take $X := \mathbb{R}$ and $T(x) := e^x$ for example). However, if X is compact or $T(X) = X$, we surely have $\hat{X} \neq \emptyset$. In this case the continuous projection $\pi_0 : \hat{X} \to X$ defined by $\pi_0\big((x_k)_{k \in \mathbb{N}_0}\big) := x_0$ makes the diagram

$$
\begin{array}{ccc}
\hat{X} & \xrightarrow{\hat{T}} & \hat{X} \\
\pi_0 \downarrow & & \downarrow \pi_0 \\
X & \xrightarrow{T} & X
\end{array}
$$

commute and clearly constitutes a semi-conjugacy if and only if $T(X) = X$. Finally, if T itself is a homeomorphism, the systems $(\hat{X}, \hat{T})$ and (X, T) are conjugate and thus indistinguishable from a dynamical point of view. The *natural extension* $(\hat{X}, \hat{T})$ may therefore be regarded as the simplest invertible dynamical system that semi-conjugates onto (X, T). The concept of natural extension admittedly looks artificial at first glance. Its usefulness, however, comes from results like the following.

Theorem 2.32. *Let T be a continuous map on a compact metric space X with $T(X) = X$. Then T is chaotic on X if and only if $\hat{T}$ is chaotic on $\hat{X}$.*

Proof. First we shall show that denseness of periodic orbits and the existence of a dense orbit carry over from T to $\hat{T}$. Observe that for $\varepsilon > 0$ we can find an $N(\varepsilon) \in \mathbb{N}$ such that

$$\hat{d}\big((x_k)_{k\in\mathbb{N}_0}, (y_k)_{k\in\mathbb{N}_0}\big) \leq \sum_{k=0}^{N(\varepsilon)} 2^{-k} \frac{d(x_k, y_k)}{1 + d(x_k, y_k)} + \frac{\varepsilon}{2}$$

for all pairs of sequences in $\hat{X}$. Assume now that Per T is dense in X and take $(x_k)_{k\in\mathbb{N}_0} \in \hat{X}$. Due to the continuity of T we can find a periodic orbit $y \in X$ satisfying $d\big(T^i(x_{N(\varepsilon)}), T^i(y)\big) < \varepsilon/4$ for all $i = 0, \ldots, N(\varepsilon)$. Let p denote the period of y. The periodic point

$$(y_k)_{k\in\mathbb{N}_0} := \big(\overline{T^{N(\varepsilon)}(y), T^{N(\varepsilon)-1}(y), \ldots T^{N(\varepsilon)-p+1}(y)}\big) \in \hat{X}$$

satisfies $\hat{d}\big((x_k)_{k\in\mathbb{N}_0}, (y_k)_{k\in\mathbb{N}_0}\big) < \varepsilon$ showing that Per $\hat{T}$ is dense in $\hat{X}$. Assume in turn that the orbit of $y \in X$ is dense, take $(x_k)_{k\in\mathbb{N}_0} \in \hat{X}$ and let $\varepsilon > 0$ be given. As before we can choose $M \in \mathbb{N}_0$ such that $d\big(T^i(x_{N(\varepsilon)}), T^{M+i}(y)\big) < \varepsilon/4$ for $i = 0, \ldots, N(\varepsilon)$. Therefore $\hat{d}\big((x_k)_{k\in\mathbb{N}_0}, \hat{T}^M(y_k)_{k\in\mathbb{N}_0}\big) < \varepsilon$ for every sequence $(y_k)_{k\in\mathbb{N}_0} \in \hat{T}^{N(\varepsilon)}\big(\pi_0^{-1}(y)\big)$. Since T is onto, $\pi_0^{-1}(\{y\})$ is not empty and therefore $\hat{X}$ contains a dense orbit.

In order to prove the theorem, assume that T be chaotic. Then X is infinite and, by our above considerations together with Exercise 2.3, $\hat{T}$ is chaotic. If in turn $\hat{T}$ is chaotic then Per T is dense in X and T is topologically transitive due to an earlier result (Lemma 2.17). Since $\#X < \infty$ would force T to be a permutation, the theorem follows. $\square$

With the technical device of natural extension at our disposal, we can now completely describe the dynamics of the map T on the solenoid Γ. Recall that the chaotic map $Q_0 : S^1 \to S^1$ with $Q_0(z) = z^2$ meets all the assumptions of Theorem 2.32. Denoting by P the projection of $X = S^1 \times D^2$ onto its mid-circle, that is $P(z, w) := z$, we have $P \circ T = Q_0 \circ P$. Therefore the rule

$$\Phi : (z, w) \mapsto (z, P \circ T^{-1}(z, w), \ldots) = \big(P \circ T^{-k}(z, w)\big)_{k\in\mathbb{N}_0}$$

defines a continuous map $\Phi : \Gamma \to \hat{S}^1$ with $\hat{Q}_0 \circ \Phi = \Phi \circ T$. This rather involved situation is best summarized by the following commuting diagram:

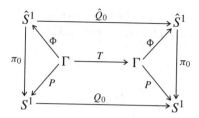

By explicitly exhibiting its inverse, we are going to show that Φ is in fact a homeomorphism and thus conjugates $(\Gamma, T|_\Gamma)$ and $(\hat{S}^1, \hat{Q}_0)$. Take any sequence $(z_k)_{k \in \mathbb{N}_0} \in \hat{S}^1$. It is easily seen that the sets $T^k(\{z_k\} \times D^2) \subseteq \{z_0\} \times D^2$ form a nested sequence of sets with diameter 10^{-k}. Therefore $\bigcap_{k \in \mathbb{N}_0} T^k(\{z_k\} \times D^2)$ consists of a single point (z_0, w^*). We now define a continuous map $\Psi : \hat{S}^1 \to \Gamma$ by $\Psi((z_k)_{k \in \mathbb{N}_0}) := (z_0, w^*)$. Since $(z, w) \in T^k(\{P \circ T^{-k}(z, w)\} \times D^2)$ for all $k \in \mathbb{N}_0$, the relation $\Psi \circ \Phi = \mathrm{id}_\Gamma$ holds; similarly we find $\Phi \circ \Psi = \mathrm{id}_{\hat{S}^1}$. In a nutshell we have

Proposition 2.33. *The systems $(\Gamma, T|_\Gamma)$ and $(\hat{S}^1, \hat{Q}_0)$ are topologically conjugate.*

This fact not only gives us a new proof that Γ is a transitive attractor but also provides a detailed description of the dynamics on that attractor. More generally, the technique of natural extension can successfully be applied to the class of *expanding attractors*. Like the solenoid, these systems admit a uniform lower bound on the growth rate of distances of close-by points. (For the solenoid we have $|Q_0'(z)| \equiv 2$ on S^1.) However, the space X, whose natural extension provides a model for the dynamics on the attractor, need not be as simple as in the case of the solenoid. As an example, we mention the so-called *Plykin attractor* which gives rise to a natural extension model $(\hat{X}, \hat{T})$ where the space X contains branching points and T^3 (rather than T) is uniformly expanding; see [20, 31] for the details.

Barely surprising, there are most interesting attractors that cannot be analysed by the method sketched here. Some of them in fact have largely defied analysis so far. A famous example in this direction comes from the family of planar maps defined by

$$H_{a,b} : \begin{cases} \mathbb{R}^2 & \to \quad \mathbb{R}^2 \\ (x_1, x_2) & \mapsto \quad (a + bx_2 - x_1^2, x_1) \end{cases} \tag{2.17}$$

and first studied by Hénon in order to mimic some effects observed for the renowned Lorenz equations ([27]). Since the definition (2.17) contains only one non-linear term, the Hénon maps are probably the simplest non-linear planar maps at all. Despite this apparent simplicity, there are still several open questions concerning that family. For example at $(a, b) = (1.4, 0.3)$ the diffeomorphism $H_{a,b}$ maps a quadrilateral Q into its interior. Consequently $\Gamma := \bigcap_{n \in \mathbb{N}_0} H_{a,b}^n(Q)$ is an attractor but up to now nobody knows whether it is transitive. There are even some doubts concerning the significance of numerically generated pictures like Figure 2.32 (cf. [27]). Nevertheless, numerical results strongly suggest that $H_{a,b}$ exhibits sensitive dependence on initial conditions for these particular parameter values (see Figure 2.32).

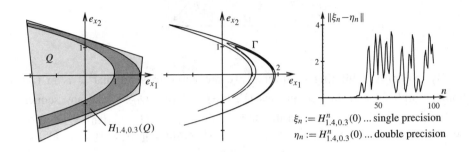

Figure 2.32. Numerically, $H_{1.4,0.3}$ exhibits sensitive dependence on initial conditions (right) on a rather complicated attractor Γ (middle).

2.7   Toral maps and shadowing

An important class of dynamical systems completely different from those considered so far is provided by *toral endomorphisms*. Historically, such systems first arose from the work of Hadamard on geodesic flows on manifolds of negative curvature and later were systematically studied by Anosov. They also provide examples of *algebraic dynamical systems*, i.e., the maps under consideration preserve an algebraic structure of the underlying space.

We define the d-dimensional torus $\mathbb{T}^d$ to be $\mathbb{R}^d/\mathbb{Z}^d$, which means that two points in $\mathbb{R}^d$ are identified if all their coordinates differ by integers. The structure of a compact abelian group may be given to $\mathbb{T}^d$ by defining an addition

$$[x] + [y] := (x + \mathbb{Z}^d) + (y + \mathbb{Z}^d) := x + y + \mathbb{Z}^d = [x + y]$$

as well as a metric

$$d([x], [y]) := \min_{k \in \mathbb{Z}^d} \|x - y + k\|.$$

The quotient map $\pi_d : \mathbb{R}^d \to \mathbb{T}^d$ with $\pi_d(x) := [x] = x + \mathbb{Z}^d$ is a continuous homomorphism from $\langle \mathbb{R}^d, + \rangle$ onto $\langle \mathbb{T}^d, + \rangle$. Observe that according to our definition the spaces $\mathbb{T}^1$ and S^1 (and more generally, $\mathbb{T}^d$ and $\prod_{i=1}^{d} S^1$) are not identical though homeomorphic (as topological spaces) and isomorphic (as groups). A geometric interpretation of the two-torus is as follows: the space $\mathbb{T}^2$ may be considered the unit square $[0, 1]^2 \subseteq \mathbb{R}^2$ where first the points $(x, 0)$ and $(x, 1)$ and then the points $(0, y)$

and $(1, y)$ are identified. This identification procedure yields the surface called *torus* in elementary geometry.

In order to define a map on $\mathbb{T}^d$ take a matrix $A \in \mathbb{R}^{d \times d}$. It is easy to see that the linear map $x \mapsto Ax$ (which we identify with A) induces a well defined map T_A on $\mathbb{T}^d$, making the diagram

$$
\begin{array}{ccc}
\mathbb{R}^d & \xrightarrow{\ A\ } & \mathbb{R}^d \\
\pi_d \downarrow & & \downarrow \pi_d \\
\mathbb{T}^d & \xrightarrow{\ T_A\ } & \mathbb{T}^d
\end{array}
$$

commute, if and only if A is an integer matrix, i.e. $A \in \mathbb{Z}^{d \times d}$. On the other hand, every *continuous* endomorphism of $\mathbb{T}^d$ is of the form T_A for some A (see [62]). Furthermore, T_A is onto if and only if $\det A \neq 0$, and it is a homeomorphism precisely if $|\det A| = 1$. In the latter case we have $T_A^{-1} = T_{A^{-1}}$. These properties are easily deduced from the above diagram which will also prove helpful for understanding the dynamics of T_A.

The case $\det A = 0$ is somewhat special, and we defer its analysis to the end of this section. As a result, it will turn out that we may assume $\det A \neq 0$ for good reasons, and therefore we shall do so for the time being.

We shall now show that the maps T_A exhibit an extremely complicated behaviour provided that A is *hyperbolic*, which means that no eigenvalue of A has modulus one. In that case we call T_A a *hyperbolic toral endomorphism*. Observe that by virtue of $DT_A|_x \equiv A$ for all $x \in \mathbb{T}^d$ any periodic orbit for T_A is certainly hyperbolic according to Definition 2.3. So the present notion of hyperbolicity in some sense extends the one introduced earlier. Since DT_A is constant, the hyperbolicity of T_A is highly uniform all over $\mathbb{T}^d$. In fact, the dynamical features of toral endomorphisms which we shall describe below substantially carry over to much more general uniformly (and even non-uniformly) hyperbolic systems which constitute an important and extensively studied field in their own right ([31]). Before giving the general result, we shall look at a specific example, namely T_A with

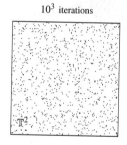

10^3 iterations

$\mathbb{T}^2$

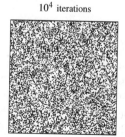

10^4 iterations

$$
A = \begin{pmatrix} 2 & 1 \\ 1 & 1 \end{pmatrix}. \tag{2.18}
$$

Figure 2.33. Orbits of the toral automorphism (2.18) may look rather stochastic.

For the eigenvalues of A we find $\lambda_{1,2} = \frac{3 \pm \sqrt{5}}{2}$. The corresponding eigenspaces are denoted by E_{λ_1} and E_{λ_2}, respectively. The dynamics of A on $\mathbb{R}^2$ is especially simple: as $n \to \infty$ we have $A^n(x) \to 0$ for $x \in E_{\lambda_2}$ and $A^{-n}(x) \to 0$ for $x \in E_{\lambda_1}$ as well as $\|A^n(x)\| \to \infty$ for $x \notin E_{\lambda_2}$ (cf. Figure 2.34). Clearly these relations imply $T_A^n(x) \to 0$ for $x \in \pi_2(E_{\lambda_2})$ and $T_A^{-n}(x) \to 0$ for $x \in \pi_2(E_{\lambda_1})$. Points in $\pi_2(E_{\lambda_1}) \cap \pi_2(E_{\lambda_2})$ thus are both forward and backward asymptotic to the origin. Observe now that this intersection is dense in $\mathbb{T}^2$. This is seen straightforwardly by noticing that the intersection of $\pi_2(E_{\lambda_{1,2}})$ with the circle $\{(0, x_2) : 0 \le x_2 < 1\}$ may be written as a (full) orbit under an irrational rotation. More formally,

$$\pi_2(E_{\lambda_{1,2}}) \cap \{(0, x_2) : 0 \le x_2 < 1\} = \left\{(0, k\frac{\pm\sqrt{5} - 1}{2}) : k \in \mathbb{Z}\right\}$$

and clearly $\frac{\pm\sqrt{5}-1}{2} \notin \mathbb{Q}$. The dynamics of T_A is thus expected to be rather complicated: every point not in $\pi_2(E_{\lambda_1}) \cap \pi_2(E_{\lambda_2})$ will considerably be pushed around; indeed, a typical orbit seems to be quite irregular (see Figure 2.33). Having noticed the strong dynamical contrast between A and T_A, the following theorem should not come as a complete surprise. (Recall that we always assume $\det A \ne 0$.)

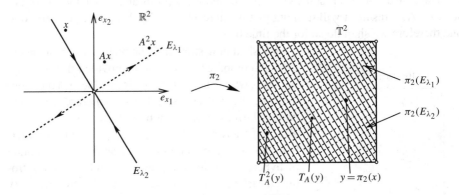

Figure 2.34. The spaces E_{λ_i} project onto dense sets under π_2.

Theorem 2.34. *Every hyperbolic toral endomorphism is chaotic.*

Proof. Let $A \in \mathbb{Z}^{d \times d}$ be a hyperbolic matrix with $\det A \ne 0$. In order to prove $\overline{\mathrm{Per}\, T_A} = \mathbb{T}^d$, take $p \in \mathbb{N}$ to be relatively prime to $|\det A|$ and observe that T_A permutes the finite set $\pi_d(\frac{1}{p}\mathbb{Z}^d)$. Since there are infinitely many such numbers p, we conclude that $\mathrm{Per}\, T_A$ is dense in $\mathbb{T}^d$. For proving topological transitivity we have to formalize our above observations. Due to the hyperbolicity of A we write $\mathbb{R}^d = E_s \oplus E_u$ with E_s and E_u being the sums of the generalized eigenspaces corresponding to eigenvalues with modulus smaller and larger than one, respectively. Assume for the moment the following two facts, the proofs of which we postpone to the end: there exists a constant $\kappa > 0$ such that $\|x - y\| \ge \kappa(\|x\| + \|y\|)$ for all $x \in E_s, y \in E_u$ and, secondly, the intersection of $\pi_d(E_s)$ and $\pi_d(E_u)$ is dense in $\mathbb{T}^d$. Taking this for granted, T_A is now shown to be topologically transitive. Let U, V be open sets in $\mathbb{T}^d$, take $u \in U \cap \pi_d(E_s) \cap \pi_d(E_u)$ and $v \in V \cap \pi_d(E_s) \cap \pi_d(E_u)$ as well as $\varepsilon > 0$. We can find

natural numbers N_1, N_2 with the following prop-
erties: $T_A^{N_1}(U)$ contains an ε-box in $\pi_d(E_u)$ and
$d(T_A^{N_1}(u), 0) < \frac{1}{2}\kappa\varepsilon$; analogously, $T_A^{-N_2}(V)$ contains
an ε-box in $\pi_d(E_s)$, and $T_A^{N_2}(w) = v$ for some w satisfying $d(w, 0) < \frac{1}{2}\kappa\varepsilon$. Since $d(T_A^{N_1}(u), w) < \kappa\varepsilon$, the
relation $T_A^{N_1}(U) \cap T_A^{-N_2}(V) \neq \emptyset$ and hence topologi-
cal transitivity follows. We can now go for the proof of
the above assumptions. The first is an easy geometric
observation. In order to prove the second one, it suffices to show that $\pi_d(E_s)$ and $\pi_d(E_u)$ both are dense
in $\mathbb{T}^d$. To this end take $x \in \mathbb{T}^d$ and $\varepsilon > 0$. Due to the
densensess of $\mathrm{Per}\, T_A$ there exists a periodic point y with
period p and $d(x, y) < \varepsilon/2$. Since in $\mathbb{R}^d$ the points

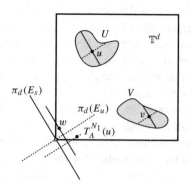

Figure 2.35. Notations used in the proof of Theorem 2.34

$A^{np} y'$ approach E_u as $n \to \infty$ for any $y' \in \pi_d^{-1}(\{y\})$,
we find $\|A^{np} y' - z'\| < \varepsilon/2$ for an appropriate $z' \in E_u$
and sufficiently large n. Consequently $d(x, \pi_d(z')) < \varepsilon$, and $\pi_d(E_u)$ is dense. If $|\det A| = 1$
the set $\pi_d(E_s)$ is dealt with similarly. Unfortunately, if $|\det A| \geq 2$ this argument will no longer
work in order to prove the topological transitivity of T_A. However, we shall close this minor
gap in the next chapter by showing that T_A is ergodic with respect to the Lebesgue measure $\lambda_{\mathbb{T}^d}$
on $\mathbb{T}^d$. $\square$

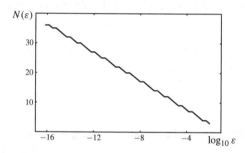

Figure 2.36. Quantifying the sensitive dependence on initial conditions of the toral automorphism (2.18)

Hyperbolic toral endomorphisms provide a large class of chaotic systems. By
means of so-called *Markov partitions*
it is possible again to relate these systems to symbolic dynamics. However,
since the details turn out to be rather
technical we shall not present them here
but refer to the literature on this topic
([20, 27, 31]).

For the rest of this section we focus
on a practical aspect of hyperbolic toral
endomorphisms. We already know that
these maps exhibit sensitive dependence
on initial conditions, and we are now go-
ing to quantify this effect. To this end fix $x \in \mathbb{T}^2$ and define for the hyperbolic toral
automorphism (2.18)

$$N(\varepsilon) := \min_{n \in \mathbb{N}} \left\{ n : d\left(T_A^n(x), T_A^n\left(x + (\varepsilon, \varepsilon)\right)\right) \geq \tfrac{1}{2} \right\}. \tag{2.19}$$

As the specific choice of the point x barely effects the result, the explicit dependency
of $N(\varepsilon)$ on x has been suppressed in the notation. The graph of the quantity $N(\varepsilon)$ in
Figure 2.36 shows how dramatic this sensitivity is for the specific toral automorphism
considered here: asymptotically, $N(\varepsilon)$ is proportional to $-\log_{10}\varepsilon$ (and not ε^{-a} for

some $a > 0$ as one might have hoped). Since locally T_A acts like a linear map with one eigenvalue of modulus greater than one this kind of behaviour of $N(\varepsilon)$ does not come as a real surprise. However, as a consequence $N(\varepsilon)$ is relatively small even for very small ε, or in other words, even the tiniest perturbation will grow to a considerable error within a frighteningly short time. Observe that the machine's precision is not better than about 10^{-16} if standard double precision floating point arithmetic is used. Consequently, if we feed a point x and a large number $N \in \mathbb{N}$ into the computer and ask the machine for $T_A^N(x)$, we must not think that the numerical result given back to us has anything to do with the correct result! From a practical point of view the system behaves unpredictably on long time scales. Computing orbits numerically therefore seems to be a waste of time and energy for such systems. However, this impression turns out to be partly misleading for hyperbolic toral endomorphisms, and it is not hard to make this assertion precise by means of the following definition.

Definiton 2.35. Let T be a continuous map on a metric space X and $\delta > 0$. A sequence $(x_n)_{n \in \mathbb{N}_0}$ in X is called a δ-*pseudo-orbit* for T if $d(T(x_n), x_{n+1}) < \delta$ for all n.

We may think of a pseudo-orbit as an orbit generated by a computer. Due to internal errors (round-off etc.) there will always be a slight discrepancy between the numerical and the true results x_{n+1} and $T(x_n)$, respectively. Can a pseudo-orbit $(x_n)_{n \in \mathbb{N}_0}$ be considered the perturbation of a true orbit nearby? In the case of the simple systems discussed here the answer definitely is positive.

Theorem 2.36. *Let T_A be a hyperbolic toral endomorphism and $\varepsilon > 0$. Then there exists a $\delta(\varepsilon) > 0$ such that every $\delta(\varepsilon)$-pseudo-orbit ε-shadows a true orbit, i.e., for every $\delta(\varepsilon)$-pseudo-orbit $(x_n)_{n \in \mathbb{N}_0}$ there is a point $\overline{x} \in \mathbb{T}^d$ such that $\sup_{n \in \mathbb{N}_0} d\left(x_n, T_A^n(\overline{x})\right) < \varepsilon$.*

Proof. Again we shall utilize the splitting $\mathbb{R}^d = E_s \oplus E_u$ introduced during the proof of Theorem 2.34. The quantity a according to

$$2a := \min\left\{|\lambda - z| : \lambda \in \sigma(A), z \in S^1\right\} = d(\sigma(A), S^1) > 0$$

will prove useful for the following. There exists a constant $C > 0$ such that for $x \in E_s$, $y \in E_u$ the bounds

$$\|A^n x\| \leq C(1-a)^n \|x\| \quad \text{and} \quad \|A^{-n} y\| \leq C(1+a)^{-n} \|y\|$$

are valid for all $n \in \mathbb{N}$. Given $\varepsilon > 0$, let $\delta(\varepsilon)$ denote the radius of the largest ball in $\mathbb{R}^d$ that fits into the open set

$$\left\{x : x = x_s + x_u, x_s \in E_s, x_u \in E_u, \max\{\|x_s\|, \|x_u\|\} < \frac{\varepsilon a}{3C}\right\}$$

and let $(x_n)_{n \in \mathbb{N}_0}$ be a $\delta(\varepsilon)$-pseudo-orbit. With appropriate $v_n \in \pi_d(E_s)$, $w_n \in \pi_d(E_u)$ we then have

$$x_n = T_A(x_{n-1}) + v_n + w_n$$

where $\max\{d(v_n, 0), d(w_n, 0)\} < \varepsilon a/(3C)$. Inductively we get

$$x_n = T_A^n(x_0) + \sum_{i=1}^{n} T_A^{n-i}(v_i) + \sum_{i=1}^{n} T_A^{n-i}(w_i)$$

for all $n \in \mathbb{N}$. Choosing $w_i^* \in E_u$ with $\pi_d(w_i^*) = w_i$ and $\|w_i^*\| < \varepsilon a/(3C)$ we observe that $\sum_{i=1}^{\infty} A^{-i} w_i^*$ converges. Define now

$$\bar{x} := x_0 + \pi_d\left(\sum_{i=1}^{\infty} A^{-i} w_i^*\right).$$

By means of a straightforward calculation we find

$$T_A^n(\bar{x}) = x_n + \pi_d\left(\sum_{i=1}^{\infty} A^{-i} w_{i+n}^*\right) - \sum_{i=1}^{n} T_A^{n-i}(v_i).$$

Combining this with the above estimates yields

$$d(T_A^n(\bar{x}), x_n) \leq C\frac{\varepsilon a}{3C} \sum_{i=1}^{\infty} (1+a)^{-i} + C\frac{\varepsilon a}{3C} \sum_{i=1}^{n} (1-a)^i < \varepsilon.$$

The point $\bar{x}$ hence has the desired property. $\square$

Although x_n will be close to $T^n(x_0)$ for a few steps only, Theorem 2.36 tells us that nevertheless there is a point $\bar{x}$ near x_0 whose orbit looks much the same as $\{x_n : n \in \mathbb{N}_0\}$. This property of T_A is usually referred to as *shadowing*: every sufficiently accurate pseudo-orbit is the "shadow" of a true orbit nearby.

One should note however that the orbit shadowed by the pseudo-orbit need not be typical in any sense. Due to the finiteness of representable numbers any orbit on any digital computer will be eventually periodic: the more powerful the machine, the longer the non-periodic sequences that may be observed. In fact, there are even more drastic examples. Let $b \in \mathbb{N}\setminus\{1\}$ denote the base of the computer's internal representation. The map $x \mapsto bx \pmod 1$ which – according to our notation – is just the endomorphism $T_{(b)}$ of $\mathbb{T}^1$ will eventually fix all machine numbers at 0 after a number of steps only depending on the machine's arithmetic. Later we shall see that contrary to that observation no nearby *typical* point of $\mathbb{T}^1$ is periodic under $T_{(b)}$. Finally, shadowing is generally not guaranteed for non-hyperbolic maps. For example, the 2δ-pseudo-orbit $(x_n)_{n\in\mathbb{N}_0}$ with $x_n := (n\delta, n\delta) \in \mathbb{T}^2$ does not ε-shadow any true orbit of the toral automorphism

$$T_A : \mathbb{T}^2 \to \mathbb{T}^2 \quad \text{with} \quad A := \begin{pmatrix} 0 & 1 \\ 1 & 0 \end{pmatrix},$$

no matter how small δ is chosen.

As announced earlier we shall close by dealing with the case $\det A = 0$. To this end introduce two subspaces of $\mathbb{R}^d$ by

$$X_1 := \bigcap_{n \geq 0} A^n(\mathbb{R}^d) \quad \text{and} \quad X_2 := \{x \in \mathbb{R}^d : A^n x = 0 \text{ for some } n \in \mathbb{N}\}$$

and recall the basic fact from linear algebra that $\mathbb{R}^d$ is the direct sum of these spaces, i.e. $\mathbb{R}^d = X_1 \oplus X_2$. Moreover, $A(X_1) = X_1$ as well as $A(X_2) \subseteq X_2$ and $A^d(X_2) = \{0\}$. Let $d_1 := \dim X_1 < d$. If $d_1 = 0$ then the dynamics of T_A is simple: all points are eventually fixed at zero. If $d_1 \geq 1$ then we may find a basis $y_1, \ldots, y_{d_1} \in \mathbb{Z}^d$ of the d_1-dimensional lattice $X_1 \cap \mathbb{Z}^d$. Furthermore let the integer matrix $B = (b_{jk}) \in \mathbb{Z}^{d_1 \times d_1}$ be defined via $A y_k = \sum_j b_{jk} y_k$; clearly $\det B \neq 0$. For any point $x \in \pi_d(X_1)$ with $\pi_d\left(\sum_{j=1}^{d_1} \xi_j y_j\right) = x$ set $\Psi(x) := \pi_{d_1}(\xi_1, \ldots, \xi_{d_1})$. The map $\Psi : \pi_d(X_1) \to \mathbb{T}^{d_1}$ is easily seen to be both a homeomorphism and a group isomorphism. Since

$$\Psi \circ T_A(x) = \Psi \circ \pi_d\left(A\left(\sum_{j=1}^{d_1} \xi_j y_j\right)\right) = \Psi \circ \pi_d\left(\sum_{j,k} b_{kj} \xi_j y_k\right) = \pi_{d_1}(B\xi) = T_B \circ \Psi(x)$$

holds for all $x \in \pi_d(X_1)$, the two dynamical systems $(\pi_d(X_1), T_A|_{\pi_d(X_1)})$ and $(\mathbb{T}^{d_1}, T_B)$ are topologically conjugate. Furthermore $T_A^n(x) \in \pi_d(X_1)$ for every $x \in \mathbb{T}^d$ and $n \geq d$; therefore it is not at all restrictive to exclusively deal with the case $\det A \neq 0$. (The cautious reader may have noticed that the matrix B above is not uniquely determined as it depends on the choice of the basis $y_1, \ldots, y_{d_1}$ of $X_1 \cap \mathbb{Z}^d$. Different choices will yield different yet topologically conjugate toral maps T_B.)

Exercises

(2.1) Show by means of examples that Sharkovsky's theorem is sharp: given a natural number p construct a continuous map on $\mathbb{R}$ having a periodic point with prime period p but none with prime period q for *any* $q \neq p$ with $p \lhd q$. Additionally, provide a continuous map on $\mathbb{R}$ the prime periods of which are precisely $\{2^n : n \in \mathbb{N}_0\}$.

(2.2) The local dynamics near a hyperbolic fixed point is easy to understand as it is completely governed by a *linear* map. Specifically, prove the following simplest version of the so-called Hartman–Grobman Theorem. Let x_0 denote a hyperbolic fixed point of $T : \mathbb{R} \to \mathbb{R}$ with $T'(x_0) = \lambda \neq 0$. Then there exist open intervals $V \subseteq U$ containing x_0 and a homeomorphism $h : U \to \mathbb{R}$ such that $h \circ T = L \circ h$ holds on V; here L denotes the linear map $x \mapsto \lambda x$ on $\mathbb{R}$. Therefore locally T is conjugate to its linearization. Is the latter also true for $\lambda = 0$?

(2.3) Let T denote a continuous map on a complete separable metric space with $T(X) = X$. Show that the following statements are equivalent:

  (i) T is topologically transitive;

  (ii) there exists a dense orbit, i.e. $\overline{O^+(x)} = X$ for some $x \in X$.

Furthermore, give examples showing that this equivalence does not hold if any of the assumptions is dropped.

(2.4) Let p denote a real polynomial of degree two or three. Show that the induced circle map N_{p,S^1} is topologically conjugate to N_{q,S^1} where in the quadratic case q equals one of the polynomials

$$q_+(x) := x^2 + 1, \quad q_\bullet(x) := x^2, \quad q_-(x) := x^2 - 1,$$

whereas in the cubic case q is to be found either among

$$c_+(x) := x^3 + x, \quad c_\bullet(x) := x^3, \quad c_-(x) := x^3 - x,$$

or else is a member of the one-parameter family

$$c_\gamma(x) := x^3 + \gamma x + 1 \quad (\gamma \in \mathbb{R}).$$

Completely describe the dynamics of N_{c_-,S^1}. Verify that the *domain of attraction* of -1, i.e. the set $\{z \in S^1 : N_{c_-,S^1}^n(z) \to -1\}$, is an open arc while the corresponding sets for $\pm i$ are countable unions of disjoint open arcs.

(2.5) For the sake of simplicity, in the exposition we have exclusively dealt with Newton maps associated to *real* polynomials. However, the generalization to *complex* coefficients and arguments is straightforward: given $p(z) = a_l z^l + \ldots + a_0$ with $l \geq 2$ and $a_k \in \mathbb{C}$, $a_l \neq 0$ we may associate the rational map

$$N_p(z) := z - \frac{p(z)}{p'(z)}$$

which is defined except for a finite number of poles. Consider the two-dimensional unit sphere $S^2 := \{x \in \mathbb{R}^3 : x_1^2 + x_2^2 + x_3^2 = 1\}$ together with the stereographic projection

$$\Phi : \left\{ \begin{array}{ccc} S^2 \backslash \{(0,0,1)\} & \to & \mathbb{C}, \\ (x_1, x_2, x_3) & \mapsto & \frac{x_1}{1-x_3} + i \frac{x_2}{1-x_3} . \end{array} \right.$$

By imitating the real case, show that N_p induces a smooth map N_{p,S^2} on the sphere such that

$$N_p(z) = \Phi \circ N_{p,S^2} \circ \Phi^{-1}(z)$$

holds whenever the right-hand side is defined.

If p is a quadratic polynomial show that N_{p,S^2} is topologically conjugate to N_{q,S^2} with q equal to

$$q_{\bullet}(z) := z^2 \quad \text{or} \quad q_{+}(z) := z^2 + 1 ;$$

completely analyse the dynamics of N_{q,S^2} for these two polynomials. In the cubic case N_{p,S^2} is – up to topological conjugacy – given by N_{c,S^2} with c belonging to the following list:

$$c_{\bullet}(z) := z^3, \quad c_{+}(z) := z^3 + 1, \quad c_{\gamma}(z) := z^3 + z + \gamma \ (\gamma \in \mathbb{C}) .$$

Explore the dynamics of N_{c_+} as detailed as you can.

(2.6) Once again consider the quadratic map $Q_c : z \mapsto z^2 - c$ with $c \in \mathbb{C}$ and assume $|c| > 2$. Call a compact set a *Cantor set* if it is perfect and no two of its points can be joined by a continuous path within that set. (This notion is the natural generalization of Definition 2.19.) Show that there exists a Cantor set Λ_c with $Q_c(\Lambda_c) = \Lambda_c$ and $Q_c^n(z) \to \infty$ for $z \notin \Lambda_c$; furthermore $(\Lambda_c, Q_c|_{\Lambda_c})$ is topologically conjugate to (Σ_2, σ).

(2.7) Let $T : X \to X$ denote a continuous map on the metric space X. Define a point x to be *recurrent* if for any neighbourhood U of x we have $T^n(x) \in U$ for some $n \in \mathbb{N}$; denote by $\mathcal{R}(T)$ the set of all recurrent points. Call x *non-wandering* if $T^{-n}(U) \cap U \neq \emptyset$ for any neighbourhood U of x and an appropriate $n \in \mathbb{N}$; analogously let $\Omega(T)$ stand for the set of all non-wandering points. Make sure that $T\big(\mathcal{R}(T)\big) \subseteq \mathcal{R}(T)$ as well as $T\big(\Omega(T)\big) \subseteq \Omega(T)$, and that $\Omega(T)$ is a closed set. Give examples showing that $\mathcal{R}(T)$ need not be closed, and all inclusions in

$$\text{Per } T \subseteq \mathcal{R}(T) \subseteq \overline{\mathcal{R}(T)} \subseteq \Omega(T)$$

may be proper. Also show that not necessarily $\Omega(T|_{\Omega(T)}) = \Omega(T)$ while trivially $\mathcal{R}(T|_{\mathcal{R}(T)}) = \mathcal{R}(T)$.

(2.8) Consider a point x in the full shift space Σ_l and define a subset of the latter as the closure of the orbit of x, that is $X := \overline{O^+(x)} = \overline{\{\sigma^n(x) : n \in \mathbb{N}_0\}}$. Clearly $\sigma(X) \subseteq X$; therefore $(X, \sigma|_X)$ is a symbolic dynamical system. As the situation for $x \in \text{Per } \sigma$ is trivial, focus on the more interesting case $x \in \mathcal{R}(\sigma) \backslash \text{Per } \sigma$. (For example, every topologically transitive subshift of finite type can be generated this way.) Show that X then is infinite with $\sigma|_X$ being topologically transitive and exhibiting sensitive dependence on initial conditions ([25]). Is $\sigma|_X$ chaotic?

(2.9) Prove that the natural extension $(\hat{X}, \hat{T})$ of a continuous map T on a metric space X is minimal in the following sense: whenever an *invertible* dynamical system (Y, S) factors onto (X, T) then it also factors onto $(\hat{X}, \hat{T})$ in such a way that the following diagram commutes:

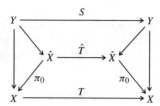

Find $(\hat{\Sigma}_{l,A}, \hat{\sigma}_A)$ where $A \in \{0, 1\}^{l \times l}$ denotes an irreducible matrix.

(2.10) A continuous map T on the metric space X is called *topologically mixing* if for any pair U, V of non-empty open sets $T^{-n}(U) \cap V \neq \emptyset$ for *all* sufficiently large $n \in \mathbb{N}$. Clearly this is a much stronger property than topological transitivity; like the latter it is preserved under semi-conjugacies.

Given an irreducible matrix $A \in \{0, 1\}^{l \times l}$, how can one decide whether the system $(\Sigma_{l,A}, \sigma_A)$ is topologically mixing? Show that there exists a natural number n_A and a decomposition of Σ_A into disjoint compact sets $\Lambda_1, \ldots, \Lambda_{n_A} \subseteq \Sigma_A$ such that $\sigma_A(\Lambda_i) = \Lambda_{i+1}$ for $i = 1, \ldots, n_A - 1$ and $\sigma_A(\Lambda_{n_A}) = \Lambda_1$; furthermore $\sigma_A^{n_A}|_{\Lambda_i}$ is topologically mixing for all i. (Commonly, $(\Lambda_i)_{i=1}^{n_A}$ is referred to as the *spectral decomposition* of $\Sigma_{l,A}$ [31].)

(2.11) For $k \in \mathbb{N}$ define the closed sets

$$B_k := \left\{ x \in \Sigma_2 : \left| \sum_{i=m}^{n} (-1)^{x_i} \right| \leq k \text{ for all } m, n \in \mathbb{N}_0 \text{ with } n \geq m \right\};$$

obviously $\{(\overline{12}), (\overline{21})\} = B_1 \subseteq B_2 \subseteq \ldots \subseteq \Sigma_2$. Since $\sigma(B_k) \subseteq B_k$ the systems $(B_k, \sigma|_{B_k})$ are symbolic dynamical systems. Show that for each k there exist an $l \in \mathbb{N}$ and an irreducible matrix $A \in \{0, 1\}^{l \times l}$ such that $(B_k, \sigma|_{B_k})$ is a factor of $(\Sigma_{l,A}, \sigma_A)$; consequently, $(B_k, \sigma|_{B_k})$ is chaotic if $k \geq 2$. However, prove by means of Exercise 2.10 that this system is *not* topologically conjugate to any subshift of finite type. (Though being a continuous image of a subshift of finite type, $(B_k, \sigma|_{B_k})$ thus itself is not of finite type; it is an example of a so-called *sofic* system [36].)

(2.12) Give a complete analysis of the dynamics of the two horseshoes depicted in Figure 2.29 – just as we did for Smale's classical horseshoe. Quantify the differences between these systems.

(2.13) Verify formula (2.14): confirm that for the billiard map $T_{\text{bill}} : (s, \alpha) \mapsto (s', \alpha')$ associated with a convex table Ω with smooth boundary $\partial\Omega$ the Jacobian is given by

$$DT_{\text{bill}} = \frac{\partial(s', \alpha')}{\partial(s, \alpha)} = \begin{pmatrix} \frac{l\kappa - \sin\alpha}{\sin\alpha'} & \frac{l}{|\partial\Omega|\sin\alpha'} \\ |\partial\Omega| \frac{l\kappa\kappa' - \kappa\sin\alpha' - \kappa'\sin\alpha}{\sin\alpha'} & \frac{l\kappa' - \sin\alpha'}{\sin\alpha'} \end{pmatrix};$$

here $|\partial\Omega|$, l, κ, κ' denotes, respectively, the length of $\partial\Omega$, the length of the billard trajectory segment between and the curvature of $\partial\Omega$ at s and s' (see Figure 2.21).

(2.14) Provide the details for the hyperbolic case of the elliptic billiard: prove that every trajectory passing between the two foci has as a caustic a hyperbola confocal with the original ellipse.

(2.15) Given $p \in \mathbb{N}$ find a $2p$-periodic point (x_p, α_p) of the map T_{bill} associated with the circular billiard which persists if the round table is perturbed to assume the shape of the stadium. While (x_p, α_p) is immediately seen to be parabolic for the circle, which means that

$$DT_{\text{bill}}^{2p}|_{(x_p, \alpha_p)} = \begin{pmatrix} 1 & * \\ 0 & 1 \end{pmatrix},$$

show that in the stadium the corresponding points are hyperbolic and unstable.

Chapter 3
Ergodic theory I. Foundations

Simply defined maps may exhibit very complicated dynamics wherein in practice the fate of individual points is completely unpredictable. We have already seen this in several illustrative examples. We also observed that the techniques necessary to rigorously analyse even such simple systems as the solenoid or toral endomorphisms tend to become quite demanding. When dealing with more complicated systems from applications, a complete mathematical analysis therefore should not be hoped for. In many cases, describing the orbit of each individual point will be too ambitious a task. Even if such a detailed description existed it would probably be of limited practical use. In applications one usually faces uncertainties due to the process of replacing real-world phenomena with simplified mathematical models, and also due to non-precise data and parameters. A reasonable analysis should be aware of these effects. From this chapter on we shall thus modify our point of view. Rather than focussing on individual orbits we are led to a *statistical* description of dynamical systems. Such a statistically motivated approach to dynamical systems theory constitutes an integral part of *ergodic theory*, which by now has developed into a multifaceted mathematical discipline. For us, the aim to analyse the asymptotic behaviour of *typical* points naturally gives a reason to introduce basic probabilistic concepts and techniques. As we shall see, the long-time behaviour of complicated systems is often most appropriately described in stochastic terms.

3.1 The statistical point of view

This short section will introduce the statistical approach to dynamical systems by means of examples. More formal definitions and considerations as well as additional examples will be discussed in subsequent sections.

When plotting the orbit of a hyperbolic toral endomorphism T_A we find very little regularity: the iterates seem to spread out over the whole space (recall for instance Figure 2.33). We shall now indicate that nevertheless there is a striking long-time regularity for this spreading. To this end we perform a statistical analysis by means of *histograms*. Such an approach has already been sketched in the Introduction and will repeatedly prove useful in the sequel.

Let us fix a not-too-small number $q \in \mathbb{N}$ and divide the space $\mathbb{T}^2$ into q^2 equal squares S_{ij}. We now take a point $x \in \mathbb{T}^2$ and draw normalized empirical histograms of the initial segment of the orbit $O^+(x)$. More formally we determine the relative frequencies

$$a_{ij}(x, n) := \frac{q^2}{n} \#\{0 \leq k < n : T_A^k(x) \in S_{ij}\} \quad (i, j, = 1, \ldots, q).$$

Although we should be careful with any conclusions drawn from numerical calculations, the result depicted in Figure 3.1 is remarkable: as n tends to infinity the relative frequencies apparently all converge to the same limit. The orbit $O^+(x)$ therefore not only spreads out over all of $\mathbb{T}^2$ but this spreading takes place uniformly in a strong statistical sense.

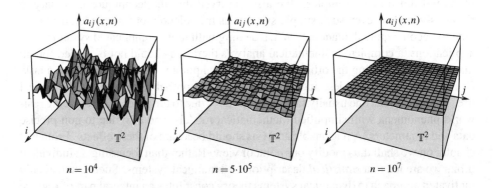

Figure 3.1. Empirical histograms for a hyperbolic toral automorphism

A lot of questions arise from these findings: Can we mathematically justify our observation? What about numerical errors, do they matter? How does the choice of x affect the results? If we had chosen x from the dense set Per T_A, then we obviously would have observed something different. Finally, what can be deduced from all this? Before generally delving into the details let us give a few prototypical answers concerning our specific example.

As we shall deduce from the general theory, the quantities $a_{ij}(x, n)$ typically converge, and

$$\lim_{n \to \infty} a_{ij}(x, n) = 1 \quad \text{for all } i, j \in \{1, \ldots, q\} \tag{3.1}$$

holds with the possible exception of a very small set of initial points x. Thus the orbit of a typical point is equally distributed all over $\mathbb{T}^2$, and empirical histograms uniformly converge towards the constant function $\mathbf{1}_{\mathbb{T}^2}$, a result strongly suggested by Figure 3.1. As we shall see, the validity of (3.1) follows from the fact that T_A is an *ergodic* map, and other properties of the dynamics (e.g. topological transitivity) may equally be deduced from this.

In the last chapter we also gave a quantitative description of T_A's sensitive dependence on initial conditions via the quantity $N(\varepsilon)$ as defined by (2.19). Roughly speaking, $N(\varepsilon)$ denotes the minimal number of iterates necessary to make points of initial distance ε drift apart to a distance of at least $\frac{1}{2}$. From Figure 3.2 an affine approximation of $N(\varepsilon)$ may be obtained which is quite accurate. For assessing the validity of numerical calculations it is most important to know how $N(\varepsilon)$ increases as ε tends towards zero. From our empirical data we find

$$\lim_{\varepsilon \to 0} \frac{N(\varepsilon)}{-\log_{10}\varepsilon} \approx 2.408 . \qquad (3.2)$$

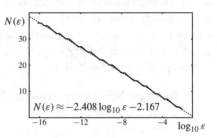

It has been pointed out earlier that this rather slow growth of $N(\varepsilon)$ as $\varepsilon \to 0$ corresponds to the long-term unpredictability of individual orbits under T_A. The statistical point of view allows a refined quantitative statement as follows. Denote by $B_n(x)$ the set of all

Figure 3.2. Approximately, there is an affine relation between $N(\varepsilon)$ and $\log_{10}\varepsilon$.

points the first n iterates of which do not deviate too much from the corresponding iterates of x, or more formally

$$B_n(x) := \left\{ y \in \mathbb{T}^2 \ : \ d\big(T_A^i(x), T_A^i(y)\big) < \frac{1}{2} \text{ for all } i = 0, \dots, n \right\}.$$

The smaller $B_n(x)$ the faster typical points near x are separated under iteration of the toral map T_A. By means of the concept of *(metric) entropy* which heavily relies on the statistical perspective we shall prove that for almost all $x \in \mathbb{T}^2$ the sets $B_n(x)$ asymptotically shrink like $\left(\frac{3+\sqrt{5}}{2}\right)^{-n}$, more precisely

$$\lim_{n \to \infty} \frac{1}{n} \log \lambda_{\mathbb{T}^2}\big(B_n(x)\big) = -\log \frac{3+\sqrt{5}}{2} .$$

Here $\lambda_{\mathbb{T}^2}$ denotes the (normalized) Lebesgue measure on $\mathbb{T}^2$ which for the sets under consideration equals their elementary area. The area of $B_n(x)$ thus shrinks exponentially as n increases. Since $\left(\log \frac{3+\sqrt{5}}{2}\right)^{-1} \approx 2.392$ this result not only agrees with (3.2) but also highlights the principal relevance of the statistical approach to dynamical systems in which we shall now embark.

In order to statistically analyse the dynamics generated by a map T on a space X we first have to decide which subsets of X are large, which ones are small, and how this size evolves under iteration of T. We therefore endow X with the structure of a measure space writing $(X, \mathcal{A}, \mu)$ instead of X. Here $\mathcal{A}$ denotes a σ-algebra, i.e. a family of subsets of X closed under countable set-theoretical operations (e.g. taking intersections, unions, complements), whereas the measure μ assigns to each $A \in \mathcal{A}$ a real number $\mu(A)$ in such a way that it may be considered the formally defined size of A. Since it is essential for the further discussion, we proceed on the assumption that the reader aiming at more than a cursory reading possesses a basic knowledge of measure theory. A few definitions and fundamental facts thereof have been compiled in Appendix A, but for a sound reference on the theme the textbooks cited there are particularly recommended.

In the sequel we need not worry too much about $\mathcal{A}$ and μ. The σ-algebra $\mathcal{A}$ will always be large enough to contain all the sets we are interested in. As far as the measure μ is concerned, all we need to know is its non-negativity and σ-additivity. We will mostly deal with normed measure spaces, which means that $\mu(X) = 1$; such spaces are also called *probability spaces*. For a better grasp of the probability space $(X, \mathcal{A}, \mu)$ it may be interpreted as follows.

(i) **Deterministic interpretation:** We think of μ as a distribution of mass on X where the mass of the whole space has been normalized to one. Given a reasonable set $A \subseteq X$ we consider $\mu(A)$ to be the mass of A, more precisely, the percentage of the total mass that is accumulated on A.

(ii) **Stochastic interpretation:** We regard X as the set of all possible outcomes ξ of an experiment governed by chance. Given $A \in \mathcal{A}$, the quantity $\mu(A)$ is just the probability $\mathbb{P}(\xi \in A)$ of the experiment to produce an outcome in A. The sets in $\mathcal{A}$ are called *events*.

A great variety of measures will show up throughout this text. Here we just mention two basic examples. On $\mathbb{R}^d$ we define $\mathcal{B}^d$ to be the smallest σ-algebra containing all d-dimensional cubes and call it the *Borel σ-algebra*. The d-dimensional *Lebesgue measure* λ^d is defined as the unique extension of the elementary d-dimensional volume to $\mathcal{B}^d$. For any space X and any point $x_0 \in X$ we can construct a measure δ_{x_0} which is concentrated at x_0, i.e., $\delta_{x_0}(A) = 1$ if and only if $x_0 \in A$, and $\delta_{x_0}(A) = 0$ otherwise. According to our above interpretations δ_{x_0} corresponds to a unit point mass situated at x_0 or to an experiment with the only possible outcome x_0, respectively. Usually δ_{x_0} is called the *Dirac measure* concentrated at x_0 (see Appendix A for details).

How can a map $T : X \to X$ be related to the measurable structure of X? Having in mind the stochastic interpretation above we would like to consider the composition $T(\xi)$ of our experiment with T as a new experiment with outcomes in X. For this to be sensible $T^{-1}(A)$ has to be an event whenever A is an event. We are therefore led to consider *measurable* maps on X. In the sequel we shall not worry much about measurability because it will be evident in most cases. A measurable map on a measure space preserves the space's structure just like a linear map does on a linear space and a continuous map does on a topological space. Let T denote a measurable map on $(X, \mathcal{A}, \mu)$. Since $T^{-1}(A) \in \mathcal{A}$ we can define

$$T\mu(A) := \mu\big(T^{-1}(A)\big) \tag{3.3}$$

for all $A \in \mathcal{A}$. It is easy to see that $T\mu$ again is a measure on $(X, \mathcal{A})$, referred to as the *induced measure* of μ under T. If we interpret μ as a mass distribution, then $T\mu$ describes the mass distribution after the space has been mixed up by T. In the light of our stochastic interpretation $T\mu$ describes the probabilistic structure of the new experiment that emerges when the old one is composed with T. A few very simple examples will illustrate this concept.

For the Dirac measure δ_{x_0} we find $T\delta_{x_0} = \delta_{T(x_0)}$ because $T\delta_{x_0}(A) = 1$ if and only if $T(x_0) \in A$. This observation may be generalized for the finite space $X := \{1, \ldots, l\}$ with $\mathcal{A} := \mathcal{P}(X)$ and $\mu := \sum_{k=1}^{l} p_k \delta_k$ where $p_k \geq 0$ and $\sum_{k=1}^{l} p_k = 1$. (For $l = 6$ this space may be considered a model for throwing a die.) If we introduce a matrix $(a_{ij}) \in \{0, 1\}^{l \times l}$ according to

$$a_{ij} := \begin{cases} 1 & \text{if } T(i) = j, \\ 0 & \text{otherwise}, \end{cases}$$

then we have $T\mu(\{k\}) = \sum_{i:T(i)=k} p_i = \sum_{i=1}^{l} p_i a_{ik}$ and therefore

$$T\mu = \sum_{k} q_k \delta_k \quad \text{with} \quad q_k = \sum_{i=1}^{l} p_i a_{ik}.$$

Therefore the assignment $\mu \mapsto T\mu$ may be represented by a matrix multiplication, and hence is *linear*. Evidently

$$a_{ij} \geq 0 \quad \text{and} \quad \sum_{k=1}^{l} a_{ik} = 1 \text{ for all } i, j;$$

later we shall say more about such non-negative matrices with rows adding up to one.

For another example consider the measure space $([0, 1], [0, 1] \cap \mathcal{B}^1, \lambda^1|_{[0,1]})$ and the map $T : x \mapsto x^2$. Since $T\lambda^1([a, b]) = \sqrt{b} - \sqrt{a} = \int_a^b \frac{dx}{2\sqrt{x}}$ the density of $T\lambda^1|_{[0,1]}$ with respect to λ^1 is just $f_1(x) := (2\sqrt{x})^{-1}$. By iteration we find $f_n(x) := 2^{-n} x^{2^{-n}-1}$ as the density of $T^n \lambda^1|_{[0,1]}$. Observe how these densities concentrate at zero (cf. Figure 3.3). Obviously, this is in perfect accordance with our intuition concerning the dynamics of T since likewise all points in $[0,1[$ are attracted towards zero.

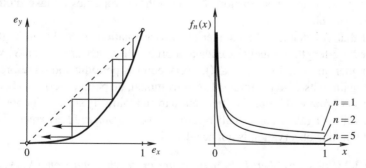

Figure 3.3. For $T : x \mapsto x^2$ orbits and densities (right) both concentrate at zero.

As a final example consider $(\mathbb{R}^1, \mathcal{B}^1, \lambda^1)$ together with the translation to the right $T : x \mapsto x + 1$. Since $T\lambda^1([a, b]) = \lambda^1([a-1, b-1]) = b - a = \lambda^1([a, b])$ we have $T\lambda^1 = \lambda^1$. Due to its fundamental importance in ergodic theory we assign a specific notion to this relationship.

Definiton 3.1.   A measurable map T on a measure space $(X, \mathcal{A}, \mu)$ is *measure-preserving* if $T\mu = \mu$. Equivalently, μ is said to be *T-invariant* in this case.

As we shall see later, invariant measures may provide a lot of information relevant to our analysis of dynamical systems. A very first impression of the intimate relation between invariant measures and dynamical properties may be gained by looking at the above examples. Clearly $T\delta_{x_0} = \delta_{x_0}$ if and only if x_0 is a *fixed point*. In the case of the finite space $X = \{1, \ldots, l\}$ we find $T\mu = \mu$ if and only if

$$p_k = \sum_{i=1}^{l} p_i a_{ik} \quad \text{for all } k \in \{1, \ldots, l\}.$$

Thus $\mu = \sum_{k=1}^{l} p_k \delta_k$ is T-invariant precisely if $(p_1, \ldots, p_l)$ is a (left) eigenvector of $(a_{ij})_{i,j=1}^{l}$. Especially, if T is a permutation of X which is irreducible in the sense that $O^+(k) = X$ for any k then $p_k = 1/l$ for all k. Recalling our stochastic interpretation this is fairly obvious: statistically a die and an irreducibly permuted die will be indistinguishable if and only if the die is *fair*, which means that all outcomes are equally likely to occur. In the case of the third example above δ_0 is invariant while $\lambda^1|_{[0,1]}$ is not. Since zero is an attracting fixed point we may hope that δ_0 is a statistically significant measure. Indeed, we shall prove later that in a reasonable sense $T^n \lambda^1|_{[0,1]} \to \delta_0$ as $n \to \infty$. The stochastic and deterministic description are thus in perfect agreement.

3.2   Invariant and ergodic measures

We are now going to develop the statistical perspective by discussing important features of measure-preserving maps. To motivate one of these properties we take another look at a simple example.

Recall that R_ϑ denotes the counter-clockwise rotation of S^1 by an angle $2\pi\vartheta$. Normalized arc-length uniquely extends to a probability measure λ_{S^1} on S^1 which is invariant under any rotation. If $\vartheta \in \mathbb{Q}$, every point is periodic and therefore returns again and again to itself in the future. If ϑ is irrational, no point is periodic, but even in this case every point will infinitely often return to an arbitrarily small neighbourhood of itself. The following theorem, usually referred to as the *Poincaré Recurrence Theorem*, points out that this observation is not accidental.

Theorem 3.2 (Poincaré). *Let T be a measure-preserving map on $(X, \mathcal{A}, \mu)$ with $\mu(X) < \infty$. For every $A \in \mathcal{A}$ the equality $\mu(A) = \mu(A_\infty)$ holds where the set $A_\infty \subseteq A$ is defined as $A_\infty := \{x \in A : T^n(x) \in A \text{ infinitely often}\}$.*

Proof. Let B denote the set of points in A which immediately leave A forever, i.e. $B :=$ $A \cap \bigcap_{n \in \mathbb{N}} T^{-n}(A^c)$. Obviously $T^{-i}(B) \cap T^{-j}(B) = \emptyset$ if $i \neq j$. Now observe that $A \backslash A_\infty \subseteq$ $\bigcup_{n \in \mathbb{N}_0} T^{-n}(B)$. Since $\mu(X) < \infty$ we must have $\mu(A \backslash A_\infty) = 0$. $\square$

The Poincaré Recurrence Theorem can be rephrased as follows: every set of positive measure will be revisited infinitely often in the future by almost all of its points. Observe that the theorem does not hold in general if the space X has infinite measure.

An important class of dynamical systems already studied in Chapter Two is provided by toral endomorphisms. We shall now take a statistical look at these systems. By $\lambda_{\mathbb{T}^d} := \pi_d \lambda^d|_{[0,1[^d}$ a probability measure is defined on $\mathbb{T}^d$ where the latter is endowed with the σ-algebra $\mathcal{B}(\mathbb{T}^d)$. Obviously $\lambda_{\mathbb{T}^d}$ is invariant under addition in $\mathbb{T}^d$, i.e. $\lambda_{\mathbb{T}^d}(x + B) = \lambda_{\mathbb{T}^d}(B)$ for all $x \in \mathbb{T}^d$ and $B \in \mathcal{B}(\mathbb{T}^d)$. In fact, $\lambda_{\mathbb{T}^d}$ can be shown to be the only reasonable probability measure with this property: it is the so-called *Haar measure* of the compact abelian group $\langle \mathbb{T}^d, + \rangle$ ([6, 47]). The relation between $\lambda_{\mathbb{T}^d}$ and toral endomorphisms is simple.

Lemma 3.3. *The toral endomorphism* $T_A : \mathbb{T}^d \to \mathbb{T}^d$ *preserves* $\lambda_{\mathbb{T}^d}$ *if and only if* $\det A \neq 0$.

Proof. If $\det A = 0$ we have $\lambda_{\mathbb{T}^d}\left(T_A^k(\mathbb{T}^d)\right) = 0$ for some k and therefore $\lambda_{\mathbb{T}^d}$ is certainly not preserved by T_A (see Section 2.7). Assume in turn $\det A \neq 0$. We first give an elementary proof for the one-dimensional case ($d = 1$): the pre-image under $T_{(k)}$ of any interval $[a, b]$ consists of $|k|$ disjoint intervals of length $\frac{b-a}{|k|}$ and hence length is preserved. More formally, $T_{(k)}^{-1}([a, b]) = \bigcup_{i=0}^{k-1} \left[\frac{a+i}{k}, \frac{b+i}{k}\right]$ if $k > 0$ and $T_{(k)}^{-1}([a, b]) = \bigcup_{i=1}^{|k|} \left[\frac{i-b}{|k|}, \frac{i-a}{|k|}\right]$ if $k < 0$. In either case $\lambda_{\mathbb{T}^d}\left(T_{(k)}^{-1}([a, b])\right) = b - a = \lambda_{\mathbb{T}^d}([a, b])$. Although it would be possible to carry over this geometric argument to higher dimensions, we prefer a formal argument for sake of brevity. Define $\mu := T_A \lambda_{\mathbb{T}^d}$ and observe $\mu(x + B) = \lambda_{\mathbb{T}^d}\left(T_A^{-1}(x + B)\right) = \lambda_{\mathbb{T}^d}\left(y + T_A^{-1}(B)\right) = \mu(B)$, where we have used the fact that T_A is onto. By the uniqueness of Haar measure $\mu = \lambda_{\mathbb{T}^d}$. $\square$

By virtue of the Poincaré Recurrence Theorem we expect most points in $\mathbb{T}^d$ to be recurrent, i.e. returning to an arbitrarily small neighbourhood infinitely often (see also Exercises 2.7 and 3.1). On the other hand we have seen that the periodic points as well as the points forward asymptotic to zero are dense in $\mathbb{T}^d$ for any hyperbolic toral endomorphism. Nevertheless, under a statistical perspective these points are rare exceptions.

From the sole existence of a finite measure preserved by T Theorem 3.2 guarantees a lot of recurrence. Clearly the conclusions might be trivial, for example if μ just describes the equidistribution on the orbit of a p-periodic point x_0, i.e. $\mu = \frac{1}{p} \sum_{i=1}^p \delta_{T^i(x_0)}$. However, in applications it is frequently observed that a specific measure quite naturally offers itself for further considerations. Sometimes significant insights may be gained from the existence of such a *natural invariant measure*. We will get an impression thereof by the subsequent discussion.

Consider a convex billiard table Ω with C^2-boundary $\partial\Omega$. Neglecting frictional and gravitational effects we associated in Section 2.5 a smooth map T_{bill} on $X :=$

$[0, 1[\times]0, \pi[$ with the motion of a billiard ball inside Ω. Moreover, relation (2.14) provided us with an expression for the differential DT_{bill} in terms of several geometric quantities. From this expression we deduce

$$\det DT_{\text{bill}}|_{(s,\alpha)} = \frac{\sin \alpha}{\sin \alpha'} . \tag{3.4}$$

If we now define a probability measure μ_{bill} on $\big(X, \mathcal{B}(X)\big)$ according to

$$\mu_{\text{bill}}(A) := \frac{1}{2} \int_A \sin \alpha \, ds \, d\alpha$$

we find for not-too-complicated sets A (rectangles, for example) by virtue of the transformation rule

$$\mu_{\text{bill}}\big(T_{\text{bill}}(A)\big) = \frac{1}{2} \int_{T_{\text{bill}}(A)} \sin \alpha' \, ds' \, d\alpha' = \frac{1}{2} \int_A \sin \alpha \, ds \, d\alpha = \mu_{\text{bill}}(A) .$$

Since T_{bill} is invertible it thus preserves μ_{bill}. As a non-trivial consequence of Theorem 3.2 we see that almost all points in X will be recurrent under T_{bill}: starting somewhere on the boundary $\partial\Omega$, the emanating billiard trajectory will not only bring us back close to our starting point eventually but we will also re-start infinitely often into nearly the same direction that we originally started off with.

When dealing with billiards we shall henceforth automatically rely upon μ_{bill} for our statistical analysis. In fact, μ_{bill} turns out to be preserved by T_{bill} even if we do not require that Ω be convex, and still if we admit a *piecewise* C^2-boundary $\partial\Omega$; a few representatives from this much larger class of admissible billiard tables are depicted in Figure 3.4. The problem with these more general situations lies in the fact that T_{bill} is a priori not defined at corners and will generally not be continuous for non-convex tables (see Figure 3.4). Although the locus of these pitfalls in X will have μ_{bill}-measure zero, some care has to be taken, and the corresponding proof will be more complicated. We omit the details (see for instance [54]) and merely summarize as follows.

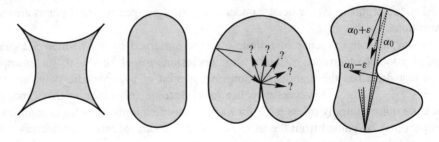

Figure 3.4. Examples of not necessarily convex tables with piecewise C^2-boundary

Proposition 3.4. *Let $\Omega \subseteq \mathbb{R}^2$ denote a bounded connected open table, the boundary of which is a finite union of piecewise C^2-curves. The associated billiard map T_{bill} can be chosen measurable, and μ_{bill} is preserved by T_{bill}.*

We are now in a position to complete our discussion of the elliptic billiard. Recall that for the case of an elliptic caustic we found the restriction of T_{bill} to an invariant curve to be topologically conjugate to a smooth homeomorphism J of S^1. Let T_X denote the point of tangency of the line joining X and $J(X)$ with the inner circle (see Figure 3.5) and define a real-valued function f according to

$$f : \begin{cases} S^1 & \to & \mathbb{R}, \\ X & \mapsto & d(X, T_X)^{-1}. \end{cases}$$

From Figure 3.5 we deduce that

$$\frac{f(X)}{f(J(X))} = \frac{d(T_X, J(X))}{d(T_X, X)} = \lim_{Y \to X} \frac{d(Q, J(X))}{d(Q, X)}$$
$$= \lim_{Y \to X} \frac{d(Q, J(X))}{d(Q, J(Y))} \frac{d(J(X), J(Y))}{d(X, Y)} = J'(X); \qquad (3.5)$$

here we have used the fact that the triangles XQY and $J(Y)QJ(X)$ are similar. If we define μ_f to be the measure on S^1 whose density with respect to λ_{S^1} is f then we have by virtue of (3.5) and the transformation rule

$$\mu_f(J^{-1}(A)) = \int_{J^{-1}(A)} f \, d\lambda_{S^1} = \int_{J^{-1}(A)} f \circ J J' \, d\lambda_{S^1} = \int_A f \, d\lambda_{S^1} = \mu_f(A).$$

The measure μ_f therefore is invariant under J. The following lemma deals with an implication of this observation.

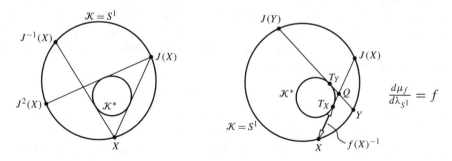

Figure 3.5. The billiard induced circle homeomorphism J preserves the measure μ_f.

Lemma 3.5. *Let T denote an orientation-preserving homeomorphism of the circle S^1, and let the finite measure μ be equivalent to λ_{S^1} and T-invariant. Then T is* *topologically conjugate to a rotation; furthermore*

$$\rho(T) = \frac{\mu([z, T(z)])}{\mu(S^1)}$$

for any $z \in S^1$; here $[a, b]$ with $a, b \in S^1$ stands for the positively oriented arc joining a and b.

Proof. It is easily verified that the map $h : S^1 \to S^1$ with

$$h(z) := e^{2\pi i \frac{\mu([1,z])}{\mu(S^1)}}$$

defines an orientation-preserving homeomorphism of the circle; furthermore the relation $\lambda_{S^1}(h(A)) = \mu(A)/\mu(S^1)$ holds for every arc $A \subseteq S^1$. Setting $R := h \circ T \circ h^{-1}$ we find

$$\lambda_{S^1}(R^{-1}(A)) = \frac{\mu(T(h^{-1}(A)))}{\mu(S^1)} = \frac{\mu(h^{-1}(A))}{\mu(S^1)} = \lambda_{S^1}(A)$$

for any arc. Therefore the circle homeomorphism R preserves λ_{S^1} and thus is a rotation. Additionally,

$$\rho(R) = \lambda_{S^1}([h(z), R \circ h(z)]) = \lambda_{S^1}(h([z, T(z)])) = \frac{\mu([z, T(z)])}{\mu(S^1)}$$

for any $z \in S^1$. $\square$

As announced earlier we thus have arrived at a complete description of the billiard inside an ellipse in the case of an elliptic caustic. When restricted to the corresponding invariant curve the billiard map is topologically conjugate to a rotation. (Exercise 3.7 asks for an explicit calculation of the corresponding rotation number.) Concerning the auxiliary map J (Figure 3.5) we may say that depending on the small circle $\mathcal{K}^*$ the broken line joining $X, J(X), J^2(X), \ldots$ either closes after the same number of steps for *all* $X \in S^1$, or densely fills the region between the two circles. This last observation is a special case of a famous theorem from projective geometry due to Poncelet ([13]).

Historically, it was in the context of differential equations that recurrence phenomena were first studied. Therefore, despite our focus on discrete dynamical systems throughout this text, let us take a short excursion to differential equations. We denote by $V : \mathbb{R}^d \to \mathbb{R}^d$ a C^1 vector field sufficiently well-behaved at infinity such that for every $x_0 \in \mathbb{R}^d$ the initial value problem

$$\dot{x} = V(x), \quad x(0) = x_0, \tag{3.6}$$

has a unique solution $x(t; x_0)$ for all $t \geq 0$. The time-t-map Φ_t defined by $\Phi_t(x_0) := x(t; x_0)$ describes the overall evolution according to (3.6) that takes place within a time interval of length t. The family $(\Phi_t)_{t \geq 0}$ is called the *flow* generated by the differential equation. Taking a not-too-complicated set $B \subseteq \mathbb{R}^d$ (a d-dimensional cube, for example) we would like to quantify the evolution of the volume $v_B(t) := \lambda^d(\Phi_t(B))$. As a standard result from analysis ([2]) we have

$$\dot{v}_B(t) = \int_{\Phi_t(B)} \operatorname{div} V \, d\lambda^d \quad \text{for } t \geq 0,$$

from which we deduce that Φ_t certainly preserves λ^d if div $V \equiv 0$. Unfortunately, Lebesgue measure is not finite. However, in some cases a bounded region $D \subseteq \mathbb{R}^d$ may be found which satisfies $\Phi_t(D) = D$ for all $t \geq 0$. In this case the analysis can be restricted to D, thus yielding a measure-preserving system on a *finite* measure space. Without any additional assumptions a wealth of recurrence phenomena is then guaranteed by Poincaré's theorem.

An important class of examples to which this last observation applies is given by autonomous *Hamiltonian systems*. By definition such systems can be written in the form

$$\dot{x}_{2k-1} = \frac{\partial H}{\partial x_{2k}}, \quad \dot{x}_{2k} = -\frac{\partial H}{\partial x_{2k-1}} \quad \text{for all } k \in \{1, \ldots, d\} \tag{3.7}$$

where $H : U \to \mathbb{R}$ denotes a C^2-function defined on some open set $U \subseteq \mathbb{R}^{2d}$. The flow generated by (3.7) preserves λ^d because

$$\text{div}\left(\frac{\partial H}{\partial x_2}, -\frac{\partial H}{\partial x_1}, \ldots, \frac{\partial H}{\partial x_{2d}}, -\frac{\partial H}{\partial x_{2d-1}} \right) = \sum_{k=1}^{d} \left(\frac{\partial^2 H}{\partial x_{2k-1} \partial x_{2k}} - \frac{\partial^2 H}{\partial x_{2k} \partial x_{2k-1}} \right) \equiv 0.$$

In many applications the *Hamiltonian function H* may be interpreted as a generalized energy exhibiting the quite natural property that $|H(x)| \to \infty$ whenever $\|x\| \to \infty$. Since H is conserved along trajectories of (3.7) by virtue of

$$\frac{d}{dt} H\big(\Phi_t(x_0)\big) = \sum_{k=1}^{2d} \frac{\partial H}{\partial x_k} \dot{x}_k = \sum_{k=1}^{d} \frac{\partial H}{\partial x_{2k-1}} \frac{\partial H}{\partial x_{2k}} + \frac{\partial H}{\partial x_{2k}} \left(-\frac{\partial H}{\partial x_{2k-1}} \right) = 0,$$

the corresponding time-t-maps fit into the framework of Poincaré's theorem when restricted to the bounded level-sets $H^{-1}([a, b])$.

It was this observation that caused considerable confusion among statistical physicists at the beginning of the twentieth century. To explain this, think of an adiabatic, i.e. perfectly insulated chamber divided into two equal cells by a removable wall. Assume that the left cell is filled with gas while the right cell is left empty (cf. Figure 3.6). If the wall is removed we expect the gas to fill the whole chamber. A first approach to this

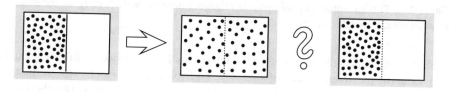

Figure 3.6. The expected return time might exceed the life-span of the universe.

process consists of modelling the gas as a large number of hard spheres (molecules) interacting elastically with one another and the walls, respectively. This model gives rise to a system of Hamiltonian type in a bounded region of a very high-dimensional space. Of course there will occur some discontinuities due to collisions but the resulting system will nevertheless be measure-preserving. The conclusion drawn from the recurrence theorem sounds paradoxical: the system will return arbitrarily close to its initial state (here: all the molecules in the left half) sometime in the future. Such a behaviour clearly contradicts our experience with physical processes like the one considered here. However, this apparent paradox can easily be resolved by noting that the theorem does not make any claim on the expected return time. Indeed, for the system under consideration the *mean return time* will be extremely large (or even infinite). With other words, returning to the initial state is extremely unlikely on short time scales – which is, after all, obvious by intuition.

By means of a simple mechanical system we shall take a closer look at the behaviour of return times. Consider a mathematical pendulum whose motion is (after an appropriate scaling) governed by the Hamiltonian

$$H_1(x_1, x_2) := -\frac{2}{\pi}\left(\cos\frac{\pi x_1}{2}\right)^2 + \frac{x_2^2}{2}.$$

As indicated in Figure 3.7 we restrict our analysis to the lens-shaped region $D_1 := \{(x_1, x_2) : H_1(x_1, x_2) \le 0, |x_1| \le 1\}$. By the above discussion the restriction of the flow to D_1 preserves $\lambda^2|_{D_1}$. The conclusion of the recurrence theorem is trivial in the

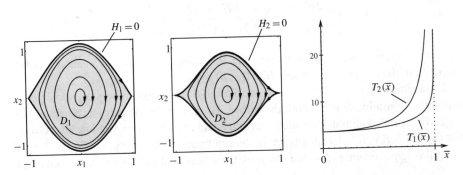

Figure 3.7. Two Hamiltonian phase portraits showing different growth rates for the length of periods near the boundary of the shaded regions

present example because all points in the interior of D_1 are periodic. But what can be said about the mean return time of a set $A \subseteq D_1$? (Notice how stochastic terminology naturally enters our discussion!) To answer this question we have to analyse the growth of periods near the boundary of D_1. For the trajectory starting at $(\bar{x}, 0)$ with $0 < \bar{x} < 1$ we can explicitly calculate the period as

$$T_1(\bar{x}) = 2\sqrt{2\pi} \int_0^{\bar{x}} \frac{d\xi}{\sqrt{\cos\pi\xi - \cos\pi\bar{x}}}$$

from which we get the asymptotic behaviour ([1])

$$T_1(\overline{x}) = -\frac{4}{\sqrt{\pi}} \log(1 - \overline{x}) + O(1) \quad \text{as } \overline{x} \to 1 .$$

Therefore T_1 has a rather weak singularity along the boundary of D_1, and we find a *finite* mean return time for every set $A \subseteq D_1$. However, such a situation need not occur for every measure-preserving system. For example, take the slightly modified Hamiltonian

$$H_2(x_1, x_2) := -\frac{\pi}{8}(1 - x_1^2)^4 + \frac{x_2^2}{2} .$$

Setting more or less as before $D_2 := \{(x_1, x_2) : H_2(x_1, x_2) \le 0, |x_1| \le 1\}$ an analogous calculation shows that

$$T_2(\overline{x}) = \frac{8}{\sqrt{\pi}} \int_0^{\overline{x}} \frac{d\xi}{\sqrt{(1 - \xi^2)^4 - (1 - \overline{x}^2)^4}} .$$

After some straightforward but lengthy analysis we deduce that

$$\lim_{\overline{x} \to 1} (1 - \overline{x}) T_2(\overline{x}) = \frac{1}{\pi\sqrt{8}} \Gamma(0.25)^2 \approx 1.479 ;$$

here Γ denotes the Euler Gamma function ([1]). Consequently, there are many subsets of D_2 for which the mean return time is *infinite*. The main reason for this unpleasant fact can be seen from the phase portrait (Figure 3.7): there is not enough mixing in the system. A point starting near the boundary of D_2 will forever stay near that boundary, moving slowly most of the time and therefore needing a very long time to return.

If there were more interaction between different parts of the space we would perhaps expect the mean return time to remain finite. In fact, a certain indecomposability condition turns out to be of fundamental importance. Although most of the terminology may be analogously introduced for flows, we now return and definitely restrict ourselves to *maps*.

Definiton 3.6. Let T be a measure preserving map on a probability space $(X, \mathcal{A}, \mu)$. We call T *ergodic* if for each $B \in \mathcal{A}$ with $T^{-1}(B) = B$ we have $\mu(B) \in \{0, 1\}$. The measure μ is then also said to be ergodic for T.

An ergodic system cannot be split into smaller pieces. If we find a set B satisfying $T^{-1}(B) = B$ then clearly $T(B) \subseteq B$ as well as $T(B^c) \subseteq B^c$. We can therefore study $T|_B$ and $T|_{B^c}$ separately. However, since T is ergodic, one of the sets B, B^c must essentially be the whole space. From this we see that neither of the Hamiltonian systems considered above is ergodic. Deciding whether a given system is ergodic will be difficult in general. Despite its technical appearance sometimes the following characterization proves useful.

Lemma 3.7. *Let T denote a measure-preserving map on the normed measure space $(X, \mathcal{A}, \mu)$. The following statements are equivalent:*

(i) *T is ergodic;*

(ii) *If $T(B_1) \subseteq B_1$, $T(B_2) \subseteq B_2$ and $\mu(B_1 \cap B_2) = 0$ for some $B_1, B_2 \in \mathcal{A}$ then $\mu(B_1)\mu(B_2) = 0$;*

(iii) *If $\mu(T^{-1}(B) \triangle B) = 0$ then $\mu(B) \in \{0, 1\}$ (here $\triangle$ denotes the symmetric difference, see Appendix A);*

(iv) *If $\mu(B) > 0$ then $\mu\left(\bigcup_{n \in \mathbb{N}_0} T^{-n}(B)\right) = 1$;*

(v) *If $\mu(B)\mu(C) > 0$ then $\mu(T^{-n}(B) \cap C) > 0$ for some $n \in \mathbb{N}_0$;*

(vi) *If $f \in L^2(\mu)$ satisfies $f \circ T = f$ $[\mu]$ then $f = \text{const.}$ $[\mu]$.*

Proof. In order to deduce (ii) from (i) define $B_i^* := \bigcup_{n \in \mathbb{N}_0} T^{-n}(B_i) \supseteq B_i$ for $i = 1, 2$. Since $T^{-1}(B_i^*) = B_i^*$ and $B_1^* \cap B_2^* \subseteq \bigcup_{n \in \mathbb{N}_0} T^{-n}(B_1 \cap B_2)$ we conclude that $\mu(B_1)\mu(B_2) = 0$, hence (ii). Assume in turn (ii) and let $\mu(T^{-1}(B) \triangle B) = 0$. The set $C := \bigcap_{n \in \mathbb{N}_0} \bigcup_{k \geq n} T^{-k}(B)$ satisfies $T(C) \subseteq C$ and $\mu(C) = \mu(B)$. Since an analogous argument applies for B^c we obtain $\mu(B)\mu(B^c) = 0$ and thus (iii). It is easy to see that $T^{-1}(B^*) \subseteq B^*$ and $\mu(T^{-1}(B^*) \triangle B^*) = 0$ for $B^* := \bigcup_{n \in \mathbb{N}_0} T^{-n}(B)$. Assuming (iii) and $\mu(B) > 0$ therefore implies $\mu(B^*) = 1$, proving (iv). Obviously (v) follows from (iv). Let f be a square-integrable function satisfying $f \circ T = f$ $[\mu]$ and set $C_{n,k} := \{x \in X : k2^{-n} \leq f(x) < (k+1)2^{-n}\}$ for $n \in \mathbb{N}, k \in \mathbb{Z}$. Clearly $\mu(\bigcup_{k \in \mathbb{Z}} C_{n,k}) = 1$ for all n. On the other hand $\mu(T^{-1}(C_{n,k}) \triangle C_{n,k}) = 0$. By (v) we get that $\mu(C_{n,k_1})\mu(C_{n,k_2}) = 0$ if $k_1 \neq k_2$. Therefore $\mu(C_{n,k_n}) = 1$ for exactly one $k_n \in \mathbb{Z}$. On $X^* := \bigcap_{n \in \mathbb{N}} C_{n,k_n}$ the function f clearly is constant. Since $\mu(X^*) = 1$ we have proved (vi). Finally, (i) follows immediately from (vi) by taking $f := \mathbf{1}_B$ for an invariant set B. $\square$

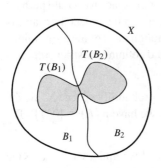

Figure 3.8. If T is ergodic then $\mu(B_1)\mu(B_2) = 0$.

We therefore have at our disposal several characterizations of ergodicity. Before applying the above lemma let us interpret some of the equivalent conditions mentioned. The meaning of condition (ii) is easily grasped: it just says that a situation as depicted in Figure 3.8 cannot occur unless at least one of the sets B_i has measure zero. From (v) we deduce that ergodicity may be considered the statistical analogue of topological transitivity: it is always possible to move from any set of positive measure to any other such set. Finally, assume that we have $f \circ T = f$ for some function $f \in L^2(\mu)$, i.e., the latter does not "feel" the action of T at all. It is easy to see that the finite measure $\mu_{|f|}$ defined by $\mu_{|f|}(A) := \int_A |f| d\mu$ is preserved by T. Clearly $\mu_{|f|} \ll \mu$, and ergodicity of T just says that the only T-invariant finite measures absolutely continuous with respect to μ are scalar multiples of μ.

We shall now take a look at two earlier examples. We have seen that dynamically the rotations R_ϑ differ considerably depending on whether ϑ is rational or not.

Theorem 3.8. *The rotation R_ϑ of S^1 is ergodic (with respect to λ_{S^1}) if and only if $\vartheta \notin \mathbb{Q}$.*

Proof. If $\vartheta = p/q$, then the function $f : S^1 \to \mathbb{C}$ with $f(z) := z^q$ satisfies $f \circ R_\vartheta = f$ without being constant. Consequently, R_ϑ is not ergodic in this case. Assume in turn $\vartheta \notin \mathbb{Q}$ and $f \circ R_\vartheta = f$ for some $f \in L^2$. By means of Fourier series $f(z) = \sum_{k \in \mathbb{Z}} c_k z^k$ we have $f \circ R_\vartheta - f = \sum_{k \in \mathbb{Z} \setminus \{0\}} c_k (e^{2\pi i k \vartheta} - 1) z^k = 0$. Since $e^{2\pi i k \vartheta} \neq 1$ for all $k \in \mathbb{Z} \setminus \{0\}$ we conclude that $c_k = 0$ for $k \neq 0$. Therefore f necessarily is a constant in L^2. $\square$

Statistically, rotations of the unit circle thus behave just as we expect them to do. Up to a certain extent this is also true for toral endomorphisms. The proof of the following result also relies on the characterization (vi) in Lemma 3.7.

Theorem 3.9. *Let the toral endomorphism T_A preserve $\lambda_{\mathbb{T}^d}$. Then T_A is ergodic if and only if no eigenvalue of A is a root of unity, i.e. if*

$$\sigma(A) \cap \{e^{2\pi i y} : y \in \mathbb{Q}\} = \emptyset.$$

Proof. The following observation is crucial: if $\sum_{k \in \mathbb{Z}^d} c_k e^{2\pi i \langle k, x \rangle}$ denotes the Fourier series of a function $f : \mathbb{T}^d \to \mathbb{C}$ then we have $f \circ T_A = \sum_{k \in \mathbb{Z}^d} c_k e^{2\pi i \langle A^T k, x \rangle}$; here A^T stands for the transpose of A. Let $e^{2\pi i \frac{p}{q}}$ be an eigenvalue of A. Then $(A^T)^q k_0 = k_0$ for some $k_0 \in \mathbb{Z}^d \setminus \{0\}$. Consequently $f(x) := \sum_{j=0}^{q-1} e^{2\pi i \langle (A^T)^j k_0, x \rangle}$ satisfies $f \circ T_A = f$ without being constant. Assume in turn $\sigma(A) \cap \{e^{2\pi i y} : y \in \mathbb{Q}\} = \emptyset$. From $f \circ T_A = f$ we deduce that $c_{(A^T)^n k} = c_k$ for all $n \in \mathbb{N}_0$. If $c_{k_1} \neq 0$ for some $k_1 \neq 0$ we have $c_l = c_{k_1} \neq 0$ for infinitely many l, contradicting $\sum_{k \in \mathbb{Z}^d} |c_k|^2 < \infty$. Therefore f necessarily is constant, and T_A is ergodic. $\square$

As a corollary, we obtain ergodicity for all hyperbolic toral endomorphisms T_A with $\det A \neq 0$. Later we shall see that this fact enables us to make precise some of our observations concerning the distribution of orbits under these maps.

For the time being we shall go back to the definition of ergodicity. Remember that our main motivation for its introduction was to rule out infinite mean return times which are not a priori barred by the recurrence theorem. Up to now we must still be uncertain whether we have chosen the right definition in order to achieve this goal. The following theorem, usually referred to as Kac's lemma, shows that we have indeed been successful. It even allows to calculate (in the mean) the *return time* to the set B defined as

$$t_B(x) := \min\{n \in \mathbb{N} : T^n(x) \in B\} \quad \text{for } x \in B,$$

where (as usual) t_B is understood to be infinite if the set on the right is empty. Due to Poincaré's theorem t_B is finite almost everywhere on B.

Theorem 3.10 (Kac). *Let T denote an ergodic map on the probability space $(X, \mathcal{A}, \mu)$. If $\mu(B) > 0$ then*

$$\frac{1}{\mu(B)} \int_B t_B \, d\mu = \frac{1}{\mu(B)} \, .$$

Proof. Clearly $t_B : B \to \mathbb{N} \cup \{\infty\}$ is measurable and $\int_B t_B d\mu = \sum_{n \in \mathbb{N}} n\mu(B_n) = \sum_{n \in \mathbb{N}} \sum_{k \geq n} \mu(B_k)$ with $B_k := \{x \in B : t_B(x) = k\}$. For every $k \in \mathbb{N}$ define $C_k := T^{-1}(B^c) \cap \ldots \cap T^{-(k-1)}(B^c) \cap T^{-k}(B)$ and observe $C_k \cap B = B_k$ as well as $T^{-1}(C_k) \cap T^{-1}(B^c) = C_{k+1}$. Due to ergodicity $\mu(\bigcup_{k \in \mathbb{N}} C_k) \geq \mu(\bigcup_{k \in \mathbb{N}} T^{-k}(B)) = 1$. Furthermore, $\mu(C_{k+1}) = \mu(C_k \backslash B) = \mu(C_k) - \mu(B_k)$ and thus $\mu(C_n) = \sum_{k \geq n}(B_k)$. From this $\int_B t_B d\mu = \sum_{n \in \mathbb{N}} \mu(C_n) = 1$ follows. $\square$

Ergodicity thus forces mean return times not only to be finite but also to behave in a very clear way: the smaller the set B is in measure, the longer it takes a point (in the mean) to return to B. In other words, an ergodic measure assigns most of the weight to the fast returning points. Our definitions therefore neatly fit together and reflect some basic dynamical properties just as we wanted them to do.

3.3   Ergodic theorems

In this section the fundamental importance of ergodicity is highlighted by means of two famous ergodic theorems and a couple of related applications. As a starting point for the subsequent discussion remember our first and simplest approach to the statistics of dynamical systems by drawing histograms. We shall now formalize this approach.

Let T be a measurable map on a probability space $(X, \mathcal{A}, \mu)$. For $A \in \mathcal{A}$ and $x \in X$ we define the relative frequency of the first n iterates of x in A as

$$h_A(x, n) := \frac{1}{n} \#\{0 \leq k < n : T^k(x) \in A\} = \frac{1}{n} \sum_{k=0}^{n-1} \mathbf{1}_A \circ T^k(x). \qquad (3.8)$$

We shall mainly be interested in the behaviour of $h_A(x, n)$ for $n \to \infty$. Does this sequence converge? If so, how does the limit depend on x and A? As a slight generalization we shall also study the asymptotic behaviour of the sums

$$S_n f(x) := \frac{1}{n} \sum_{k=0}^{n-1} f \circ T^k(x) \quad \text{for } f \in L^1(\mu). \qquad (3.9)$$

For another motivation for studying expressions like (3.9) recall that we repeatedly interpreted $(X, \mathcal{A}, \mu)$ as the probabilistic structure describing the outcome of a single experiment ξ. If $(\xi_n)_{n \in \mathbb{N}_0}$ is a sequence of *independent* experiments all governed by the same law μ, then the standard Laws of Large Numbers guarantee the convergence of $\frac{1}{n} \sum_{k=0}^{n-1} f(\xi_k)$ to the mean value $\mathbb{E}f = \int_X f d\mu$ in one sense or another. Unfortunately, the outcomes $x, T(x), T^2(x), \ldots$ are definitely *not* independent: they are coupled by the map T. Studying the asymptotic properties of $S_n f$ therefore incorporates the question, whether orbits under T may – at least asymptotically – be regarded as sequences of outcomes of independent experiments. The following simple example shows that the behaviour of $S_n f$ need not be obvious from the outset.

Let X be the finite space $\{1, \ldots, 6\}$ with $\mathcal{A} := \mathcal{P}(X)$ and $\mu := \frac{1}{6} \#$ and consider the measure-preserving maps

$$T_1 : \begin{cases} 1 & 2 & 3 & 4 & 5 & 6 \\ \downarrow & \downarrow & \downarrow & \downarrow & \downarrow & \downarrow \\ 2 & 3 & 4 & 5 & 6 & 1 \end{cases} \quad \text{and} \quad T_2 := T_1 \circ T_1 \, .$$

As far as T_1 is concerned it is easily seen that for any $f : X \to \mathbb{C}$

$$\lim_{n \to \infty} S_n f(x) = \frac{1}{6} \sum_{i=1}^{6} f(i) = \int_X f \, d\mu$$

which does not depend on the choice of x. For T_2 things are slightly more complicated. Setting $B := \{1, 2, 3\}$, $O := \{1, 3, 5\}$ and $E := \{2, 4, 6\}$ for example we find

$$\lim_{n \to \infty} h_B(x, n) = \begin{cases} \frac{2}{3} & \text{if } x \in O \, , \\ \frac{1}{3} & \text{if } x \in E \, , \end{cases}$$

so this limit does depend on x. Furthermore it also depends on the chosen set: taking $C := \{1, 3\}$ we obtain

$$\lim_{n \to \infty} h_C(x, n) = \begin{cases} \frac{2}{3} & \text{if } x \in O \, , \\ 0 & \text{if } x \in E \, . \end{cases}$$

In this simple example, of course, every quantity of interest can be explicitly calculated. However, in the general situation the behaviour of $S_n f$ needs some further analysis.

Two tools turn out to be crucial for such an analysis: the concept of *conditional expectations* and the σ-algebra of *almost invariant sets*. Basic definitions concerning conditional expectations may be found in Appendix A. For us, this concept only serves as a technical device which we use for mathematical convenience. Lack of knowledge in this field should not cause serious problems in comprehending the essence of the theorems below. For practical use the concept could be outlined as follows. Given a sub-σ-algebra $\mathcal{B}$ of $\mathcal{A}$, conditional expectation $\mathbb{E}(\cdot | \mathcal{B})$ yields a positive linear projection operator from $L^1(X, \mathcal{A}, \mu)$ onto $L^1(X, \mathcal{B}, \mu|_{\mathcal{B}})$ which satisfies

$$\int_B \mathbb{E}(f | \mathcal{B}) \, d\mu = \int_B f \, d\mu \quad \text{for all } B \in \mathcal{B}, \, f \in L^1(X, \mathcal{A}, \mu) \, .$$

In more probabilistic terms, $\mathbb{E}(\cdot | \mathcal{B})$ describes the expectation of $f(\xi)$ with the information $\mathcal{B}$ available, with $\mathbb{E}(\cdot | \{\emptyset, X\})$ being just the usual unconditioned expectation $\mathbb{E}(\cdot)$ with no significant information about ξ available, and $\mathbb{E}(\cdot | \mathcal{A})$ being the identity on $L^1(X, \mathcal{A}, \mu)$.

The second tool is introduced as follows: for every measure-preserving map T the family of sets $\mathcal{A}_T := \{A \in \mathcal{A} : \mu(T^{-1}(A) \triangle A) = 0\}$ is a sub-σ-algebra of $\mathcal{A}$, usually referred to as the *σ-algebra of almost invariant sets*. The reader may wish to check that a function $f \in L^1(\mu)$ satisfies $f \circ T = f$ if and only if it is $\mathcal{A}_T$-measurable.

Armed with these concepts we are now in a position to formulate a combination of two famous theorems.

Theorem 3.11. *Let T be a measure-preserving map on the probability space $(X, \mathcal{A}, \mu)$ and $f \in L^1(\mu)$. There exists a function $f^* \in L^1(\mu)$ with $f^* \circ T = f^*$ and $\int_X f^* d\mu = \int_X f d\mu$ such that*

(i) $\lim_{n \to \infty} S_n f(x) = f^*(x)$ *for almost all points $x \in X$ ("Birkhoff Ergodic Theorem");*

(ii) $S_n f \to f^*$ *in L^1, that is, $\int_X |S_n f - f^*| d\mu \to 0$ as $n \to \infty$ ("von Neumann Ergodic Theorem").*

Proof. Our strategy is to show that $f^* = \mathbb{E}(f | \mathcal{A}_T)$ from which the first two properties follow. Let us begin with a preliminary consideration. Take $g \in L^1$ and define a sequence of L^1-functions according to $G_n(x) := \max_{1 \le k \le n} \sum_{i=0}^{k-1} g \circ T^i(x)$. Obviously, $\left(G_n(x)\right)_{n \in \mathbb{N}}$ is non-decreasing and thus either diverges to $+\infty$ or otherwise converges. Define the set $B := \{x : \sup_{n \in \mathbb{N}} G_n(x) < \infty\} \in \mathcal{A}$. From the easily proved relation $G_{n+1} - G_n \circ T = g - \min\{0, G_n \circ T\}$ we deduce that $B \in \mathcal{A}_T$. On B^c the sequence $\left(G_n(x)\right)_{n \in \mathbb{N}}$ diverges, and

$$0 \le \int_{B^c} (G_{n+1} - G_n) \, d\mu = \int_{B^c} (G_{n+1} - G_n \circ T) \, d\mu \;\to\; \int_{B^c} g \, d\mu = \int_{B^c} \mathbb{E}(f | \mathcal{A}_T) d\mu$$

by dominated convergence and the definition of conditional expectation, respectively. Thus if $\mathbb{E}(g | \mathcal{A}_T) < 0$ on X we must have $\mu(B^c) = 0$. On B the relation $\overline{\lim}_{n \to \infty} \frac{1}{n} \sum_{k=0}^{n-1} g \circ T^k \le 0$ holds.

Now we tackle the Birkhoff Ergodic Theorem. Take $f \in L^1$, $\varepsilon > 0$ and set $g := f - \mathbb{E}(f | \mathcal{A}_T) - \varepsilon$. Since $\mathbb{E}(g | \mathcal{A}_T) = -\varepsilon < 0$ we have

$$\overline{\lim}_{n \to \infty} \frac{1}{n} \sum_{k=0}^{n-1} g \circ T^k(x) = \overline{\lim}_{n \to \infty} \left(S_n f(x) - \mathbb{E}(f | \mathcal{A}_T) - \varepsilon\right) \le 0$$

for almost every $x \in X$. An analogous argument for $\mathbb{E}(f | \mathcal{A}_T) - f - \varepsilon$ finally shows that $S_n f(x) \to \mathbb{E}(f | \mathcal{A}_T)$ almost everywhere. In order to prove (ii) assume $f \in L^1$ to be bounded. In this case $S_n f$ is bounded and converges to $\mathbb{E}(f | \mathcal{A}_T)$ almost everywhere by (i). Therefore $\|S_n f - \mathbb{E}(f | \mathcal{A}_T)\| = \int_X |S_n f - \mathbb{E}(f | \mathcal{A}_T)| d\mu \to 0$ by dominated convergence. Given an arbitrary $f \in L^1$ and $\varepsilon > 0$ we may choose a bounded f_ε in L^1 such that $\|f - f_\varepsilon\| < \varepsilon$. But then

$$\|S_n f - \mathbb{E}(f | \mathcal{A}_T)\| \le 2\varepsilon + \|S_n f_\varepsilon - \mathbb{E}(f_\varepsilon | \mathcal{A}_T)\|$$

and $\overline{\lim}_{n \to \infty} \|S_n f - \mathbb{E}(f | \mathcal{A}_T)\| \le 2\varepsilon$. Since $\varepsilon > 0$ was arbitrary, (ii) follows. □

Let us make a few comments on this theorem. If T is ergodic, each set in $\mathcal{A}_T$ has measure zero or one. Therefore $\mathbb{E}(f | \mathcal{A}_T) = \mathbb{E}(f) = \int_X f d\mu$ and $\lim_{n \to \infty} S_n f(x) = \int_X f d\mu = $ const. in that case. Physicists usually rephrase the Birkhoff Ergodic Theorem by saying that the *time average* $\lim_{n \to \infty} S_n f$ and the *space average* $\int_X f d\mu$

coincide. Especially, this coincidence implies that $\lim_{n \to \infty} h_A(x, n) = \mu(A)$ almost everywhere in the ergodic case. If we remember Kac's lemma, then this last result should not come as a surprise. In non-technical language it simply says that the orbit of a typical point is distributed according to μ: the measure is concentrated exactly where the orbits of most of the points concentrate.

In our simple example above, T_1 is ergodic and therefore

$$\lim_{n \to \infty} S_n f(x) = \int_X f \, d\mu = \frac{1}{6} \sum_{i=1}^{6} f(i)$$

for every $x \in X$. While this result has always been obvious we can now give an analogous result for T_2. The difficulties we had with this map clearly came from the fact that it is not ergodic. However, the σ-algebra $\mathscr{A}_{T_2}$ equals $\{\emptyset, O, E, X\}$, and the Birkhoff Ergodic Theorem yields

$$\lim_{n \to \infty} S_n f(x) = \mathbb{E}(f | \mathscr{A}_{T_2}) = \frac{1}{3} \sum_{i \in O} f(i) \, \mathbf{1}_O(x) + \frac{1}{3} \sum_{i \in E} f(i) \, \mathbf{1}_E(x) \qquad (3.10)$$

for every $x \in X$. By means of the conditional measures on O and E,

$$\mu_O(B) := \frac{\mu(B \cap O)}{\mu(O)} = \frac{1}{3}\#(B \cap O) \quad \text{and} \quad \mu_E(B) := \frac{\mu(B \cap E)}{\mu(E)} = \frac{1}{3}\#(B \cap E),$$

respectively, (3.10) may be given the more concise form

$$\lim_{n \to \infty} S_n f(x) = \int_X f \, d\mu_O \, \mathbf{1}_O(x) + \int_X f \, d\mu_E \, \mathbf{1}_E(x). \qquad (3.11)$$

The measures μ_O and μ_E are ergodic for the restrictions $T_2|_O$ and $T_2|_E$, respectively. According to $\mu = \mu(O)\mu_O + \mu(E)\mu_E$, the original measure μ is a weighted sum of two ergodic measures. An analogous though more technical result turns out to hold under very general conditions. Such an *ergodic decomposition* of invariant measures provides an important tool in advanced ergodic theory ([44, 62]). Here we shall content ourselves with only a few elementary observations.

For any measurable map T we denote by $\mathscr{M}_T(X, \mathscr{A})$ the set of all T-invariant probability measures on $(X, \mathscr{A})$. If $\mu, \nu \in \mathscr{M}_T(X, \mathscr{A})$ then $t\mu + (1-t)\nu \in \mathscr{M}_T(X, \mathscr{A})$ for all $t \in [0, 1]$, hence $\mathscr{M}_T(X, \mathscr{A})$ is a (possibly empty) convex set. Recall that an *extreme point* of a convex set C is defined as a point which does not lie on any line-segment in C with endpoints different from that point.

Lemma 3.12. *The extreme points in $\mathscr{M}_T(X, \mathscr{A})$ are precisely the ergodic measures for T.*

Proof. Let μ be ergodic and assume $\mu = t\mu_1 + (1-t)\mu_2$ with T-invariant measures $\mu_1, \mu_2 \in \mathscr{M}_T(X, \mathscr{A})$ and $t \in [0, 1]$. Obviously $\mu_1 \ll \mu$ if $t > 0$. Consider the Radon–Nikodym derivative $f_1 := \frac{d\mu_1}{d\mu}$ and define $A := \{x : f_1(x) > 1\}$. Since both $\mu_1\big(A\backslash T^{-1}(A)\big) = \mu_1\big(T^{-1}(A)\backslash A\big)$ and $\mu\big(A\backslash T^{-1}(A)\big) = \mu\big(T^{-1}(A)\backslash A\big)$ we must have $\mu\big(A\triangle T^{-1}(A)\big) = 0$; therefore $\mu(A) \in \{0, 1\}$. In fact, we have $\mu(A) = 0$ because $\mu(A) = 1$ would imply $\mu_1(X) > 1$. By an analogous argument $\mu(\{x : f_1(x) < 1\}) = 0$ and thus $\mu_1 = \mu_2$ which shows that μ is extremal. If on the other hand μ is not ergodic then $T^{-1}(A) = A$ for some A with $0 < \mu(A) < 1$. The conditional measures $\mu_1(\cdot) := \mu(\cdot \cap A)/\mu(A)$ and $\mu_2(\cdot) := \mu(\cdot \cap A^c)/\mu(A^c)$ are points in $\mathcal{M}_T(X, \mathcal{A})$ and $\mu = \mu(A)\mu_1 + \big(1 - \mu(A)\big)\mu_2$, clearly contradicting the extremality of μ. □

Building invariant measures from ergodic ones therefore can be interpreted as reconstructing a convex set from its extreme points, which is a standard problem in convex analysis ([28, 44]). Due to their extremality, ergodic measures are quite incompatible as the following lemma shows.

Lemma 3.13. *Let $\mu_1 \neq \mu_2$ be two ergodic measures in $\mathcal{M}_T(X, \mathcal{A})$. Then μ_1 and μ_2 are mutually singular, that is, $\mu_1(A^c) = \mu_2(A) = 0$ for some set $A \in \mathcal{A}$.*

Proof. By virtue of the Lebesgue Decomposition Theorem A.10 there exist two uniquely determined measures α, β such that $\mu_1 = \alpha + \beta$ with $\alpha \ll \mu_2$ and $\beta \perp \mu_2$. Since $T\alpha \ll \mu_2$ and $T\beta \perp \mu_2$ these measures are both T-invariant. From Lemma 3.12 we deduce that $\alpha(X) \in \{0, 1\}$. If $\alpha(X) = 1$ then $\mu_1 \ll \mu_2$ and hence $\mu_1 = \mu_2$ as in the proof of Lemma 3.12. This contradiction implies $\alpha(X) = 0$, and thus $\mu_1 \perp \mu_2$. □

As a final remark we should point out that in general the set on which $S_n f(x)$ does not converge to $\mathbb{E}(f|\mathcal{A}_T)(x)$ is not empty. For example, consider the ergodic toral endomorphism $T_{(2)}$ on $\mathbb{T}^1$ and $f : \mathbb{T}^1 \to \mathbb{R}$ with $f(x) := \cos 2\pi x$. Choosing any eventually fixed point x^* we find that

$$\lim_{n \to \infty} S_n f(x^*) = f(0) = 1 \neq 0 = \int_{\mathbb{T}^1} f \, d\lambda_{\mathbb{T}^1}.$$

Hence $S_n f(x)$ does not converge to zero on a dense set. However, the Birkhoff Ergodic Theorem assures that this exceptional set cannot have positive measure.

As a first application of Theorem 3.11 let us deal with the *Strong Law of Large Numbers*, which by now is a classical topic from probability theory. Since we aim at a probabilistic view on dynamical systems the mere existence of such an application seems promising. Recall that we interpreted a probability measure μ on $(X, \mathcal{A})$ as the governing mechanism of an experiment ξ of chance with outcomes in X, that is, the probability $\mathbb{P}(\xi \in A)$ equals $\mu(A)$. Taking a numerical function $f : X \to \mathbb{C}$ we intuitively expect that we may find an approximation for the mean value $\int_X f \, d\mu$ of $f(\xi)$ by performing a sufficiently large number of *independent* experiments and then calculating the empirical average. In other words, we suppose that

$$\frac{1}{n} \sum_{k=0}^{n-1} f(\xi_k) \to \int_X f \, d\mu \quad \text{as } n \to \infty, \tag{3.12}$$

provided that the ξ_k all obey the law μ and are independent. As presented below the Strong Law of Large Numbers makes this assertion precise.

A crucial point in discussing (3.12) concerns the reasonable formalization of a sequence $(\xi_k)_{k\in\mathbb{N}_0}$ of independent, identically distributed experiments. A usual way of dealing with such a sequence is to consider it an experiment of chance in the product space $Y := \prod_{k\in\mathbb{N}_0} X$, i.e. the space of all infinite sequences in X. A probability measure ν on Y incorporating the assumed independence should satisfy

$$\nu(A_0 \times \ldots \times A_m \times X \times \ldots) = \mathbb{P}(\xi_0 \in A_0, \ldots, \xi_m \in A_m)$$
$$= \mathbb{P}(\xi_0 \in A_0) \cdot \ldots \cdot \mathbb{P}(\xi_m \in A_m) \qquad (3.13)$$
$$= \mu(A_0) \cdot \ldots \cdot \mu(A_m)$$

for all $m \in \mathbb{N}_0$ and $A_k \in \mathcal{A}$. It turns out that (3.13) suffices to uniquely determine ν on $(Y, \bigotimes_{k\in\mathbb{N}_0}\mathcal{A})$ where by definition $\bigotimes_{k\in\mathbb{N}_0}\mathcal{A}$ denotes the smallest σ-algebra containing all possible events showing up in (3.13); see Appendix A for some formal details. From a practical point of view the easy-to-grasp relation (3.13) is sufficient for our use of the new probabilistic model $(Y, \bigotimes_{k\in\mathbb{N}_0}\mathcal{A}, \nu)$. If $X = \{1, \ldots l\}$ we rediscover thé space Σ_l, though now the emphasis is on the measure-theoretic rather than on the topological point of view.

What is the benefit of the above construction? Recalling our experience with symbolic dynamics in the previous chapter it seems natural to have a closer look at the map

$$\sigma : \begin{cases} \prod_{k\in\mathbb{N}_0} X & \to & \prod_{k\in\mathbb{N}_0} X \\ (x_k)_{k\in\mathbb{N}_0} & \mapsto & (x_{k+1})_{k\in\mathbb{N}_0} \end{cases}$$

commonly termed the (one-sided) *Bernoulli shift* on $(Y, \bigotimes_{k\in\mathbb{N}_0}\mathcal{A}, \nu)$.

Lemma 3.14. *The Bernoulli shift is ergodic.*

Proof. Since

$$\nu\big(\sigma^{-1}(A_0 \times \ldots \times A_m \times X \times \ldots)\big) = \nu(X \times A_0 \times \ldots \times A_m \times X \times \ldots)$$
$$= \nu(A_0 \times \ldots \times A_m \times X \times \ldots)$$

the shift σ preserves ν. Let now $\sigma^{-1}(A) = A$ and $\varepsilon > 0$. As the family of cylinders generates $\bigotimes_{k\in\mathbb{N}_0}\mathcal{A}$ there exists a finite union B of cylinder sets such that $\nu(A \,\Delta\, B) < \varepsilon$. Since

$$\nu(\sigma^{-m}(B) \cap B^c) = \nu(B)\nu(B^c) = \nu(\sigma^{-m}(B^c) \cap B)$$

for sufficiently large $m \in \mathbb{N}$ we find

$$2\nu(B)\big(1 - \nu(B)\big) = \nu(\sigma^{-m}(B) \,\Delta\, B)$$
$$\leq \nu\big(\sigma^{-m}(B) \,\Delta\, \sigma^{-m}(A)\big) + \nu(\sigma^{-m}(A)\Delta A) + \nu(A \,\Delta\, B) \leq 2\varepsilon$$

and thus $\nu(A)\big(1 - \nu(A)\big) < 2\varepsilon + \varepsilon^2$. Since ε has been arbitrary $\nu(A) \in \{0, 1\}$. $\square$

In fact, Bernoulli shifts exhibit statistical properties much stronger than ergodicity (see Exercise 3.15) but Lemma 3.14 is all we need to make (3.12) precise.

Theorem 3.15 (Strong Law of Large Numbers). *Given $f \in L^1(\mu)$ the relation*

$$\lim_{n \to \infty} \frac{1}{n} \sum_{k=0}^{n-1} f(\xi_k) = \int_X f \, d\mu$$

holds almost surely, that is, for ν-almost all sequences $(\xi_k)_{k \in \mathbb{N}_0}$.

Proof. Define the measurable projection π_0 from $Y = \prod_{k \in \mathbb{N}_0} X$ onto its zeroth factor via $\pi_0(x) = \pi_0((x_k)_{k \in \mathbb{N}_0}) := x_0$. Then $F := f \circ \pi_0$ is integrable on $(Y, \bigotimes_{k \in \mathbb{N}_0} \mathcal{A}, \nu)$ with $\int_Y F \, d\nu = \int_X f \, d\mu$. On the other hand $\sum_{k=0}^{n-1} F \circ \sigma^k(x) = \sum_{k=0}^{n-1} f(x_k)$ for all $n \in \mathbb{N}$. The Birkhoff Ergodic Theorem thus yields the desired result. □

Readers who have seen purely probabilistic proofs of Theorem 3.15 will perhaps appreciate the dynamical argument given here for being short and transparent. It must however not be overlooked that rather advanced techniques have been made use of during the proof of the powerful Ergodic Theorem 3.11.

The second example we are going to discuss briefly provides another interesting application of the ergodic theorem but also sheds some light on potential limitations. Fixing a natural number $b \geq 2$ we can assign to every $x \in [0, 1[$ a b-adic expansion

$$x = \sum_{i=1}^{\infty} x_i b^{-i} \quad \text{with } x_i \in \{0, \ldots, b-1\}. \tag{3.14}$$

This expansion is unique except for a countable set. A natural question in this context is whether some patterns of the digits $0, \ldots, b-1$ are more likely to occur than others. Observing that for $T_{(b)} : \mathbb{T}^1 \to \mathbb{T}^1$ and $x \in [0, 1[$

$$\frac{1}{n} \#\{1 \leq i \leq n : x_i = k\} = h_A(n, \pi_1(x)) \quad \text{where } A := \pi_1\left(\left[\frac{k}{b}, \frac{k+1}{b}\right[\right)$$

we deduce from the Birkhoff Ergodic Theorem the following result often referred to as *Borel's Theorem on Normal Numbers* and first proved by means of number theoretical methods.

Theorem 3.16 (Borel). *Almost every number x in $[0, 1[$ is normal with respect to every base, i.e., for every $b \in \mathbb{N} \setminus \{1\}$ the relation*

$$\lim_{n \to \infty} \frac{1}{n} \#\{1 \leq i \leq n : x_i = k\} = \frac{1}{b}$$

holds for all $k \in \{0, \ldots, b-1\}$ where $(x_i)_{i \in \mathbb{N}}$ denotes the b-adic expansion (3.14) of x.

For any base b therefore the digits of a typical real number are equally distributed, all of them appearing with the same asymptotic relative frequency $1/b$. The simplicity of the argument is contrasted by the fact that rather sophisticated tools are required for explicitly constructing numbers which are normal with respect to *every* base. (Various techniques for *special* bases, however, have extensively been studied [22, 51].) Observe that Theorem 3.16 itself does not explicitly give any normal number. We conclude from this observation that despite their important implications the general ergodic theorems may leave some interesting questions open, which have to be answered by means of improved or even completely different methods. In the sequel we are going to discuss a few examples related to this fact.

Recall that the rotation R_ϑ of S^1 is ergodic if and only if ϑ is irrational (Theorem 3.8). Therefore $\lim_{n\to\infty} h_A(z, n) = \lambda_{S^1}(A)$ for λ^1-almost every point $z \in S^1$, and we may wonder whether there are any exceptional points at all. Indeed, we are going to show that there are none. The proof will not make use of the ergodic theorem but rather exploit the specific structure of S^1.

Theorem 3.17 (Weyl). *For every bounded Riemann-integrable function $f : S^1 \to \mathbb{C}$ and every irrational rotation R_ϑ of S^1 the relation*

$$\lim_{n\to\infty} S_n f(z) = \int_{S^1} f \, d\lambda_{S^1} = \int_0^1 f(e^{2\pi i x}) \, dx$$

holds for every $z \in S^1$.

Proof. We first prove the theorem for monomials $f_l(z) := z^l$ ($l \in \mathbb{Z}$). Since the claim is obviously true if $l = 0$, we take $l \neq 0$ and observe

$$\left| S_n f_l(z) \right| = \left| \frac{z^l}{n} \sum_{k=0}^{n-1} e^{2\pi i l k \vartheta} \right| \leq \frac{2}{n \left| e^{2\pi i l \vartheta} - 1 \right|} \to 0 = \int_{S^1} f_l \, d\lambda_{S^1} .$$

Therefore the result holds for polynomials on S^1 and even for continuous functions by virtue of the Weierstrass Approximation Theorem. Finally, assume that f is a real-valued function exhibiting a finite number of steps, and take $\varepsilon > 0$. We can find continuous functions g_ε and h_ε with

$$g_\varepsilon \leq f \leq h_\varepsilon \quad \text{and} \quad \int_{S^1} f - \varepsilon \leq \int_{S^1} g_\varepsilon \leq \int_{S^1} h_\varepsilon \leq \int_{S^1} f + \varepsilon .$$

From this we deduce $\overline{\lim}_{n\to\infty} |S_n f(z) - \int_{S^1} f \, d\lambda_{S^1}| \leq \varepsilon$. Since real and imaginary part of $f : S^1 \to \mathbb{C}$ may be considered separately and $\varepsilon > 0$ as well as z have been arbitrary the result follows. $\square$

As an obvious corollary we observe that $\lim_{n\to\infty} h_A(z, n) = \lambda_{S^1}(A)$ for every $z \in S^1$ and every arc A. In case of an irrational rotation the orbit of every point therefore not only fills the circle densely but even is equally distributed. This last result is frequently referred to as *Weyl's Equidistribution Theorem*.

Before taking a look at the analogous result for higher dimensions let us give an immediate application. We already know that every natural number k will eventually appear as the initial segment of the decimal representation of 2^n. Now we are going to show that nevertheless some digits are more likely to occur than others. Recall that the decimal representation of 2^n begins with k if and only if

$$R_{\log_{10} 2}^n(1) \in \pi\left([\log_{10} k, \log_{10}(k+1)[\right) =: A_k.$$

From Weyl's theorem we conclude that

$$\lim_{n\to\infty} h_{A_k}(1, n) = \lambda_{S^1}(A_k) = \log_{10}(k+1) - \log_{10} k. \tag{3.15}$$

Observe that we would not have been able to guarantee this relation by solely applying the Birkhoff Ergodic Theorem. The larger k is, the less likely 2^n is to begin with k. For example, we expect the first digit of 2^n to be 7 more often than 8.

As can be seen from the table in Figure 3.9, relation (3.15) perfectly agrees with the numerical data for large n. For small n, however, 8 is found considerably more often then 7. This is due to the fact that the numerical value of $\log_{10} 2 \approx 0.301$ is quite close to $\frac{3}{10}$ and consequently $R_{\log_{10} 2}$ initially resembles a rational rotation with period ten. Indeed, when looking unbiased at the table from Figure 1.19 one seemingly observes a periodic phenomenon. It takes some time for the relative frequencies to evolve as predicted by (3.15). Since $\log_{10} 2 + \log_{10} 5 = 1$, a similar effect is to be expected for powers of five. Actually, it is not until $n = 50$ that 8 first appears as the leading digit of 5^n.

k	$h_{A_k}(1, 10)$	$h_{A_k}(1, 10^2)$	$h_{A_k}(1, 10^3)$	$h_{A_k}(1, 10^4)$	$\lambda_{S^1}(A_k)$
1	0.3	0.30	0.301	0.3010	0.3010
2	0.2	0.17	0.176	0.1761	0.1761
3	0.1	0.13	0.125	0.1249	0.1249
4	0.1	0.10	0.097	0.0970	0.0969
5	0.1	0.07	0.079	0.0791	0.0792
6	0.1	0.07	0.069	0.0670	0.0669
7	0.0	0.06	0.056	0.0579	0.0580
8	0.1	0.05	0.052	0.0512	0.0512
9	0.0	0.05	0.045	0.0458	0.0458

Figure 3.9. Relative frequencies of the first digits of 2^n and their limit (right column)

We shall now give a higher-dimensional version of the Equidistribution Theorem 3.17. To this end we define for every $\vartheta \in \mathbb{R}^d$ the ϑ-translation R_ϑ of $\mathbb{T}^d$ by $R_\vartheta(x) := x + \pi_d(\vartheta)$. As a continuous analogue we set $\psi_t(x; w) := R_{wt}(x) = x + \pi_d(wt)$ where $w \in \mathbb{R}^d$ is interpreted as a velocity vector. In order to describe the dynamics of R_ϑ and $\psi_t(\cdot; w)$ we make use of a notion from number theory. The complex numbers $\gamma_1, \ldots, \gamma_d$ are said to be *independent over* $\mathbb{Q}$ if $\sum_{i=1}^d \gamma_i k_i = 0$ with $k_i \in \mathbb{Q}$ necessarily implies that $k_i = 0$ for all i. By means of Fourier series the translation $R_{(\vartheta_1, \ldots, \vartheta_d)}$ may be seen to be ergodic if and only if $1, \vartheta_1, \ldots, \vartheta_d$ are independent over $\mathbb{Q}$; if $d = 1$ this simply reduces to $\vartheta_1 \notin \mathbb{Q}$. In addition, $\psi_t\big(\cdot; (w_1, \ldots, w_d)\big)$ is ergodic for at least one $t > 0$ if and only if $w_1, \ldots, w_d$ are independent over $\mathbb{Q}$. Since the proof of the following theorem runs along exactly the same lines as in the one-dimensional case we omit the details.

Theorem 3.18. *Let $f : \mathbb{T}^d \to \mathbb{C}$ be a bounded Riemann-integrable function.*

(i) *If $1, \vartheta_1, \ldots, \vartheta_d$ are independent over $\mathbb{Q}$, then under $R_{(\vartheta_1, \ldots, \vartheta_d)}$*

$$\lim_{n \to \infty} S_n f(x) = \int_{\mathbb{T}^d} f \, d\lambda_{\mathbb{T}^d}$$

for every $x \in \mathbb{T}^d$.

(ii) *If $w_1, \ldots, w_d$ are independent over $\mathbb{Q}$, then the relation*

$$\lim_{T \to \infty} \frac{1}{T} \int_0^T f\big(\psi_t(x; (w_1, \ldots, w_d))\big) dt = \int_{\mathbb{T}^d} f \, d\lambda_{\mathbb{T}^d}$$

holds for every $x \in \mathbb{T}^d$.

The application of this theorem to the rectangular billiard is straightforward. Remember that we assigned to this system a straight motion on the rectangle with side-lengths $2a$ and $2b$, respectively (cf. Figure 1.2). Scaling this rectangle onto the unit square we obtain the continuous translation $\psi_t(\cdot; w)$ where $w = v/(2ab) \cdot (b \cos \alpha, a \sin \alpha) \in \mathbb{R}^2$. We already know that the motion starting at an arbitrary point will be periodic if and only if $a/b \tan \alpha \in \mathbb{Q} \cup \{\pm\infty\}$. By means of Theorem 3.18 we can now completely describe the rectangle billiard for the case $a/b \tan \alpha \in \mathbb{R} \backslash \mathbb{Q}$: the trajectory of any point fills the rectangle densely, and moreover it is equally distributed all over the billiard table. Obviously, assertion (1.1) from the Introduction concerning the long-term behaviour of relative frequencies in this case is just an equivalent form of the latter statement.

Surely, such an exhaustive analysis would be desirable also for tables other than rectangles. In case of the circular billiard we can again use translations on $\mathbb{T}^2$ to obtain a satisfactory description of the dynamics. More specifically, fix $R > 0$ and $\alpha \in]0, \frac{\pi}{2}[$ and consider the map $F : \mathbb{T}^2 \to \mathbb{R}^2$ defined as

$$F(x_1, x_2) := R\sqrt{\cos^2 \alpha + (2x_2 - 1)^2 \sin^2 \alpha} \, \big(\cos \varphi(x_1, x_2), \sin \varphi(x_1, x_2)\big) \quad (3.16)$$

with the angle $\varphi(x_1, x_2)$ given by

$$\varphi(x_1, x_2) := \alpha + 2(\pi x_1 - \alpha x_2) \quad (3.17)$$
$$+ \operatorname{sign}(2x_2 - 1) \arccos \frac{\cos \alpha}{\sqrt{\cos^2 \alpha + (2x_2 - 1)^2 \sin^2 \alpha}}.$$

A careful check confirms that F is indeed a continuous map. Moreover, $F(\mathbb{T}^2)$ is an annulus in the plane centered at the origin with radii $R|\cos \alpha|$ and R, respectively

(Figure 3.10). Since the rather cumbersome computational details should not distract us from the essence of the argument, we leave it as an exercise to derive the ingenious formulae (3.16) and (3.17) as well as the facts used below (Exercise 3.8). What we are going to investigate is how F acts on the trajectories of translations on $\mathbb{T}^2$ with velocity $w = (1, \pi/\alpha)$. Taking for instance $z := (\zeta, 0) \in \mathbb{T}^2$, what can be said about $F(\psi_t(z; w))$? If we define a sequence n_k of unit vectors according to

$$n_k := \big(\cos(2\pi\zeta + (2k+1)\alpha),\ \sin(2\pi\zeta + (2k+1)\alpha)\big) \in \mathbb{R}^2 \quad (k \in \mathbb{N}_0)$$

we observe that the relation

$$\big\langle F\big(\psi_t(z; w)\big), n_k \big\rangle = R\cos\alpha = \text{const.}$$

holds for $k < \pi t/\alpha < k+1$. Therefore the point $F(\psi_t(z; w))$ travels along the straight line joining the points P_k and P_{k+1} where

$$P_k := \big(R\cos(2\pi\zeta + 2k\alpha),\ R\sin(2\pi\zeta + 2k\alpha)\big) \in \mathbb{R}^2$$

(see Figure 3.10). Moreover, this movement takes place with constant speed because

$$\left\| \frac{d}{dt} F\big(\psi_t(z; w)\big) \right\|_2 = \frac{2\pi R \sin\alpha}{\alpha} = \text{const.}$$

by virtue of an elementary though lengthy calculation. We deduce from these observations that the curve $\big\{ F\big(\psi_t(z; w)\big) : t \geq 0 \big\}$ is just the billiard trajectory emanating from P_0 with an angle α. This not only confirms our earlier results on periodic and dense trajectories, it also enables us to determine the long-term behaviour of the relative frequencies if α/π is irrational. Let A be a subset of the annulus $F(\mathbb{T}^2)$ such that $1_A \circ F$ is Riemann-integrable. Then

$$\lim_{T \to \infty} \frac{1}{T} \int_0^T 1_A \circ F\big(\psi_t(z; w)\big)\, dt = \lambda_{\mathbb{T}^2}\big(F^{-1}(A)\big) = F\lambda_{\mathbb{T}^2}(A)$$

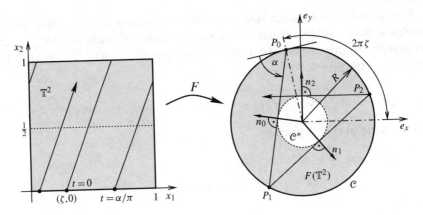

Figure 3.10. The circular billiard may be considered a translation on $\mathbb{T}^2$.

from which we obtain via the defining relation (3.16)

$$\lim_{T\to\infty} \frac{1}{T}\int_0^T \mathbf{1}_A\big(F(\psi_t(z; w))\big)\, dt = \frac{1}{2\pi R \sin\alpha} \iint_A \frac{r\, dr\, d\varphi}{\sqrt{r^2 - R^2 \cos^2\alpha}}. \qquad (3.18)$$

In (3.18) the integral on the right has been transformed to polar coordinates because $F\lambda_{\mathbb{T}^2}$ is invariant with respect to any rotation of the annulus $F(\mathbb{T}^2)$. Via formula (3.18) which has already been announced in the Introduction we have arrived – just as for the rectangular table – at a complete description of the circular billiard by means of translations on a torus.

It seems natural to ask for a similar analysis for other shapes of the billiard table. However, already our rather tricky treatment of the elliptic billiard could warn us that this might be too ambitious a task. Therefore a more statistically oriented analysis may be wished for, which for instance addresses the question of ergodicity. Recall that we associated a piecewise smooth map T_{bill} on $[0, 1[\times]0, \pi[$ with the billiard within a region $\Omega \subseteq \mathbb{R}^2$ with (piecewise) smooth boundary; in addition, we found a natural probability measure μ_{bill} invariant under T_{bill}. Is T_{bill} ergodic with respect to that measure? For the rectangular billiard the answer trivially is negative because every trajectory can attain at most four different directions of motion. More interestingly, the elliptic billiard is not ergodic either: by means of the two families of caustics one can easily construct T_{bill}-invariant sets having measure neither zero nor one. Obviously the latter argument generally disproves ergodicity for billiards which admit a sufficient number of caustics. A theorem due to Lazutkin asserts that there always exists a family of caustics $(\mathcal{E}_i^*)_{i\in I}$ for which the corresponding sets $M(\mathcal{E}_i^*) \subseteq [0, 1[\times]0, \pi[$ fill a set of positive measure if only the table is convex and has a sufficiently smooth boundary (see [54] for details and references). Such billiards therefore are not ergodic. On the other hand the stadium (which is convex but has no caustics as its boundary is only C^1), the dispersing billiard as well as the other tables depicted in Figure 3.11 provide examples of ergodic billiard systems. Although a proof of this statement definitely lies beyond the scope of this book, its general significance should be quite clear by now: the dynamics of such billiard systems cannot be satisfyingly analysed without adopting the statistical point of view.

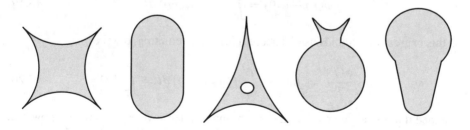

Figure 3.11. For these tables the associated billiard map T_{bill} is ergodic.

A final remark on billiard systems concerns the behaviour of billiards inside triangles or, more generally, polygons. Despite their geometrical simplicity these systems have not yet yielded to a definite comprehensive understanding: there are surprisingly many open questions in this field. As just one fact stressing the difficulties with such billiards we mention that the dichotomy observed for billiard trajectories in the rectangle – either closed or equally distributed – does not even hold for triangles. On the contrary, there exist triangles which exhibit non-periodic trajectories that are not even dense on the table ([12, 54]).

As another and somewhat different application of translations let us now deal with the problem of *mean motion*. Recall that in the Introduction we set out to describe – at least asymptotically and on average – the rotation of the endpoint of a chain consisting of d rotating rods. To put this more formally, we wrote in the complex plane

$$z(t) = a_1 e^{i\omega_1 t} + \ldots + a_d e^{i\omega_d t} =: r(t) e^{i\varphi(t)}$$

where $a_k \in \mathbb{C}\backslash\{0\}$ and $\omega_k \in \mathbb{R}$. Clearly $\varphi(t)$ is not defined if $z(t) = 0$. Ignoring this technical difficulty for the moment and assuming φ to be continuous whenever it is defined we find

$$\dot{\varphi} = \Re\left(\frac{\dot{z}}{iz}\right) = \Re\left(\frac{\sum_k a_k \omega_k e^{i\omega_k t}}{\sum_k a_k e^{i\omega_k t}}\right).$$

We now write $a_k = |a_k| e^{2\pi i \alpha_k}$ for $k = 1, \ldots, d$; furthermore we define an $L^1(\lambda_{\mathbb{T}^d})$-function $f : \mathbb{T}^d \to \mathbb{R}$ as

$$f(x_1, \ldots, x_d) := \Re\left(\frac{\sum_k |a_k| \omega_k e^{2\pi i x_k}}{\sum_k |a_k| e^{2\pi i x_k}}\right)$$

and also introduce the quantities

$$\alpha := \pi_d(\alpha_1, \ldots, \alpha_d) \in \mathbb{T}^d, \quad \omega := \frac{1}{2\pi}(\omega_1, \ldots, \omega_d) \in \mathbb{R}^d$$

for the sake of brevity. If $z(t) \neq 0$ for $0 \leq t \leq T$ we find

$$\varphi(T) - \varphi(0) = \int_0^T f(\psi_t(\alpha; \omega)) \, dt . \tag{3.19}$$

As this expression reminds us of Theorem 3.18 it is tempting to write

$$\omega_\infty = \lim_{T \to \infty} \frac{\varphi(T)}{T} = \lim_{T \to \infty} \frac{1}{T} \int_0^T f(\psi_t(\alpha; \omega)) \, dt = \int_{\mathbb{T}^d} f \, d\lambda_{\mathbb{T}^n} , \tag{3.20}$$

provided that the angular velocities ω_k are independent over $\mathbb{Q}$. There are, however, two technical difficulties which have to be taken care of. First, f is not bounded so that Theorem 3.18 does not directly apply; and falling back on the Birkhoff Ergodic Theorem merely yields (3.20) for *almost all* α. Secondly, as it stands, (3.19) will only hold if the endpoint of the chain never passes through the origin. Fortunately, both difficulties can be overcome, and (3.20) is in fact true for *all* α if only $\omega_1, \ldots, \omega_d$ are independent over $\mathbb{Q}$ (see [64] for the details). Observe that λ^d-almost every angular velocity vector ω drawn from $\mathbb{R}^d$ will satisfy the condition of independence.

Equation (3.20) may be evaluated to give

$$\omega_\infty = \int_{\mathbb{T}^d} f \, d\lambda_{\mathbb{T}^d} = \sum_{k=1}^{d} \omega_k \, \Re\left(\int_{[0,1]^d} \frac{|a_k| e^{2\pi i x_k}}{\sum_l |a_l| e^{2\pi i x_l}} \, dx_1 \ldots dx_d\right) = \sum_{k=1}^{n} \omega_k \, p_k \, ,$$

where the coefficients p_k are defined by

$$p_k := \Re\left(\int_{[0,1]^d} \frac{|a_k| e^{2\pi i x_k}}{\sum_l |a_l| e^{2\pi i x_l}} \, dx_1 \ldots dx_d\right)$$

$$= \lambda_{\mathbb{T}^d}\left(\{x \in \mathbb{T}^d \; : \; |a_k| > |\sum_{l \neq k} |a_l| e^{2\pi i x_l}|\}\right);$$

here the last equality follows from the elementary fact that for $b \in \mathbb{C} \backslash S^1$

$$\int_0^1 \frac{e^{2\pi i t} \, dt}{e^{2\pi i t} + b} = \begin{cases} 1 & \text{if } |b| < 1, \\ 0 & \text{if } |b| > 1. \end{cases}$$

Clearly $p_k \geq 0$ and $\sum_{k=1}^{d} p_k = 1$. We therefore have shown that the average angular velocity ω_∞ is a barycentric mean of the angular velocities ω_k for almost every choice of the latter. Evidently, the weights p_k of this mean only depend on the geometry, i.e. $p_k = p_k(|a_1|, \ldots, |a_d|)$. The special case discussed by Lagrange is easily analysed: if $|a_{k_0}| > \sum_{l \neq k_0} |a_l|$ for some k_0 then $p_{k_0} = 1$ and $\omega_\infty = \omega_{k_0}$. In this case the average movement solely depends on the rotation of the dominant rod.

Having seen several interesting applications of translations on a torus let us close by briefly sketching a final one which is most important for both theoretical considerations and explicit calculations. Recall that we termed a system of $2d$ ordinary differential equations *Hamiltonian* if it may be written as

$$\dot{x}_{2k-1} = \frac{\partial H}{\partial x_{2k}}, \quad \dot{x}_{2k} = -\frac{\partial H}{\partial x_{2k-1}} \qquad (k = 1, \ldots, d) \qquad (3.21)$$

with the (autonomous) Hamiltonian function $H : U \to \mathbb{R}$ defined on some open set $U \subseteq \mathbb{R}^{2d}$. (The usual notion of a Hamiltonian system is much more general, as e.g. in [2], but the notion adopted here suffices for the subsequent discussion.) Defining for any two C^1-functions $f, g : U \to \mathbb{R}$ their *Poisson bracket* $\{f, g\}$ as

$$\{f, g\} := \sum_{k=1}^{d} \left(\frac{\partial f}{\partial x_{2k-1}} \frac{\partial g}{\partial x_{2k}} - \frac{\partial f}{\partial x_{2k}} \frac{\partial g}{\partial x_{2k-1}}\right),$$

it is easily seen that along trajectories of (3.21)

$$\frac{d}{dt} f(x_1(t), \ldots, x_{2d}(t)) = \sum_{k=1}^{d} \left(\frac{\partial f}{\partial x_{2k-1}} \dot{x}_{2k-1} + \frac{\partial f}{\partial x_{2k}} \dot{x}_{2k} \right) = \{f, H\}.$$

Therefore, under the evolution according to (3.21) the value of f is conserved precisely if $\{f, H\} = 0$.

Consider now the particular Hamiltonian system derived from

$$H(x_1, \ldots, x_{2d}) := \frac{1}{2} \sum_{k=1}^{2d} \alpha_k^2 x_k^2 \tag{3.22}$$

where $\alpha_k > 0$ denote fixed parameters. In mechanical terms (3.22) describes the simultaneous though uncoupled motion of d undamped oscillators ([2]). Introducing new coordinates $(I_k, \varphi_k)_{k=1}^{d}$ according to

$$I_k := \pi \frac{\alpha_{2k-1}}{\alpha_{2k}} x_{2k-1}^2 + \pi \frac{\alpha_{2k}}{\alpha_{2k-1}} x_{2k}^2,$$

$$\varphi_k := \frac{1}{2\pi} \arg(\alpha_{2k-1} x_{2k-1} + i\, \alpha_{2k} x_{2k}),$$

we find for the Hamiltonian in these new quantities

$$H = \frac{1}{2\pi} \sum_{k=1}^{d} \alpha_{2k-1} \alpha_{2k}\, I_k.$$

Clearly, the angles φ_k are only determined up to additive integers. With respect to the new coordinates the equations of motion read

$$\dot{I}_k = 0 = \frac{\partial H}{\partial \varphi_k}, \quad \dot{\varphi}_k = -\frac{\alpha_{2k-1}\alpha_{2k}}{2\pi} = -\frac{\partial H}{\partial I_k} \quad (k = 1, \ldots, d). \tag{3.23}$$

The transformed system thus is Hamiltonian again with the quantities I_k being constant along any trajectory. If we solve (3.23) subject to the initial conditions $I_k(0) = I_{k,0}$, $\varphi_k(0) = \varphi_{k,0}$ then the solution always remains within the set

$$M_0 := \{x \in \mathbb{R}^{2d} : I_1(x) = I_{1,0}, \ldots, I_d(x) = I_{d,0}\}.$$

If $I_{k,0} > 0$ for all k then M_0 equals $\mathbb{T}^d$ via the coordinates $\varphi_1, \ldots, \varphi_d$, and when restricted to M_0 the flow generated by (3.23) is just a translation on this set, more formally

$$\Phi_t|_{M_0} \equiv \psi_t(\cdot\,; w) \quad \text{with } w := -\frac{1}{2\pi}(\alpha_1 \alpha_2, \ldots, \alpha_{2d-1}\alpha_{2d}).$$

Moreover, up to a set of λ^d-measure zero (corresponding to motions with one or several oscillators at rest which means that $I_{k,0} = 0$ for some k) the whole space $\mathbb{R}^{2d}$ is a disjoint union of the sets M_0. Thus the overall picture is easily grasped: essentially the whole space is filled with *invariant tori* on each of which the flow is just a translation.

The simple-minded example (3.22) itself does not deserve much interest. However, the geometric picture brought about may also be found in a more general context. Assume that we have – by ingenuity or whatever method – found d functions J_k : $U \to \mathbb{R}$ which remain constant along any trajectory of (3.21). Moreover, assume that these functions are independent in the sense that their gradients are pointwise linearly independent, and $\{J_k, J_l\} = 0$ for all k, l. Traditionally, the quantities J_k are called *first integrals*, and a Hamiltonian system exhibiting d independent first integrals is termed *completely integrable*. In the above example we can simply take $J_k := I_k$ as $\{I_k, I_l\} = 0$ for all k, l. In general, the famous Liouville–Arnold Theorem ([2, 31]) asserts that one can analogously find new coordinates $(I_k, \varphi_k)_{k=1}^d$ such that in these new coordinates the system is again Hamiltonian, but the transformed Hamiltonian function solely depends on the quantities $I_k = I_k(J_1, \ldots, J_d)$, so that

$$\dot{I}_k = \frac{\partial H}{\partial \varphi_k} = 0 \,, \quad \dot{\varphi}_k = -\frac{\partial H}{\partial I_k} = \text{const} \,.$$

Moreover, if the set $M_0 := \{x \in U : J_1(x) = J_{1,0}, \ldots, J_d(x) = J_{d,0}\}$ is compact and connected then it equals $\mathbb{T}^d$ up to a smooth change of coordinates. As above, when restricted to M_0 the flow is just a translation.

We have seen that translations on tori constitute a crucial dynamical pattern for Hamiltonian systems. As one might guess, completely integrable systems are rare. In quite a number of situations (most prominently perhaps in celestial mechanics) the system under consideration may, however, be regarded as a slightly perturbed version of a completely integrable system. It is thus both natural and important to ask which of the latter's properties will persist under small perturbations. This major problem is settled by the celebrated KAM theory, a fascinating subject that deserves, as the extensive field of Hamiltonian dynamics in general, study in its own right ([2, 31, 37]).

3.4 Aspects of mixing

In this section we shall deepen our statistical understanding of dynamical systems. First we are going to define two mixing properties which naturally strengthen the concept of ergodicity. Complementing this rather qualitative approach we shall then discuss metric entropy which is among the most important tools for quantifying the irregularity of dynamical systems. Moreover, entropy vitally relies on the statistical approach without which its meaning could hardly be appreciated. As always our emphasis is on specific examples.

3.4.1   Mixing properties

When we first introduced the concept of ergodicity we were mainly concerned with very long return times in certain measure-preserving systems. We thought of an ergodic map as being indecomposable and thus mixing up the whole space. Having seen some of the various aspects of ergodicity we shall briefly discuss whether an ergodic map really generates a mixture in the sense of common language. In order to give a precise meaning to the term *mixture* think of a glass of water, formalized as X, which contains a drop of ink $B \subseteq X$. Let the action of $T : X \to X$ consist of stirring once with a spoon (cf. Figure 3.12). What kind of behaviour do we expect for the ink? Fix an arbitrary reference set $A \subseteq X$. Then, after sufficiently many stirrings, the percentage of ink in A will more or less equal the percentage of ink in the whole glass.

Figure 3.12. The true meaning of *mixing* (Definition 3.19)

Definiton 3.19. A measure-preserving map T on $(X, \mathcal{A}, \mu)$ is called *mixing* if

$$\lim_{n \to \infty} \mu(T^{-n}(A) \cap B) = \mu(A)\mu(B) \quad \text{for all } A, B \in \mathcal{A}.$$

Recall our stochastic interpretation of $(X, \mathcal{A}, \mu)$ which leads to the following probabilistic interpretation of mixing. Assume that μ is the governing law for an experiment of chance ξ_0 with outcomes in X, that is, $\mathbb{P}(\xi_0 \in A) = \mu(A)$. For $n \in \mathbb{N}_0$ the composite experiments $\xi_n := T^n(\xi_0)$ also obey that law. If T is mixing, then the joint probability $\mathbb{P}(\xi_n \in A, \xi_0 \in B)$ approaches the product $\mathbb{P}(\xi_n \in A)\mathbb{P}(\xi_0 \in B)$ so that the experiments ξ_0 (the "present") and ξ_n (the "future" after n iterations of T) asymptotically become *independent*.

In the literature a lot of different mixing properties are studied. What we call mixing here is usually termed *strongly* mixing. We refer to the references, especially [42], for these finer specifications. Since ergodicity of T may equivalently be expressed as

$$\lim_{n \to \infty} \frac{1}{n} \sum_{i=0}^{n-1} \mu(T^{-i}(A) \cap B) = \mu(A)\mu(B) \quad \text{for all } A, B \in \mathcal{A},$$

every mixing map is ergodic. The converse, however, need not be true. Bearing in mind Figure 3.12 this should not come as a surprise. Take for example an irrational rotation R_ϑ of S^1. No set will ever be spread out over S^1 by R_ϑ in the way the drop of ink is spread throughout the whole glass of water. More formally, let A and B denote the upper and lower half of S^1, respectively, and notice that

$$\underline{\lim}_{n \to \infty} \lambda_{S^1}(R_\vartheta^{-n}(A) \cap B) = 0 \quad \text{and} \quad \overline{\lim}_{n \to \infty} \lambda_{S^1}(R_\vartheta^{-n}(A) \cap B) = \frac{1}{2},$$

which corresponds to the observation that the intersection of the two semi-circles $R_\vartheta^{-n}(A)$ and B will infinitely often amount to an arbitrarily large and as often to an arbitrarily small percentage of B.

To present an example of a mixing map we take a look at the homomorphisms $T_{(k)}$ of $\mathbb{T}^1$ with $|k| \geq 2$. Due to an earlier result these maps are ergodic. It suffices to verify the mixing property for the special sets $A := \pi_1([\frac{a}{|k|^l}, \frac{a+1}{|k|^l}])$, $B := \pi_1([\frac{b}{|k|^m}, \frac{b+1}{|k|^m}])$ with $a, b, l, m \in \mathbb{N}$. A straightforward calculation shows that $\lambda_{\mathbb{T}^1}(T_{(k)}^{-n}(A) \cap B) = |k|^{-(l+m)} = \lambda_{\mathbb{T}^1}(A)\lambda_{\mathbb{T}^1}(B)$ for all $n \geq m$. A look at the graphs of $T_{(k)}^n$ gives a good impression of how arbitrary intervals are uniformly smeared over $\mathbb{T}^1$ (see Figure 3.13). In fact, $T_{(k)}$ exhibits a property which turns out to be even stronger than mixing: for any small interval $A \subseteq \mathbb{T}^1$ we have $T_{(k)}^n(A) = \mathbb{T}^1$ for n sufficiently large. Especially, the relation $\lim_{n \to \infty} \lambda_{\mathbb{T}^1}(T_{(k)}^n(A)) = 1$ holds for all intervals with positive length. In order to conveniently formalize this type of behaviour we shall utilize the following simple observation.

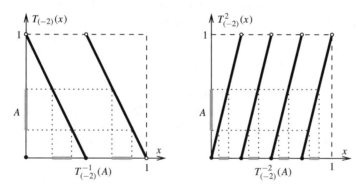

Figure 3.13. Pre-images of intervals under the endomorphism $T_{(-2)}$ spread over $\mathbb{T}^1$.

Lemma 3.20. *Let T be a measure-preserving map on $(X, \mathcal{A}, \mu)$ with $T(A) \in \mathcal{A}$ for all $A \in \mathcal{A}$. Then $\lim_{n \to \infty} \mu(T^n(A)) = 1$ for all $A \in \mathcal{A}$ with $\mu(A) > 0$ precisely if the σ-algebra $\bigcap_{n \geq 0} T^{-n}\mathcal{A}$ only contains sets of measure zero or one.*

Proof. Assume that $\mathcal{A}_\infty := \bigcap_{n \geq 0} T^{-n}\mathcal{A}$ exclusively contains sets of trivial measure and let $\mu(A) > 0$. The sets $B_n := T^{-n}(T^n(A)) \in T^{-n}\mathcal{A}$ satisfy $B_n \subseteq B_{n+1}$, and $B_0 = A$. Therefore $\bigcup_{n \geq 0} B_n \in \mathcal{A}_\infty$ and thus $1 = \mu(\bigcup_{n \geq 0} B_n) = \lim_{n \to \infty} \mu(T^n(A))$. Conversely, if $C \in \mathcal{A}_\infty$ then $C = T^{-n}(A_n)$ for appropriate $A_n \in \mathcal{A}$. Assuming that $\mu(C) > 0$ we have $1 = \lim_{n \to \infty} \mu(T^n(C)) \leq \lim_{n \to \infty} \mu(A_n) \leq 1$, hence $\mu(C) = 1$. $\square$

As a consequence of Lemma 3.20 we can label a strong stochastic property of dynamical systems.

Definiton 3.21. A measure-preserving map T on the probability space $(X, \mathcal{A}, \mu)$ is called *exact* if $\mu(A) \in \{0, 1\}$ for every $A \in \bigcap_{n \geq 0} T^{-n}\mathcal{A}$.

An exact map necessarily is ergodic because every invariant set is contained in $\bigcap_{n \geq 0} T^{-n} \mathcal{A}$. However, from the definition it is not at all obvious that it is mixing too. In the next section we shall prove this fact by means of the methods developed there. Observe that no invertible map can ever be exact since

$$\mu\big(T^n(A)\big) = \mu\big(T^{-n}(T^n(A))\big) = \mu(A) = \text{const}.$$

In order to get a feeling for the dynamical properties just introduced, three different maps on $\mathbb{T}^2$ are investigated numerically.

(i) With $\vartheta := (\sqrt{2}, \sqrt{3}) \in \mathbb{R}^2$ the translation R_ϑ on $\mathbb{T}^2$ is ergodic. Analogously to the one-dimensional case R_ϑ can be shown to be not mixing.

(ii) The hyperbolic toral automorphism T_{A_1} with $A_1 := \begin{pmatrix} 2 & 1 \\ 1 & 1 \end{pmatrix}$ turns out to be mixing (we shall prove this later). Due to its invertibility T_{A_1} cannot be exact. In fact, T_{A_1} exhibits the strongest mixing property possible for invertible maps: it is an example of a *Kolmogorov-automorphism* (or *K-system*, for short, [3, 42, 63]).

(iii) The hyperbolic toral endomorphism T_{A_2} with $A_2 := \begin{pmatrix} 4 & 1 \\ 2 & 4 \end{pmatrix}$ is exact. This, too, will be proved in the next chapter.

In Figure 3.14 the action of the three maps above is displayed. On a small square filled with 10^4 points the maps are iterated a few times. Depending on the dynamical properties of the respective map the iterates of these points exhibit more or less regularity. Observe that in the case of the exact map T_{A_2} it does not take more than thirty iterations to produce a rather stochastic pattern.

3.4.2   The concept of entropy

Remember that breaking up a space into smaller pieces and roughly describing the dynamics by coding was one of our motivations for studying symbolic dynamics. As we shall see, the concept of entropy naturally grows from quantifying the effect of partitioning the space. Let $\alpha = \{X_1, \ldots, X_m\}$ denote a finite measurable partition of the normed measure space $(X, \mathcal{A}, \mu)$, i.e. $\mu\big(\bigcup_{i=1}^m X_i\big) = 1$ with $\mu(X_i \cap X_j) = 0$ for $i \neq j$ and $\mu(X_i) > 0$ for all i. If $\beta = \{Y_1, \ldots, Y_l\}$ is another partition we define the common refinement of α and β as

$$\alpha \vee \beta := \big\{X_i \cap Y_j : 1 \leq i \leq m,\ 1 \leq j \leq l,\ \mu(X_i \cap Y_j) > 0\big\}.$$

For any measurable map T on X, the family of sets

$$T^{-1}\alpha := \big\{T^{-1}(X_i) : 1 \leq i \leq m,\ \mu\big(T^{-1}(X_i)\big) > 0\big\}$$

yields a finite partition, too. We call β *finer* than α, in symbols $\alpha \prec \beta$, if every element of α is – possibly up to a set of measure zero – a union of elements of β; obviously

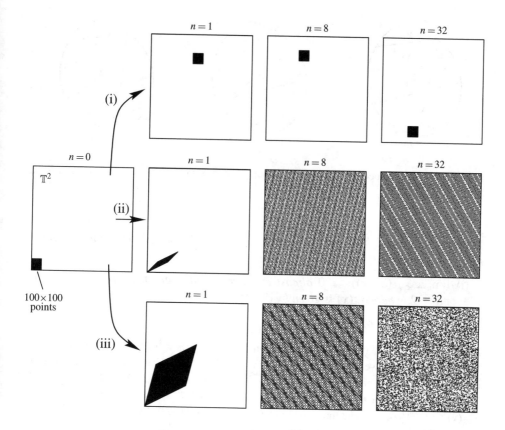

Figure 3.14. Visualizing ergodicity, mixing, and exactness by iteration of 10^4 points on $\mathbb{T}^2$

$\alpha \prec \beta$ precisely if $\alpha \vee \beta = \beta$ up to sets of measure zero. Given a partition α we may assign to each $x \in X$ a set $\alpha(x)$ from α such that $x \in \alpha(x)$. This choice is unique almost everywhere. Now consider a measure-preserving map T and define for $n \in \mathbb{N}_0$

$$\alpha_n(x) := \bigvee_{i=0}^{n} T^{-i}\alpha(x) = \left\{ y \in X \: : \: \alpha\left(T^i(y)\right) = \alpha\left(T^i(x)\right) \text{ for all } 0 \le i \le n \right\}.$$

$$(3.24)$$

Intuitively, we see that the points in $\alpha_n(x)$ share the dynamical fate of x relative to the partition α for at least n steps. For fixed $x \in X$ the sequence $\left(\mu(\alpha_n(x))\right)_{n \in \mathbb{N}_0}$ is non-increasing – the faster it decreases, the more complicated is the dynamics taking place in the vicinity of x (cf. Figure 3.15). We thus expect the asymptotic behaviour of $\mu(\alpha_n(x))$ to provide some information about the dynamics near x. Given our knowledge of the meaning of ergodicity the following result is barely surprising.

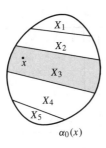
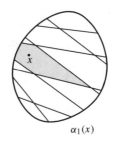
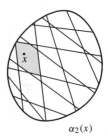

$\alpha_0(x)$ $\alpha_1(x)$ $\alpha_2(x)$

Figure 3.15. With respect to the partition α each point in $\alpha_n(x)$ has the same history as x for at least n iterations of T.

Lemma 3.22. *Let T be a measure-preserving map on $(X, \mathcal{A}, \mu)$. The following statements are equivalent:*

(i) $\lim_{n\to\infty} \mu(\alpha_n(x)) = 0$ *almost everywhere for every finite partition α with $\#\alpha \geq 2$, where $\alpha_n(x)$ is given by (3.24).*

(ii) T^k *is ergodic for every $k \in \mathbb{N}$.*

Proof. Assume $T^{-m_0}(A) = A$ for some $m_0 \in \mathbb{N}$ with $0 < \mu(A) < 1$. Setting $\alpha := \bigvee_{i=0}^{m_0-1} T^{-i}\{A, A^c\}$ we see that $\alpha_n(x) = \alpha(x)$ for all $n \geq 0$. Hence (i) is not satisfied. If in turn (ii) holds, define $\alpha_\infty(x) := \bigcap_{n\in\mathbb{N}_0} \alpha_n(x) \in \mathcal{A}$. If $\mu(\alpha_\infty(x)) > 0$ then $\alpha_\infty(x) \cap T^{-m}(\alpha_\infty(x)) \neq \emptyset$ for some $m \in \mathbb{N}$. For any $y \in \alpha_\infty(x) \cap T^{-m}(\alpha_\infty(x))$ we have $\alpha_\infty(x) = \alpha_\infty(y) \subseteq T^{-m}(\alpha_\infty(x))$ and thus $\mu(\alpha_\infty(x) \triangle T^{-m}(\alpha_\infty(x))) = 0$. By the ergodicity of T^m it is impossible that $\mu(\alpha_\infty(x)) > 0$; hence we deduce $\lim_{n\to\infty} \mu(\alpha_n(x)) = \mu(\alpha_\infty(x)) = 0$. $\square$

In order to take a closer look at the behaviour of $\mu(\alpha_n(x))$ we quote a slight generalization of the ergodic theorems of Birkhoff and von Neumann which we shall use later. Since its proof is rather technical we shall not execute it here, but refer to the literature ([38]).

Proposition 3.23. *Let T denote a measure-preserving map on $(X, \mathcal{A}, \mu)$ and $(f_n)_{n\in\mathbb{N}}$ a sequence of L^1-functions converging to f (in L^1). Then*

$$\lim_{n\to\infty} \frac{1}{n} \sum_{i=0}^{n-1} f_{n-i} \circ T^i = \mathbb{E}(f|\mathcal{A}_T) \tag{3.25}$$

holds in L^1. If in addition $f_n \to f$ almost everywhere, then (3.25) holds almost everywhere, too.

If the sets $\alpha_n(x)$ shrink at all we are interested in the speed by which they do so. Therefore we define for a given partition α a sequence of non-negative L^1-functions

$$a_n(x) := -\frac{\log \mu(\alpha_n(x))}{n} \qquad (n \in \mathbb{N}). \tag{3.26}$$

Assume for the moment that $\underline{\lim}_{n\to\infty} a_n(x_0) = a > 0$ for some $x_0 \in X$. In this case

$$\mu\big(\alpha_n(x_0)\big) \leq e^{-n(a-\varepsilon_n)}$$

with $\lim_{n\to\infty} \varepsilon_n = 0$, that is, the size of $\alpha_n(x_0)$ tends to zero *exponentially*. The map T therefore dramatically separates typical points near x_0. If there is a positive limit for a_n at least somewhere in the space, then the overall dynamics of T will tend to be rather complicated. The question arises whether the sequence $(a_n)_{n\in\mathbb{N}}$ does exhibit any limit at all. In order to satisfyingly answer this important question we shall introduce one more technical tool.

Given two finite partitions $\alpha = \{X_1, \ldots, X_m\}$ and $\beta = \{Y_1, \ldots, Y_l\}$ we define the *information function* of α *relative to* (or *conditional on*) β by

$$I(\alpha|\beta) := -\sum_{i=1}^{m}\sum_{j=1}^{l} \log \frac{\mu(X_i \cap Y_j)}{\mu(Y_j)} \mathbf{1}_{X_i \cap Y_j}. \tag{3.27}$$

Moreover, we set $I(\alpha) := I(\alpha|\{X\}) = -\sum_i \log \mu(X_i)\mathbf{1}_{X_i}$. Recalling our stochastic interpretation of $(X, \mathcal{A}, \mu)$ we may interpret $I(\alpha|\beta)$ as follows. Suppose our experiment of chance happened to produce an outcome in $Y_j \in \beta$. Then $I(\alpha|\beta)$ measures the extra information we gain if we additionally learn in which $X_i \in \alpha$ the outcome is to be found. Equivalently, $I(\alpha|\beta)$ quantifies the uncertainty of the possible outcome's position with respect to α. Consequently $I(\beta|\beta) = 0$, and the mean information gained amounts to $\frac{1}{\mu(Y_j)} \int_{Y_j} I(\alpha|\beta)\,d\mu$. This quantity is maximized if all the sets $X_i \cap Y_j$ have equal measure; then all possible outcomes with respect to α are equally likely. Though also an important tool in their own right ([42]), information functions mainly serve as a technical device here. A few basic results about this tool are collected in the following lemma.

Lemma 3.24. *For any measure-preserving map T on $(X, \mathcal{A}, \mu)$ and finite partitions $\alpha, \alpha_1, \alpha_2, \beta$ the following properties hold:*

(i) $I(\alpha_1 \vee \alpha_2|\beta) = I(\alpha_1|\alpha_2 \vee \beta) + I(\alpha_2|\beta)$;

(ii) $I(T^{-1}\alpha|T^{-1}\beta) = I(\alpha|\beta) \circ T$;

(iii) $\int_X I(\alpha_1|\beta)\,d\mu \leq \int_X I(\alpha_2|\beta)\,d\mu$ if $\alpha_1 \prec \alpha_2$;

(iv) $\int_X I(\beta|\alpha_1)\,d\mu \geq \int_X I(\beta|\alpha_2)\,d\mu$ if $\alpha_1 \prec \alpha_2$;

(v) $\int_X I(\alpha)\,d\mu \leq \log \#\alpha$;

(vi) $\int_X I(T^{-1}\alpha|T^{-1}\beta)\,d\mu = \int_X I(\alpha|\beta)\,d\mu$.

Proof. In order to prove (i) let $\alpha_1 = \{X_1^{(1)}, \ldots, X_{m_1}^{(1)}\}$, $\alpha_2 = \{X_1^{(2)}, \ldots, X_{m_2}^{(2)}\}$ and $\beta = \{Y_1, \ldots, Y_l\}$ and observe that

$$I(\alpha_1 \vee \alpha_2 | \beta) = -\sum_{i,j,k} \log \frac{\mu(X_i^{(1)} \cap X_j^{(2)} \cap Y_k)}{\mu(Y_k)} \mathbf{1}_{X_i^{(1)} \cap X_j^{(2)} \cap Y_k}$$

$$= -\sum_{i,j,k} \log \frac{\mu(X_i^{(1)} \cap X_j^{(2)} \cap Y_k)}{\mu(X_j^{(2)} \cap Y_k)} \mathbf{1}_{X_i^{(1)} \cap X_j^{(2)} \cap Y_k} - \sum_{j,k} \log \frac{\mu(X_j^{(2)} \cap Y_k)}{\mu(Y_k)} \mathbf{1}_{X_j^{(2)} \cap Y_k}$$

$$= I(\alpha_1 | \alpha_2 \vee \beta) + I(\alpha_2 | \beta).$$

Property (ii) trivially follows from (3.27). Concerning (iii) we deduce from (i) with $\alpha_1 \vee \alpha_2 = \alpha_2$ that $I(\alpha_2 | \beta) = I(\alpha_2 | \alpha_1 \vee \beta) + I(\alpha_1 | \beta)$. Since by definition I is non-negative we only need to integrate $I(\alpha_2 | \beta) \geq I(\alpha_1 | \beta)$ in order to obtain (iii). The proof of (iv) is slightly more tedious because an analogous inequality between information functions does *not* hold in general. However, notice that by $\alpha_1 \prec \alpha_2$ we have $X_i^{(1)} = \bigcup_{k \in M_i} X_k^{(2)}$ for all i with appropriate disjoint sets of indices $M_i \subseteq \{1, \ldots, m_2\}$. Setting $\varphi(x) := -x \log x$ if $x > 0$ and $\varphi(0) := 0$ for the sake of brevity, we observe that

$$\int_X I(\beta | \alpha_2) \, d\mu = \sum_{j,k} \varphi\left(\frac{\mu(Y_j \cap X_k^{(2)})}{\mu(X_k^{(2)})}\right) \mu(X_k^{(2)})$$

$$= \sum_{i,j} \left(\sum_{k \in M_i} \varphi\left(\frac{\mu(Y_j \cap X_k^{(2)})}{\mu(X_k^{(2)})}\right) \frac{\mu(X_k^{(2)})}{\mu(X_i^{(1)})} \right) \mu(X_i^{(1)})$$

$$\leq \sum_{i,j} \varphi\left(\frac{\mu(Y_j \cap X_i^{(1)})}{\mu(X_i^{(1)})}\right) \mu(X_i^{(1)}) = \int_X I(\beta | \alpha_1) \, d\mu;$$

here we have made use of the fact that

$$\sum_k \varphi(a_k) b_k \leq \varphi\left(\sum_k a_k b_k\right) \quad \text{if } a_k, b_k \geq 0 \text{ and } \sum_k b_k = 1,$$

i.e., φ is concave on $[0, +\infty[$. This latter property also yields

$$\int_X I(\alpha_1) \, d\mu = \sum_i \varphi\left(\mu(X_i^{(1)})\right) \leq m_1 \varphi(m_1^{-1}) = \log m_1$$

hence (v). Finally (vi) follows from (ii) by integration. $\square$

We are now in a position to answer the question concerning the limit of (3.26). By making use of information functions, the formal argument is quite elegant and straightforward.

Theorem 3.25 (Shannon, McMillan, Breiman). *Let T denote a measure-preserving map on a normed measure space $(X, \mathcal{A}, \mu)$ and α a finite partition. Then the limit*

$$a(x) := \lim_{n \to \infty} a_n(x) = -\lim_{n \to \infty} \frac{\log \mu\big(\alpha_n(x)\big)}{n}$$

exists almost everywhere and in L^1; moreover, $a \circ T = a \, [\mu]$.

Proof. From the definitions and Lemma 3.24 we obtain

$$na_n = I\left(\bigvee_{k=0}^{n} T^{-k}\alpha\right) = \sum_{k=0}^{n-1} I\left(\alpha| \bigvee_{j=1}^{n-k} T^{-j}\alpha\right) \circ T^k + I(\alpha) \circ T^n.$$

The sum at the right can be given a form to which Proposition 3.23 applies by setting $f_k := I\left(\alpha| \bigvee_{j=1}^{k} T^{-j}\alpha\right)$. By means of the increasing sequence of σ-algebras $\mathcal{A}_k := A_\sigma\left(\bigvee_{j=1}^{k} T^{-j}\alpha\right)$ we may state that $f_k = -\sum_{i=1}^{m} \log \mathbb{E}(\mathbf{1}_{X_i}|\mathcal{A}_k)\,\mathbf{1}_{X_i}$; here $\alpha = \{X_1, \ldots, X_m\}$. Setting $\mathcal{A}_\infty := A_\sigma(\mathcal{A}_k : k \in \mathbb{N}) = A_\sigma(T^{-k}\alpha : k \in \mathbb{N})$ we deduce from the Increasing Martingale Theorem (see Appendix A) that f_k converges almost everywhere to $f := -\sum_{i=1}^{m} \log \mathbb{E}(\mathbf{1}_{X_i}|\mathcal{A}_\infty)\mathbf{1}_{X_i}$. Thus an application of Proposition 3.23 will yield the desired result once we have shown that $f_k \to f$ also in L^1. To this end it suffices to prove that $f^* := \sup_{k\in\mathbb{N}} f_k$ is an element of L^1. Setting $F(t) := \mu(\{x : f^*(x) > t\})$ we have as a basic fact of integration theory (see Appendix A)

$$\int_X f^* \, d\mu = \int_0^{+\infty} F(t) \, dt.$$

Furthermore

$$F(t) = \sum_{i=1}^{m} \mu(X_i \cap \{x : \inf_k \mathbb{E}(\mathbf{1}_{X_i}|\mathcal{A}_k) < e^{-t}\}) = \sum_{i=1}^{m} \sum_{k=1}^{\infty} \mu(X_i \cap A_{i,k})$$

where the sets $A_{i,k}$ are defined as

$$A_{i,k} := \left\{x : \mathbb{E}(\mathbf{1}_{X_i}|\mathcal{A}_l) \geq e^{-t} \text{ for } 1 \leq l < k, \mathbb{E}(\mathbf{1}_{X_i}|\mathcal{A}_k) < e^{-t}\right\}.$$

Due to this definition $A_{i,k} \in \mathcal{A}_k$ as well as $\sum_k \mu(A_{i,k}) \leq 1$. Additionally, the estimate $\mu(X_i \cap A_{i,k}) \leq e^{-t}\mu(A_{i,k})$ holds. Thus $F(t) \leq me^{-t}$ and $\int_X f^* d\mu \leq m$ which shows that $f^* \in L^1(\mu)$ and hence proves the theorem. $\square$

The limit $a(x)$ which we found to be of considerable interest therefore exists almost everywhere. When dealing with specific examples, however, an explicit calculation of $a(x)$ may be difficult, so that one often is content to determine an average value for a. Since $a_n \to a$ in L^1 we have

$$h_\mu(T, \alpha) := \int_X a \, d\mu = \lim_{n\to\infty} \int_X a_n d\mu = \lim_{n\to\infty} -\frac{1}{n} \sum_{B\in\bigvee_0^n T^{-k}\alpha} \mu(B) \log \mu(B). \quad (3.28)$$

Obviously $h_\mu(T, \alpha) \geq 0$; if T is ergodic then the limit a is constant almost everywhere, and $a_n(x) \to h_\mu(T, \alpha)$. According to our discussion above we expect the map T to exhibit complicated and difficult-to-predict dynamics at least on parts of the space if $h_\mu(T, \alpha) > 0$ for some partition α.

Definiton 3.26. Let T denote a measure-preserving map on $(X, \mathcal{A}, \mu)$. For any finite partition α the quantity $h_\mu(T, \alpha)$ is called the *(metric) entropy* of T with respect to α. The *entropy* $h_\mu(T)$ of T is defined as

$$h_\mu(T) := \sup_\alpha h_\mu(T, \alpha)$$

where the supremum is taken over all finite partitions.

Clearly, entropy is non-negative, the case of infinite $h_\mu(T)$ not being excluded a priori. From a statistical point of view the property $h_\mu(T) > 0$ may justly be interpreted as T behaving chaotically ([11, 31]). Indeed, a widespread notion defines a continuous map T on a compact metric space X to be chaotic if $h_\mu(T) > 0$ for at least one T-invariant probability measure on $(X, \mathcal{B}(X))$. As we shall see, this statistical interpretation of irregularity perfectly matches our deterministic notion of chaos for all the systems considered hitherto (e.g. toral endomorphisms and subshifts of finite type). For continuous maps on the unit interval a famous theorem due to Misiurewicz asserts that this statistical notion of chaos rigorously coincides with the terminology mentioned in Section 2.3 (see e.g. [11]) which in turn yields many insights comparable to our usage of the term *chaos* according to Definition 2.21.

Calculating $h_\mu(T)$ directly from the definitions admittedly seems impossible. We therefore quote two auxiliary results before turning towards examples. Assertion (ii) in the following lemma is often referred to as the *Kolmogorov–Sinai Theorem*.

Lemma 3.27. *Let T denote a measure-preserving map on $(X, \mathcal{A}, \mu)$.*

(i) *If $(\beta_n)_{n \in \mathbb{N}_0}$ is an increasing sequence of finite partitions, which means that $\beta_0 \prec \beta_1 \prec \ldots$, and $A_\sigma(\beta_n : n \in \mathbb{N}_0) = \mathcal{A}$ then $h_\mu(T) = \lim_{n \to \infty} h_\mu(T, \beta_n)$.*

(ii) *If the finite partition α generates $\mathcal{A}$, i.e. $A_\sigma(T^{-n}\alpha : n \in \mathbb{N}_0) = \mathcal{A}$, then $h_\mu(T) = h_\mu(T, \alpha)$.*

Proof. As a preliminary result we are going to prove that

$$h_\mu(T, \alpha) \le h_\mu(T, \beta) + \int_X I(\alpha|\beta)\, d\mu \tag{3.29}$$

holds for arbitrary finite partitions α, β. Indeed, from Lemma 3.24 we deduce that

$$\int_X I\left(\bigvee_{i=0}^n T^{-i}\alpha\right) d\mu \le \int_X I\left(\bigvee_{i=0}^n T^{-i}\alpha \vee \bigvee_{i=0}^n T^{-i}\beta\right) d\mu$$

$$= \int_X I\left(\alpha \mid \bigvee_{i=1}^n T^{-i}\alpha \vee \bigvee_{i=0}^n T^{-i}\beta\right) d\mu + \int_X I\left(\bigvee_{i=1}^n T^{-i}\alpha \mid \bigvee_{i=0}^n T^{-i}\beta\right) d\mu + \int_X I\left(\bigvee_{i=0}^n T^{-i}\beta\right) d\mu$$

$$\le \int_X I(\alpha|\beta)\, d\mu + \int_X I\left(\bigvee_{i=0}^{n-1} T^{-i}\alpha \mid \bigvee_{i=0}^{n-1} T^{-i}\beta\right) d\mu + \int_X I\left(\bigvee_{i=0}^n T^{-i}\beta\right) d\mu .$$

Inductively we obtain

$$\frac{1}{n} \int_X I\left(\bigvee_{i=0}^n T^{-i}\alpha\right) d\mu \le \frac{n+1}{n} \int_X I(\alpha|\beta)\, d\mu + \frac{1}{n} \int_X I\left(\bigvee_{i=0}^n T^{-i}\beta\right) d\mu ,$$

and hence a passage to the limit yields (3.29). Furthermore we notice that for $\alpha = \{X_1, \ldots, X_m\}$, $\beta = \{Y_1, \ldots, Y_l\}$

$$\int_X I(\alpha|\beta)\, d\mu = -\sum_{i,j} \log \frac{\mu(X_i \cap Y_j)}{\mu(Y_j)} \mu(X_i \cap Y_j) = \sum_i \int_X \varphi\big(\mathbb{E}(\mathbf{1}_{X_i}|A_\sigma(\beta))\big)\, d\mu$$

with $\varphi(x) = -x \log x$ (if $x > 0$) and $\varphi(0) = 0$ as in the proof of Lemma 3.24.

With these preparations let us now tackle (i). Since the finite σ-algebras $A_\sigma(\beta_n)$ monotonically increase towards A the Increasing Martingale Theorem implies that $\mathbb{E}(1_{X_i}|A_\sigma(\beta_n)) \to 1_{X_i}$ almost everywhere for all i. As $0 \le \mathbb{E}(1_{X_i}|A_\sigma(\beta_n)) \le 1$ and $\varphi(x) \le e$ for $x \in [0,1]$ we have $\lim_{n\to\infty} \int_X I(\alpha|\beta_n)\,d\mu = 0$ by virtue of the Dominated Convergence Theorem. Therefore by (3.29)

$$h_\mu(T,\alpha) \le \underline{\lim}_{n\to\infty} h_\mu(T,\beta_n) + \lim_{n\to\infty}\int_X I(\alpha|\beta_n)\,d\mu = \underline{\lim}_{n\to\infty} h_\mu(T,\beta_n)$$

for any finite partition α and thus $h_\mu(T) \le \underline{\lim}_{n\to\infty} h_\mu(T,\beta_n)$. On the other hand we have $h_\mu(T) \ge h_\mu(T,\beta_n)$ for all n and thus $h_\mu(T) \ge \overline{\lim}_{n\to\infty} h_\mu(T,\beta_n)$ from which (i) follows. The proof of (ii) is a rather straightforward application of (i). Defining $\beta_n := \bigvee_{i=0}^n T^{-i}\alpha$ we find $h_\mu(T) = \lim_{n\to\infty} h_\mu(T,\beta_n)$ by virtue of (i). But since $\bigvee_{i=0}^k T^{-i}\beta_n = \beta_{n+k}$ clearly $h_\mu(T,\beta_n) = h_\mu(T,\alpha)$ for all $n \in \mathbb{N}_0$. $\square$

The following result is barely surprising. Iterates of maps with zero entropy will also have zero entropy, while iterates of maps with positive entropy grow more and more irregular.

Lemma 3.28. *For any measure-preserving map T on (X, A, μ) and $k \in \mathbb{N}_0$ the relation $h_\mu(T^k) = kh_\mu(T)$ holds. If T is invertible then $h_\mu(T^k) = |k|h_\mu(T)$ for every $k \in \mathbb{Z}$.*

Proof. Let α denote a finite partition. The case $k = 0$ is easily dealt with because $\frac{1}{n}I(\bigvee_{i=0}^n \mathrm{id}_X^{-i}\alpha) = \frac{1}{n}I(\alpha) \to 0$ implying $h_\mu(\mathrm{id}_X,\alpha) = 0$ and also $h_\mu(\mathrm{id}_X) = 0$. If $k > 1$ we may define $\beta := \bigvee_{i=0}^{k-1} T^{-i}\alpha$ and observe $\bigvee_{i=0}^n T^{-ik}\beta = \bigvee_{i=0}^{(n+1)k-1} T^{-i}\alpha$. Consequently

$$\frac{1}{n}I\left(\bigvee_{i=0}^n T^{-ik}\beta\right) = \frac{(n+1)k-1}{n} \cdot \frac{1}{(n+1)k-1} I\left(\bigvee_{i=0}^{(n+1)k-1} T^{-i}\alpha\right)$$

from which we deduce $h_\mu(T^k) \ge h_\mu(T^k,\beta) = kh_\mu(T,\alpha)$ and also $h_\mu(T^k) \ge kh_\mu(T)$. On the other hand, by virtue of Lemma 3.24 we have

$$\frac{1}{n}\int_X I\left(\bigvee_{i=0}^n T^{-ik}\alpha\right)d\mu \le \frac{k}{kn}\int_X I\left(\bigvee_{i=0}^{kn} T^{-i}\alpha\right)d\mu$$

which yields $h_\mu(T^k,\alpha) \le kh_\mu(T,\alpha) \le kh_\mu(T)$ and therefore also the reverse inequality $h_\mu(T^k) \le kh_\mu(T)$. If T is invertible we notice that

$$I\left(\bigvee_{i=0}^n (T^{-1})^{-i}\alpha\right) = I\left(T^n\left(\bigvee_{i=0}^n T^{-i}\alpha\right)\right) = I\left(\bigvee_{i=0}^n T^{-i}\alpha\right) \circ T^{-n}.$$

Since T^{-1} is also measure-preserving Lemma 3.24 implies $h_\mu(T^{-1},\alpha) = h_\mu(T,\alpha)$ from which the remaining assertion follows. $\square$

Let us now review a few of our earlier examples in the light of entropy. First recall that the translations $R_{(\vartheta_1,\dots,\vartheta_d)}$ of $\mathbb{T}^d$ are certainly not chaotic (though possibly ergodic with respect to $\lambda_{\mathbb{T}^d}$). In order to calculate $h_{\lambda_{\mathbb{T}^d}}(R_{(\vartheta_1,...,\vartheta_d)})$ we make use of partitions α_m of $\mathbb{T}^d$ into cubes of side-length $1/m$, more formally

$$\alpha_m := \left\{ \left[\frac{i_1}{m}, \frac{i_1+1}{m}\right[\times \ldots \times \left[\frac{i_d}{m}, \frac{i_d+1}{m}\right[: 0 \le i_1, \ldots, i_d < m \right\} \quad (m \in \mathbb{N}).$$

It is easy to see that the refined partition $\bigvee_{i=0}^{n} R_{(\vartheta_1,...,\vartheta_d)}^{-i} \alpha_m$ does not contain more than $(n+1)^d m^d$ elements. Therefore by virtue of Lemma 3.24 for all $m \in \mathbb{N}$

$$h_{\lambda_{\mathbb{T}^d}}(R_{(\vartheta_1,...,\vartheta_d)}, \alpha_m) \le \lim_{n \to \infty} \frac{1}{n} \log\big((n+1)^d m^d\big) = 0.$$

Furthermore $\alpha_1 \prec \alpha_2 \prec \alpha_{2^2} \prec \ldots$ and $\mathcal{B}_{\mathbb{T}^d} = A_\sigma(\alpha_{2^m} : m \in \mathbb{N}_0)$. Therefore Lemma 3.27 implies $h_{\lambda_{\mathbb{T}^d}}(R_{(\vartheta_1,...,\vartheta_d)}) = 0$ which is exactly the result we expected. If $(\vartheta_1, \ldots, \vartheta_d) \in \mathbb{Q}^d$ we could equally have invoked Lemma 3.28 because in this case $R_{(\vartheta_1,...,\vartheta_d)}^p = \mathrm{id}_{\mathbb{T}^d}$ for some p.

After having reasonably defined σ_A-invariant measures we shall later calculate the entropy of subshifts of finite type (see Exercise 4.7); also shall we find $h_{\mu^*}(F_4)$ where μ^* denotes the unique F_4-invariant probability measure which is absolutely continuous with respect to λ^1. (At present we are not even sure that μ^* exists; cf. Exercise 4.8). Right now we shall deal with another class of chaotic systems, namely hyperbolic toral endomorphisms which we know to be mixing with respect to $\lambda_{\mathbb{T}^d}$. Let us first focus on the case $d = 1$ where these maps are given by $T_{(k)} : x \mapsto kx \pmod 1$ with $|k| \ge 2$. Taking $\alpha_{|k|} = \left\{\left[\frac{i}{|k|}, \frac{i+1}{|k|}\right[: 0 \le i < |k|\right\}$ as before we see that all elements in $\bigvee_{i=0}^{n} T_{(k)}^{-i} \alpha_{|k|} = \alpha_{|k|^{n+1}}$ are intervals of the same length $|k|^{-(n+1)}$. Therefore $I\big(\bigvee_{i=0}^{n} T_{(k)}^{-i} \alpha_{|k|}\big) = (n+1) \log |k|$, and

$$h_{\lambda_{\mathbb{T}^1}}(T_{(k)}, \alpha_{|k|}) = \log |k| = h_{\lambda_{\mathbb{T}^1}}(T_{(k)}, \alpha_{|k|^l}) \quad \text{for all } l \in \mathbb{N}.$$

But $A_\sigma(\alpha_{|k|^l} : l \in \mathbb{N}) = \mathcal{B}_{\mathbb{T}^d}$ and thus

$$h_{\lambda_{\mathbb{T}^1}}(T_{(k)}) = \log |k|.$$

Unfortunately, the argument is considerably more involved for $d \ge 2$. As the necessary techniques are not available to us here we refer to [62] for a rigorous treatment. The final result, however, is written down easily if we adhere to our convention that the spectrum of a matrix may repeatedly contain an eigenvalue, according to its multiplicity (cf. Appendix A).

Theorem 3.29. *For* $\det A \ne 0$ *the toral endomorphism* T_A *has entropy*

$$h_{\lambda_{\mathbb{T}^d}}(T_A) = \sum_{z \in \sigma(A)} \log\left(\max\{1, |z|\}\right) = \sum_{z \in \sigma(A):|z| \ge 1} \log |z|.$$

In the hyperbolic case we therefore always have $h_{\lambda_{\mathbb{T}^d}}(T_A) > 0$. Again this perfectly meets our expectations. For example we have

$$h_{\lambda_{\mathbb{T}^2}}(T_A) = \log \frac{3 + \sqrt{5}}{2} \approx 0.962 \quad \text{for } A = \begin{pmatrix} 2 & 1 \\ 1 & 1 \end{pmatrix}.$$

By virtue of Theorem 3.25 we obtain for any non-trivial partition α and almost every $x \in \mathbb{T}^2$

$$\lambda_{\mathbb{T}^2}(\alpha_n(x)) = \left(\frac{3 + \sqrt{5}}{2}\right)^{-n(1-\varepsilon_n)}$$

where $\varepsilon_n \to 0$ as $n \to \infty$. At the very beginning of this chapter numerical experiments suggested that areas of points sharing roughly the same fate under iteration of T_A for at least n steps shrink exponentially. Having developed the statistical point of view we are now able to give a satisfactory quantitative description of this phenomenon by means of entropy.

Let us finally say a few words about billiards. Recall that there is an associated billiard map T_{bill} for any bounded connected table with piecewise C^2 boundary; moreover T_{bill} preserves a reasonable probability measure μ_{bill}. We already know that T_{bill} is not ergodic for the elliptic and rectangular billiard. Hardly surprising, $h_{\mu_{\text{bill}}}(T_{\text{bill}}) = 0$ for these systems. More generally, $h_{\mu_{\text{bill}}}(T_{\text{bill}}) = 0$ holds for all polygonal billiards, thus especially for triangles for which the problem of ergodicity has not completely been clarified yet ([54]). Contrary to these statistically tame systems the stadium, the dispersing billiard and the other billiards depicted in Figure 3.11 as well are mixing and have positive entropy. In fact, rather general design principles ensuring $h_{\mu_{\text{bill}}}(T_{\text{bill}}) > 0$ have been devised ([15]). Individual trajectories of such billiards typically behave unpredictably even for a moderate number of reflections. By their simple geometric definition these billiard systems provide excellent examples showing the necessity for both a geometric as well as a statistical description of chaos. In this context it is interesting to note that the initiation of ergodic theory is commonly attributed to Boltzmann who introduced the statistical point of view when studying the model of a gas as a huge number of elastically colliding hard spheres. His seminal insights grew out of the postulation of *molecular chaos* (sic!) for such a billiard-like system ([21, 32, 59]).

Historically, the notion of entropy was introduced by Kolmogorov in order to solve an isomorphism problem (see Exercise 4.9). As can be imagined even on the basis of our brief discussion, entropy offers itself for many fascinating and illuminating calculations. The multifarious aspects of entropy, its relations to periodic points, number theory, classification of certain shifts, Hausdorff dimension etc., definitely deserve separate study. We refer to the literature for all these interesting topics ([31, 38, 42, 50, 62, 63]).

Exercises

(3.1) Let X denote a separable metric space and $T : X \to X$ a continuous map. Prove that every finite T-invariant measure is supported on the set $\Omega(T)$ of non-wandering points (Exercise 2.7), i.e. $\mu(X \backslash \Omega(T)) = 0$. Show that this conclusion may fail if μ is not finite.

(3.2) Again let T denote a continuous map on a separable metric space X. In addition, assume that the probability measure μ assigns positive measure to every non-empty open set and that T is ergodic with respect to μ. Prove that the orbit of almost every point is dense, that is

$$\mu\big(\{x \in X : \overline{O^+(x)} = X\}\big) = 1 .$$

(3.3) Define a map T on $[0,1]$ according to

$$T(x) := \begin{cases} 0 & \text{if } x = 0, \\ \frac{1}{x} - \lfloor \frac{1}{x} \rfloor & \text{if } x \neq 0. \end{cases}$$

Show that the probability measure μ on $[0, 1] \cap \mathcal{B}^1$ defined as

$$\frac{d\mu}{d\lambda^1}(x) := \frac{1}{(1+x)\log 2}\, \mathbf{1}_{[0,1]}(x)$$

is T-invariant. (Frequently, μ is referred to as the *Gauss measure*.) Clearly, all points in $[0, 1] \cap \mathbb{Q}$ are eventually fixed. To every $x \in [0, 1]\backslash\mathbb{Q}$ assign a sequence of natural numbers $r_n(x) := \lfloor (T^{n-1}(x))^{-1} \rfloor$. Define $\rho_n(x) \in \mathbb{Q}^+$ by

$$\rho_n(x) := [r_1(x), \ldots, r_n(x)] := \cfrac{1}{r_1(x) + \cfrac{1}{r_2(x) + \cfrac{1}{\cdots + \cfrac{1}{r_{n-1}(x) + \cfrac{1}{r_n(x)}}}}}$$

and verify the relation $\lim_{n \to \infty} \rho_n(x) = x$ for all $x \in [0, 1]\backslash\mathbb{Q}$. Deduce from this and $r_{n+1}(x) = r_n(T(x))$ that T acts as a shift on the *continued fraction expansion* of irrational numbers. Can you verify that T is in fact ergodic on $([0, 1], [0, 1] \cap \mathcal{B}^1, \mu)$?

(3.4) With the notion of the preceding exercise it is shown in [38] that T is exact on the probability space $([0, 1], [0, 1] \cap \mathcal{B}^1, \mu)$, its entropy being $h_\mu(T) = \pi^2/(6 \log 2)$. Taking these facts for granted show that $\rho_n(x)$ typically converges towards x *exponentially*, i.e.

$$\lim_{n \to \infty} \frac{1}{n} \log |x - \rho_n(x)| = c < 0$$

for almost all x; determine c. Give examples showing that for individual points the asymptotic rate of convergence may differ considerably from this typical value.

(3.5) Let T be a measure-preserving map on the probability space $(X, \mathcal{A}, \mu)$. The natural extension $\hat{T} : \hat{X} \to \hat{X}$ clearly is $\hat{\mathcal{A}}$-measurable with $\hat{\mathcal{A}} := \hat{X} \cap \bigotimes_{n \in \mathbb{N}_0} \mathcal{A}$. By definition, the σ-algebra $\bigotimes_{n \in \mathbb{N}_0} \mathcal{A}$ is generated by the family of all finite cylinders

$C[A_0, \ldots, A_m] := \{(x_n)_{n \in \mathbb{N}_0} \in \hat{X} : x_i \in A_i \text{ for } i = 0, \ldots, m\} \subseteq X^{\mathbb{N}_0};$ see also Appendix A. Define a normed measure $\hat{\mu}$ on $\hat{\mathcal{A}}$ by

$$\hat{\mu}(C[A_0, \ldots, A_m]) := \mu(A_m).$$

Show that $\hat{\mu}$ is preserved by $\hat{T}$. In addition, prove that $\hat{T}$ is ergodic on $(\hat{X}, \hat{\mathcal{A}}, \hat{\mu})$ if and only if T is ergodic on $(X, \mathcal{A}, \mu)$. Does $h_{\hat{\mu}}(\hat{T}) = h_\mu(T)$ hold in general?

(3.6) Let T denote a measure preserving map on $(X, \mathcal{A}, \mu)$ and assume that the map $h : X \to Y$ is $\mathcal{A}$-$\mathcal{B}$-measurable where $\mathcal{B}$ denotes a σ-algebra in Y. Demonstrate that the measure $h\mu$ on $\mathcal{B}$ is preserved by any $\mathcal{B}$-measurable map $S : Y \to Y$ with $S \circ h = h \circ T$. As an application consider the maps $P_k : z \mapsto z^k$ on S^1 that preserve λ_{S^1} if $k \in \mathbb{Z} \setminus \{0\}$ and δ_1 if $k = 0$. Show that there is a unique polynomial $p_k : [-1, 1] \to [-1, 1]$ satisfying $p_k \circ h = h \circ P_k$ with $h(z) := \Re z$. Obviously $p_0 \equiv 1$, $p_1(x) = x$ and $p_{-k} = p_k$. Verify the recurrence relation $p_{k+1}(x) = 2x\, p_k(x) - p_{k-1}(x)$ for $k \in \mathbb{Z}$. (The polynomial p_k is called the $|k|$-th *Chebyshev polynomial*.) Determine the measure induced by h and show that for $|k| \geq 2$ all p_k are exact with respect to that measure.

(3.7) Given $a, b \in \mathbb{R}^+$ with $a > b$ and $\alpha \in \,]0, \pi[$ consider the billiard map T_{bill} associated with the ellipse $\mathcal{E}$ with principal axes $2a$ and $2b$, respectively. Let a billiard trajectory emanate from $(0, b)$ with angle α (see Figure 3.16); assume $a/b \sin \alpha < 1$ so that the caustic $\mathcal{E}^*$ is an ellipse. As we have seen in the text, $T_{\text{bill}}|_{M(\mathcal{E}^*)}$ is topologically conjugate to a circle rotation. Explicitly calculate $\rho(T_{\text{bill}}|_{M(\mathcal{E}^*)})$ as a function of a/b and α. (For a convenient representation of the result a quick glance into a reference of elliptic integrals, e.g. [1], might prove helpful.)

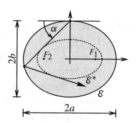

Figure 3.16.   Calculate $\rho(T_{\text{bill}}|_{M(\mathcal{E}^*)})$!

(3.8) Derive formulae (3.16) and (3.17) which correlate the circular billiard with translations on $\mathbb{T}^2$. Given $R > 0$ and $\alpha \in \,]0, \frac{\pi}{2}[$ first look for a polar coordinates representation of the straight lines touching the circle $x^2 + y^2 = R^2 \cos^2 \alpha$. Then make the image of the translated point move with constant velocity.

(3.9) For $\alpha \in \mathbb{R}$ a smooth curve x_α in the square $[-1, 1]^2 \subseteq \mathbb{R}^2$ is defined via $x_\alpha(t) := (\cos t, \cos \alpha t)$. Show that x_α is closed if and only if α is rational; furthermore, identify x_α as the graph of a polynomial if $\alpha \in \mathbb{Z}$. For irrational α and any rectangle $R \subseteq [-1, 1]^2$ verify by explicit calculation that the asymptotic relative frequency

$$\lim_{T \to \infty} \frac{1}{T} \int_0^T 1_R(x_\alpha(t))\, dt$$

is positive; deduce that x_α is dense in $[-1, 1]^2$ in this case.

(3.10) Consider the problem of mean motion with three rods and assume that their lengths $|a_1|, |a_2|, |a_3|$ generate a non-trivial triangle. (There is thus *no* dominant rod.) Confirm that

$$\omega_\infty = \frac{\alpha_1 \omega_1 + \alpha_2 \omega_2 + \alpha_3 \omega_3}{\pi}$$

where α_i denotes the triangle's angle vis-à-vis the side with length $|a_i|$.

(3.11) Consider the one-dimensional motion of two equal
masses between two walls according to Figure 3.17: both
masses move with constant velocity unless $x_1 = 0$,
$x_1 = x_2$ or $x_2 = L$; in each of these cases a completely
elastic collision (no loss of kinetic energy) takes place.
Let $x_1(t)$, $x_2(t)$ denote the position at time t of the left
and right mass, respectively. Furthermore assume that

$$x_i(0) = x_{i,0}, \ \dot{x}_i(0) = v_i \quad (i = 1, 2).$$

Show that for λ^2-almost every initial velocity $(v_1, v_2) \in$
$\mathbb{R}^2$ the asymptotic relative frequencies

$$F_i(x) := \lim_{T \to \infty} \frac{1}{T} \int_0^T \mathbf{1}_{[0,x]}(x_i(t)) \, dt \quad (i = 1, 2)$$

exist for all $x \in [0, L]$. Can you generalize your results
to an arbitrary number of masses?

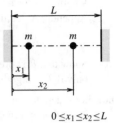

$0 \le x_1 \le x_2 \le L$

Figure 3.17. A mechanical system equivalent to a billiard

(3.12) Given a measure-preserving map T on $(X, \mathcal{A}, \mu)$ define a linear map U_T on $L^2(\mu)$ by
$U_T(f) := f \circ T$. Show that U_T is an isometry, i.e. $\langle U_T(f), U_T(g) \rangle = \langle f, g \rangle$ for all
$f, g \in L^2(\mu)$. A complex number λ is called an *eigenvalue* of T if $U_T(f) = \lambda f$ for some
non-zero $f \in L^2(\mu)$, which correspondingly is called an *eigenfunction*. Denoting by
$\sigma_p(T)$ the set of all eigenvalues of T show that $\sigma_p(T)$ is a subgroup of S^1 (with complex
multiplication); eigenfunctions corresponding to different eigenvalues are orthogonal.
Clearly, T is ergodic if and only if the eigenvalue 1 is simple. Calculate $\sigma_p(R_\vartheta)$ and
explain how this observation suits Theorem 3.8. Can you also characterize the property
of T to be *mixing* in terms of the associated isometry U_T?

(3.13) With the notion of the previous exercise two measure-preserving maps T_i on
$(X_i, \mathcal{A}_i, \mu_i)_{i=1,2}$ are called *spectrally isomorphic* if $W \circ U_{T_1} = U_{T_2} \circ W$ for some
linear isomorphism $W : L^2(\mu_1) \to L^2(\mu_2)$ which additionally preserves the inner
products, that is $\langle W(f), W(g) \rangle_2 = \langle f, g \rangle_1$ for all $f, g \in L^2(\mu_1)$; here $\langle ., . \rangle_i$ stands
for the inner product in $L^2(\mu_i)$. A dynamical property of a measure-preserving map is a
spectral invariant if it is shared by any two spectrally isomorphic maps. Given a measure-
preserving map T confirm that ergodicity and mixing of T are spectral invariants, as is
$\sigma_p(T)$. Prove or disprove the following statement: two rotations of S^1 are spectrally
isomorphic if and only if they are topologically conjugate.

The simple observations of the last two exercises mark the entrance to the classical field
of *spectral analysis* of dynamical systems, the aim of which is to gain insights into the
dynamics of T by carefully analysing the associated isometry U_T ([62]).

(3.14) Let T denote a measure-preserving map on $(X, \mathcal{A}, \mu)$ and fix $B \in \mathcal{A}$ with $\mu(B) > 0$. In
the context of Kac's lemma (Theorem 3.10) we came to define for $x \in B$ the return time
to B as

$$t_B(x) := \begin{cases} \min\{n \in \mathbb{N} : T^n(x) \in B\} & \text{if } \{n \in \mathbb{N} : T^n(x) \in B\} \neq \emptyset, \\ \infty & \text{otherwise}, \end{cases}$$

which is measurable and finite almost everywhere. Define now the *induced map* T_B

according to

$$T_B(x) := \begin{cases} T^{t_B}(x) & \text{if } t_B(x) \in \mathbb{N}, \\ x & \text{if } t_B(x) = \infty. \end{cases}$$

Show that T_B is a measure-preserving map on $(B, B \cap \mathcal{A}, \mu_B)$ where $\mu_B := \frac{1}{\mu(B)} \mu|_{B \cap \mathcal{A}}$ designates the conditional measure on $B \cap \mathcal{A}$. Verify that T_B is ergodic if T is; if $\mu\left(\bigcup_{n=0}^{\infty} T^{-n}(B)\right) = 1$ the converse also holds.

(3.15) Prove that for any probability space $(X, \mathcal{A}, \mu)$ the one-sided Bernoulli shift σ on $\bigotimes_{n \in \mathbb{N}_0} (X, \mathcal{A}, \mu)$ is exact.

Chapter 4
Ergodic theory II. Applications

Many interesting aspects of dynamics can be described and quantified by means of invariant (ergodic, exact) measures. We have already observed this in the light of some specific systems. However, until now we have completely avoided several questions which obviously are crucial in this context: Given a map on a certain space, does there exist an invariant measure at all? If it does, is this measure meaningful for our understanding of the dynamics? And if so, how can we find or at least approximate that measure? If we want ergodic theory to be applicable to real-world problems, we will definitely have to face these questions. As throughout the whole text we shall not deal with the most general situation but rather develop some important and powerful tools within the scope of special situations. The abstract notions of ergodic theory introduced in the previous chapter will become quite familiar in this context. By dealing with special maps on the unit interval as well as elementary stochastic processes we shall finally gain a unified perspective on deterministic and stochastic aspects of dynamical systems.

4.1   The Frobenius–Perron operator

Let T denote a measurable map on the probability space $(X, \mathcal{A}, \mu)$ and take $f \in L^1(\mu)$ with $f \geq 0$ and $\int_X f \, d\mu = 1$. The measure ν defined by $\frac{d\nu}{d\mu} := f$ is a probability measure on $(X, \mathcal{A})$. Recall our stochastic interpretation of such a measure: ν may be considered the governing law of an experiment of chance ξ with outcomes in X. What is the corresponding law for $T(\xi)$? We have already answered this question: the experiment $T(\xi)$ is governed by $T\nu$. Since measures defined via densities are especially convenient to deal with we may ask: Does $T\nu$ have a density with respect to μ and, if so, what does it look like? In the sequel we shall always assume the map T to be *non-singular* by which we mean that $T\mu \ll \mu$. Equivalently, T is non-singular if $\mu(T^{-1}(A)) = 0$ whenever $\mu(A) = 0$, so that, intuitively speaking, T cannot annihilate mass by moving it to places which are "invisible" to μ. Obviously any measure-preserving map is non-singular. On familiar spaces there are several easy-to-check conditions implying non-singularity ([13]), and we shall henceforth not bother much about this property. If T is non-singular then $T\nu \ll \mu$ so that $T\nu$ has a unique μ-density $g \in L^1(\mu)$ with $\int_A g \, d\mu = \int_{T^{-1}(A)} f \, d\mu$ for all $A \in \mathcal{A}$. This relation suggests the following definition which is crucial for most of the present chapter.

Definiton 4.1.   Let T denote a non-singular map on $(X, \mathcal{A}, \mu)$. The unique linear operator $P_T : L^1(\mu) \to L^1(\mu)$ satisfying

$$\int_A P_T f \, d\mu = \int_{T^{-1}(A)} f \, d\mu \quad \text{for all } A \in \mathcal{A}, \, f \in L^1(\mu) \tag{4.1}$$

is called the *Frobenius–Perron operator* associated with T.

If the random variable ξ is distributed according to the density f, then the transformed random variable $T(\xi)$ is distributed according to $P_T f$. Before listing some of the basic properties of the Frobenius–Perron operator let us have a look at an example.

Take $([0, 1], [0, 1] \cap \mathcal{B}^1, \lambda^1|_{[0,1)})$ as the probability space and assume the map $T : [0, 1] \to [0, 1]$ to be non-singular. Then by (4.1)

$$\int_0^x P_T f(y) \, dy = \int_{T^{-1}([0,x])} f(y) \, dy \quad \text{for } 0 \le x \le 1 \text{ and } f \in L^1(\lambda^1|_{[0,1)}),$$

from which we deduce via differentiation the almost everywhere relation

$$P_T f(x) = \frac{d}{dx} \int_{T^{-1}([0,x])} f(y) \, dy. \tag{4.2}$$

Implicitly, we have already used this formula before in order to obtain the relation $P_T^n f(x) = 2^{-n} x^{-1+2^{-n}} f(x^{2^{-n}})$ for $T(x) := x^2$ (recall the calculations accompanying Figure 3.3). Plugging our old friend F_4 into (4.2) we find

$$P_{F_4} f(x) = \frac{1}{4\sqrt{1-x}} \left(f\left(\frac{1 - \sqrt{1-x}}{2} \right) + f\left(\frac{1 + \sqrt{1-x}}{2} \right) \right).$$

In anticipation of our future investigations it is instructive to study the effects iteration has on P_{F_4}. As can be seen from Figure 4.1 the iterates $P_{F_4}^n f$ seem to converge rapidly towards a density f^* depicted by a dashed line. The specific choice of f turns out to

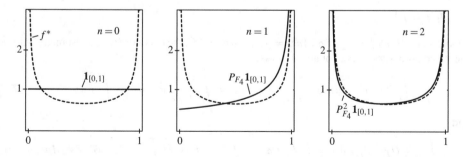

Figure 4.1. The sequence $(P_{F_4}^n \mathbf{1}_{[0,1]})_{n \in \mathbb{N}_0}$ converges to f^* rather quickly.

be rather irrelevant for this observation. At the moment we have no explicit description of f^* at our disposal. Nevertheless, it seems as if we had already met f^* before when studying the long-time behaviour of F_4 by means of histograms (cf. Figure 1.13). We therefore expect the asymptotic properties of the Frobenius–Perron operator to give some insight into the long-time structure of typical orbits. One should also notice how – in case of F_4 – the fast convergence of densities is strongly contrasted by the erratic and unpredictable behaviour of individual points (recall for instance Figure 1.14).

We gather the most important elementary properties of the Frobenius–Perron operator in the following proposition.

Proposition 4.2. *For the Frobenius–Perron operator* $P_T : L^1(\mu) \to L^1(\mu)$ *associated with the non-singular map* T *on* $(X, \mathcal{A}, \mu)$ *the following properties hold:*

(i) P_T *is positive, i.e.* $P_T f \geq 0$ *whenever* $f \geq 0$;

(ii) $\|P_T f\| = \|f\|$ *if* $f \geq 0$;

(iii) $P_{T^n} = P_T^n$ *for all* $n \in \mathbb{N}_0$;

(iv) *Define a linear operator* $U_T : L^\infty \to L^\infty$ *by* $U_T g := g \circ T$; *then* $P_T^* = U_T$, *which means that* $\int_X P_T f\, g\, d\mu = \int_X f\, U_T g\, d\mu$ *holds for all* $f \in L^1(\mu)$, $g \in L^\infty(\mu)$.

As a consequence of (iii), iterating T corresponds to iterating P_T, a fact we have already used above. The linear operator U_T is sometimes called the *Koopman operator* induced by T; according to (iv) it is just the adjoint of P_T. Evidently, the Koopman operator is also positive, and $\|U_T f\|_\infty \leq \|f\|_\infty$ for all $f \in L^\infty(\mu)$. If T is measure-preserving rather than merely non-singular, then U_T may be considered an operator on $L^1(\mu)$, too. In this case P_T and U_T are even more closely related (see also Exercises 3.12 and 3.13).

Lemma 4.3. *Let* T *be measure-preserving on* $(X, \mathcal{A}, \mu)$. *Then* $U_T : f \mapsto f \circ T$ *is a positive linear isometry of* $L^1(\mu)$. *Furthermore*

$$P_T \circ U_T(f) = f \quad \text{and} \quad U_T \circ P_T(f) = \mathbb{E}(f|T^{-1}\mathcal{A})$$

for all $f \in L^1(\mu)$.

Proof. Since $\|U_T f\| = \int_X |f \circ T|\, d\mu = \int_X |f|\, d\mu = \|f\|$ the operator U_T is an isometry with the stated properties. Since for any $A \in \mathcal{A}$

$$\int_A P_T \circ U_T f\, d\mu = \int_{T^{-1}(A)} U_T f\, d\mu = \int_{T^{-1}(A)} f \circ T\, d\mu = \int_A f\, d\mu$$

and

$$\int_{T^{-1}(A)} U_T \circ P_T f\, d\mu = \int_A P_T f\, d\mu = \int_{T^{-1}(A)} f\, d\mu = \int_{T^{-1}(A)} \mathbb{E}(f|T^{-1}\mathcal{A})\, d\mu$$

the remaining assertions follow. $\square$

The first two properties in Proposition 4.2 are going to prove most important in the sequel. A standard notion refers to that fact.

Definiton 4.4. A linear operator $P : L^1(\mu) \to L^1(\mu)$ is called a *Markov operator* if it is positive, and if $\|Pf\| = \|f\|$ whenever $f \geq 0$.

Clearly, every Frobenius–Perron operator also is a Markov operator. However, Definition 4.4 should not be considered the mere introduction of a new name. Markov operators provide a rich class of operators relevant to many areas of mathematics ([14, 28]). In particular, they play an important role in probability theory whereof we shall get a modest impression later on. By classifying the Frobenius–Perron operator as a special Markov operator we therefore subsume our considerations in a probabilistic setting. After all, it is this subsumption by means of which we shall finally take a unified view on dynamical systems.

Right from Definition 4.4 we obtain

$$\|Pf\| = \int_X |Pf|\, d\mu \leq \int_X P|f|\, d\mu = \int_X |f|\, d\mu = \|f\|;$$

thus any Markov operator P is non-expansive and satisfies $\|P\| = 1$. We call P *deterministic* if its adjoint P^* maps indicator functions to indicator functions. Since $P_T^* = U_T$ and $U_T 1_A = 1_{T^{-1}(A)}$, every Frobenius–Perron operator is deterministic. Notice that in the measure-preserving case U_T (understood as an operator on L^1) is itself a Markov operator with the additional feature that it preserves products, i.e. $U_T(fg) = U_T(f)U_T(g)$.

In order to exhibit a non-deterministic Markov operator we consider a finite set of different non-singular maps $T_1, \ldots, T_l$ on $(X, \mathcal{A}, \mu)$. Suppose we are watching an experiment of chance ξ in X which is governed by v, that is $\mathbb{P}(\xi \in A) = v(A)$. Afterwards we *independently* choose and apply one of the maps $T_1, \ldots, T_l$ where T_i is chosen with probability $p_i \geq 0$, $\sum_{i=1}^l p_i = 1$. For notational convenience we model this procedure of selection by means of a random quantity η on $\{1, \ldots, l\}$ with $\mathbb{P}(\eta = i) = p_i$. What can be said about the stochastic output $T_\eta(\xi)$? By independence we obtain

$$T_\eta v(A) := \mathbb{P}(T_\eta(\xi) \in A) = \sum_{i=1}^l \mathbb{P}(T_\eta(\xi) \in A, \eta = i)$$

$$= \sum_{i=1}^l p_i \mathbb{P}(T_i(\xi) \in A) = \sum_{i=1}^l p_i T_i v(A).$$

If $\frac{dv}{d\mu} = f \in L^1(\mu)$ we thus are naturally led to study the assignment

$$f \mapsto Pf := \frac{dT_\eta v}{d\mu} = \sum_{i=1}^l p_i P_{T_i} f.$$

This linear operator P is easily seen to be Markov; but if $\sum_{i=1}^{l} p_i^2 < 1$ it is not deterministic and thus certainly not the Frobenius–Perron operator of any map on X. It may be conceived even from this simple example that the concepts introduced so far may help to combine deterministic and stochastic aspects of dynamics. Before having a closer look at the Frobenius–Perron operator we shall briefly discuss two more examples to underpin this last statement.

Let I denote a finite or countable space and endow $\big(I, \mathcal{P}(I)\big)$ with the counting measure #. A function $f : I \to \mathbb{R}$ is integrable with respect to that measure if and only if $\sum_{i \in I} |f(i)| < \infty$. A continuous linear operator P on $L^1(\#)$ is given by $(Pf)(j) = \sum_{i \in I} f(i) p_{ij}$ with some (possibly infinite) quadratic matrix $(p_{ij})_{i,j \in I}$ which satisfies $\sup_{i \in I} \sum_{j \in I} |p_{ij}| < \infty$. It is easily seen that $P = (p_{ij})$ is Markov if and only if $p_{ij} \geq 0$ and $\sum_{j \in I} p_{ij} = 1$ for all $i \in I$. In the latter case the matrix $(p_{ij})_{i,j \in I}$ is called *stochastic*. We shall later interpret stochastic matrices in terms of Markov chains and also find them naturally related to dynamical systems.

As a second example, consider the measure space $(\mathbb{R}, \mathcal{B}^1, \lambda^1)$ and define for $t > 0$ an operator $P_t : L^1(\lambda^1) \to L^1(\lambda^1)$ according to

$$P_t f(x) := \frac{1}{\sqrt{4\pi t}} \int_{\mathbb{R}} f(y) e^{-\frac{(x-y)^2}{4t}} dy \quad \text{for } f \in L^1(\lambda^1). \tag{4.3}$$

Clearly, P_t is a positive linear operator; by integration we see that we have in fact defined a family of Markov operators. From Figure 4.2 one can get an impression of how this family works. Furthermore, a straightforward but cumbersome calculation yields the semigroup-property $P_t(P_s f) = P_{t+s} f$ for all $t, s > 0$. Since for any set $A \in \mathcal{B}^1$ with positive measure $P_t \mathbf{1}_A > 0$ we see that P_t cannot be the Frobenius–Perron operator of any measurable map T; otherwise there would exist a bounded interval B with $T\lambda^1(B) > 0$. But then $\int_{\mathbb{R}} P_t \mathbf{1}_{T^{-1}(B)} \mathbf{1}_{B^c} = \int_{\mathbb{R}} \mathbf{1}_{T^{-1}(B)} \mathbf{1}_{T^{-1}(B^c)} = 0$, implying that $\lambda^1(B^c) = 0$ which obviously contradicts $\lambda^1(B) < \infty$. (One could equivalently argue that P_t cannot be deterministic.)

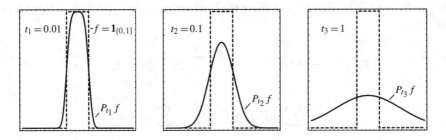

Figure 4.2. The effect of the Markov operators P_t according to (4.3)

So, after all, where does the family $(P_t)_{t>0}$ come from? In order to answer this question we fix $f \in L^1(\lambda^1)$ and write $u(t, x) := P_t f(x)$. It is not too difficult to see that u is twice differentiable and that $\frac{\partial u}{\partial t} = \frac{\partial^2 u}{\partial x^2}$ holds for $t > 0$. Moreover, if f is continuous and vanishes outside a bounded interval we obtain from

$$\|P_t f - f\| = \int_{\mathbb{R}} |P_t f(x) - f(x)| \, dx \leq \frac{1}{\sqrt{2\pi}} \int_{\mathbb{R}} e^{-\frac{y^2}{2}} \int_{\mathbb{R}} |f(x + y\sqrt{2t}) - f(x)| \, dx \, dy$$

the relation $\lim_{t \to 0} \|P_t f - f\| = 0$ by applying the Dominated Convergence Theorem to the right-hand side. Since continuous functions with bounded support are dense in $L^1(\lambda^1)$ we have $\lim_{t \to 0} P_t f = f$ for all $f \in L^1(\lambda^1)$. In this latter sense, $P_t f$ provides a solution to the initial value problem for the *heat equation*

$$\frac{\partial u}{\partial t} = \frac{\partial^2 u}{\partial x^2} \quad \text{for } t > 0, \quad u(0, x) = f(x). \tag{4.4}$$

In fact, one can prove that $P_t f$ is the unique solution to that problem in any reasonable sense ([45]). We thus have found a physical phenomenon which is described by a semigroup of Markov operators: the conduction of heat as the prototypical example of a *diffusive process*. On a macroscopic level such processes can be modelled by partial differential equations, the simplest of which is given by the above heat equation (4.4). On that level the problem looks completely deterministic. This however need not be true any longer if we adopt a microscopic point of view. Although we cannot develop the latter in detail here we will nevertheless give a stochastic interpretation of the above family $(P_t)_{t>0}$. To this end assume f to be a λ^1-density on $\mathbb{R}$, i.e. $f \geq 0$ and $\int_{\mathbb{R}} f(x) \, dx = 1$. We shall think of f as the governing law of a particle's position ξ on the real line which means that for all $a \in \mathbb{R}$

$$\mathbb{P}(\xi \leq a) = \int_{-\infty}^{a} f(x) \, dx.$$

Let η_t denote another random quantity which obeys a normal distribution with mean zero and variance $2t$ and which is independent of ξ. Then

$$\mathbb{P}(\xi + \eta_t \leq a) = \int_{-\infty}^{a} \frac{1}{\sqrt{4\pi t}} \int_{-\infty}^{\infty} f(y) e^{-\frac{(x-y)^2}{4t}} \, dy \, dx = \int_{-\infty}^{a} P_t f(x) \, dx,$$

so that $P_t f$ governs the position of the particle after it has been subjected to the random translation η_t. Since by the semigroup-property $P_t = (P_{t/n})^n$ for every $n \in \mathbb{N}$ we may iterate this argument and hence interpret $P_t f$ as a description of the particle's position after n independent, normally distributed translations $\eta_{t/n}$. According to this interpretation the family of Markov operators $(P_t)_{t>0}$ describes a very complicated random motion of a particle on the real line which is usually referred to as *Brownian motion* and which has found numerous applications in physics and economics (see also Section 5.4). It has been well known to physicists for a long time that both the macroscopic and the microscopic point of view may be given experimental evidence ([43]). It is thus highly satisfactory that the corresponding mathematical descriptions also fit together.

Having seen a few aspects of the versatility of Markov operators let us now return to the study of our main tool, the Frobenius–Perron operator. In passing from $T : X \to X$ to $P_T : L^1 \to L^1$ we have replaced the generally non-linear map T by the linear operator P_T. In principle, an analysis of P_T need not be any simpler because $L^1(\mu)$ will typically be an infinite-dimensional space. However, having at our disposal the powerful tools of functional analysis, some dynamical features of T will be more easily investigated by looking at P_T.

Theorem 4.5. *Let P_T denote the Frobenius–Perron operator associated with the non-singular map T on $(X, \mathcal{A}, \mu)$.*

(i) *$P_T \mathbf{1}_X = \mathbf{1}_X$ if and only if T preserves μ. More generally, the measure ν^* with $\frac{d\nu^*}{d\mu} = f^*$ is T-invariant precisely if $P_T f^* = f^*$.*

(ii) *For all $f \in L^1(\mu)$ the sequence $\frac{1}{n} \sum_{i=0}^{n-1} P_T^i f$ converges weakly to the constant function $\int_X f \, d\mu$ if and only if T is ergodic.*

(iii) *For all $f \in L^1(\mu)$ the sequence $P_T^n f$ converges weakly to $\int_X f \, d\mu$ if and only if T is mixing.*

(iv) *For all $f \in L^1(\mu)$ the sequence $P_T^n f$ converges (strongly) to $\int_X f \, d\mu$ if and only if T is exact.*

Proof. Statement (i) is just a reformulation of the definitions. As we know, T is ergodic precisely if $\frac{1}{n} \sum_{i=0}^{n-1} \mu(T^{-i}(A) \cap B) \to \mu(A)\mu(B)$ for all $A, B \in \mathcal{A}$. By the denseness of stepfunctions and $P_T^* = U_T$ assertion (ii) follows. An analogous argument proves (iii). In order to establish the equivalence (iv) first assume that $P_T^n f \to \int_X f \, d\mu$ for all $f \in L^1(\mu)$ and take $A \in \mathcal{A}_\infty := \bigcap_{n \in \mathbb{N}_0} T^{-n} \mathcal{A}$ with $\mu(A) > 0$. Clearly, T is measure-preserving, and for $n \in \mathbb{N}$ we have $A = T^{-n}(A_n)$ with $A_n \in \mathcal{A}$, $\mu(A_n) = \mu(A)$. Setting $f := \mu(A)^{-1} \mathbf{1}_A$ and fixing $\varepsilon > 0$ we find $\| P_T^N f - \mathbf{1}_X \| < \varepsilon$ for an appropriate N. Hence by virtue of Lemma 4.3

$$2\big(1 - \mu(A)\big) = \| \mu(A)^{-1} \mathbf{1}_{A_N} - \mathbf{1}_X \| = \| P_T^N f - \mathbf{1}_X \| < \varepsilon \, .$$

Since ε was arbitrary we obtain $\mu(A) = 1$, thus T is exact. Conversely, given $f \in L^1(\mu)$ Lemma 4.3 yields $P_T^n f \circ T^n = \mathbb{E}(f|T^{-n}\mathcal{A})$ for every $n \in \mathbb{N}$, the right side of which converges in $L^1(\mu)$ to the constant $\mathbb{E}(f|\mathcal{A}_\infty) = \int_X f \, d\mu$ by virtue of the Decreasing Martingale Theorem. Therefore

$$\| P_T^n f - \int_X f \, d\mu \| = \int_X |P_T^n f - \int f| \, d\mu = \int_X |P_T^n f \circ T^n - \int f| \, d\mu \to 0$$

as $n \to \infty$ which proves (iv). $\square$

The above theorem characterizes the mixing properties of T by means of certain stability properties of P_T. The map T is measure-preserving if and only if the constant density $\mathbf{1}_X$ is fixed by P_T and, more specifically, T is ergodic, mixing or exact, precisely if $\mathbf{1}_X$ is weakly Cesaro attracting, weakly attracting or (strongly) attracting for densities, respectively. In order to give an application of this result we shall prove some assertions concerning the dynamics of toral endomorphisms which we cited in the preceding chapter. For example, we claimed that the hyperbolic automorphism T_{A_1} of $\mathbb{T}^2$ with $A_1 = \begin{pmatrix} 2 & 1 \\ 1 & 1 \end{pmatrix}$ is mixing.

Theorem 4.6. *A toral endomorphism is mixing if and only if it is ergodic.*

Proof. Ergodicity clearly is necessary. Assume that T_A is ergodic, i.e., no eigenvalue of A is a root of unity (Theorem 3.9). We have to prove that

$$\int_{\mathbb{T}^d} P^n_{T_A} f \, g \, d\lambda_{\mathbb{T}^d} \to \int_{\mathbb{T}^d} f \, d\lambda_{\mathbb{T}^d} \int_{\mathbb{T}^d} g \, d\lambda_{\mathbb{T}^d} \quad \text{for all } f \in L^1, \, g \in L^\infty.$$

It suffices to verify this relation for dense subsets in L^1 and L^∞, respectively. Setting $f(x) := e^{2\pi i \langle k,x \rangle}$, $g(x) := e^{2\pi i \langle l,x \rangle}$ with $k, l \in \mathbb{Z}^d$ we obtain

$$\int_{\mathbb{T}^d} P^n_{T_A} f \, g \, d\lambda_{\mathbb{T}^d} = \int_{\mathbb{T}^d} e^{2\pi i \langle k + (A^T)^n l, x \rangle} \, d\lambda_{\mathbb{T}^d} = \begin{cases} 1 & \text{if } k + (A^T)^n l = 0, \\ 0 & \text{otherwise}. \end{cases}$$

If $l \neq 0$ the relation $k + (A^T)^n l = 0$ holds for at most one n; if $l = 0$ it trivially holds if and only if $k = 0$. In any case $\lim_{n \to \infty} \int_{\mathbb{T}^d} P^n_{T_A} f \, g \, d\lambda_{\mathbb{T}^d} = \int_{\mathbb{T}^d} f \, d\lambda_{\mathbb{T}^d} \int_{\mathbb{T}^d} g \, d\lambda_{\mathbb{T}^d}$ for these special functions. Denseness of trigonometric polynomials therefore completes the proof. $\square$

In addition, we claimed that T_{A_2} with $A_2 = \begin{pmatrix} 4 & 1 \\ 2 & 4 \end{pmatrix}$ is exact. Before giving a general result about exactness of toral maps let us inspect the one-dimensional case which is especially simple. To this end fix $k \in \mathbb{N} \setminus \{1\}$. For the endomorphism $T_{(k)}$ of $\mathbb{T}^1$ the Frobenius–Perron operator reads

$$P_{T_{(k)}} f(x) = \frac{1}{k} \sum_{i=0}^{k-1} f\left(\frac{i+x}{k}\right). \tag{4.5}$$

Iterating this relation yields

$$P^n_{T_{(k)}} f(x) = \frac{1}{k^n} \sum_{i=0}^{k^n-1} f\left(\frac{i+x}{k^n}\right).$$

Here the right-hand side is a Riemann sum, and therefore $P^n_{T_{(k)}} f \to \int_{\mathbb{T}^1} f \, d\lambda_{\mathbb{T}^1}$ uniformly on $\mathbb{T}^1$ if f is continuous. Since continuous functions are dense in $L^1(\lambda_{\mathbb{T}^1})$ we have shown that the maps $T_{(k)}$ are in fact exact. Unfortunately, this elegant argument does not carry over to higher dimensions in a straightforward manner. We shall thus restrict ourselves to a special case.

Theorem 4.7. *If all eigenvalues of $A \in \mathbb{Z}^{d \times d}$ have modulus greater than one, then the toral endomorphism T_A is exact.*

Proof. Let us first derive an explicit formula for P_{T_A}. For the sake of lucidity we shall use the same symbols for points in $\mathbb{R}^d$ and $\mathbb{T}^d$ here. Since the inverse image of any d-dimensional interval consists of $|\det A|$ sets of equal measure, a straightforward calculation shows that

$$P_{T_A} f(x) = \frac{1}{|\det A|} \sum_{k \in A^{-1}\mathbb{Z}^d \cap [0,1[^d} f(k + A^{-1}x),$$

a formula generalizing (4.5). Again, this may be iterated to give

$$P_{T_A}^n f(x) = \frac{1}{|\det A|^n} \sum_{k \in A^{-n}\mathbb{Z}^d \cap [0,1[^d} f(k + A^{-n}x).$$

The diameter of the cells in the lattice $A^{-n}\mathbb{Z}^d$ tends to zero because all eigenvalues of A have modulus greater than one. Therefore the right-hand side again is a Riemann sum, and we deduce $P_{T_A}^n f \to \int_{\mathbb{T}^d} f \, d\lambda_{\mathbb{T}^d}$ in $L^1(\lambda_{\mathbb{T}^d})$ as before. $\square$

Note that all we needed for the proof was that the diameter of the cells in $A^{-n}\mathbb{Z}^d$ becomes small. The eigenvalue condition was imposed in order to assure this type of shrinking. One might wonder whether other such conditions exist. For example, exactness also holds for the hyperbolic matrices

$$A_3 := \begin{pmatrix} 3 & -1 \\ -1 & 1 \end{pmatrix} \quad \text{and} \quad A_4 := \begin{pmatrix} 2 & 0 & 0 \\ 0 & 7 & 1 \\ 0 & 3 & 1 \end{pmatrix}$$

which both have an eigenvalue inside the unit circle. On the other hand the slightly modified matrix

$$A_5 := \begin{pmatrix} 2 & 0 & 0 \\ 0 & 7 & 2 \\ 0 & 3 & 1 \end{pmatrix}$$

still is hyperbolic but T_{A_5} definitely is not exact because $\|P_{T_{A_5}} f - 1_{\mathbb{T}^3}\| = \|f - 1_{\mathbb{T}^3}\|$ for all $f \in L^1(\lambda_{\mathbb{T}^3})$ which do not depend on the first coordinate.

We close this section by reviewing an earlier example in the light of the Frobenius–Perron operator. In Section 3.4 the dynamics of the toral maps

$$\text{(i) } R_{(\sqrt{2},\sqrt{3})}, \quad \text{(ii) } T_{A_1} \quad \text{and} \quad \text{(iii) } T_{A_2}$$

has been visualized. We now know that T_{A_1} is mixing and T_{A_2} is exact. Correspondingly, $1_{\mathbb{T}^2}$ is weakly and strongly attracting under the respective $P_{T_{A_i}}$ for all densities in $L^1(\lambda_{\mathbb{T}^2})$. As far as $R_{(\sqrt{2},\sqrt{3})}$ is concerned we have

$$P_{R_{(\sqrt{2},\sqrt{3})}} f = f \circ R_{(\sqrt{2},\sqrt{3})}^{-1} = f \circ R_{(-\sqrt{2},-\sqrt{3})}$$

so that $P_{R_{(\sqrt{2},\sqrt{3})}}$ merely shifts all densities without changing their shape. Figure 4.3 graphically underpins these facts.

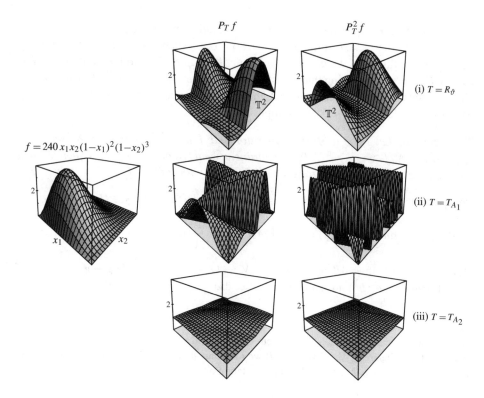

Figure 4.3. Visualizing the Frobenius–Perron operator for an ergodic, mixing and exact map on $\mathbb{T}^2$, respectively (see also Figure 3.14)

4.2 Asymptotic behaviour of densities

We will now tackle the problem of finding and approximating invariant measures. Recall that a measure with density f^* is T-invariant if and only if $P_T f^* = f^*$, which just says that f^* is an eigenvector of P_T corresponding to the eigenvalue one. There are at least two strategies for dealing with this eigenvalue problem. The first approach consists of studying the spectral properties of Frobenius–Perron or, more generally, Markov operators. In fact, a rich spectral theory of these operators has been developed ([28, 35]). Due to the great variety of different situations that may be covered by the concept of Markov operators it need not be straightforward to adapt the abstract results to specific applications. We shall therefore mention just one important fact from the general theory. As a second strategy for the eigenvalue problem, one could more specifically focus on familiar spaces and exploit their additional structure. This approach is closer to classical analysis, and we shall cite a famous theorem as well as some applications in this direction.

Given a Markov operator P the following situation may be regarded to be the most desirable one: there exists a unique density $f^* \in L^1$ such that $P^n f \to f^*$ for every density f. In this case we call P *asymptotically stable*. Due to the continuity of P clearly $Pf^* = f^*$, implying that f^* is the unique invariant density of P towards which all other densities are attracted. In fact, the mere existence of an invariant density may be ensured under much weaker assumptions.

Theorem 4.8. *Let P be a Markov operator and f an L^1-function. If the sequence of averages $A_n f := \frac{1}{n} \sum_{i=0}^{n-1} P^i f$ contains a weakly convergent subsequence, then $(A_n f)_{n \in \mathbb{N}}$ itself converges (strongly) to a limit f^* with $Pf^* = f^*$.*

Proof. Denote by f^* the weak limit of $A_{n_k} f$. Since for any functional $y \in (L^1)'$

$$\langle Pf^*, y \rangle = \lim_{k \to \infty} \langle PA_{n_k} f, y \rangle = \lim_{k \to \infty} \left[\langle A_{n_k} f, y \rangle + \frac{1}{n_k} \left(\langle P^{n_k} f, y \rangle - \langle f, y \rangle \right) \right] = \langle f^*, y \rangle,$$

clearly $Pf^* = f^*$. If we define the subspace $V := (P - \mathrm{id}_{L^1})(L^1) = \{ Pg - g : g \in L^1 \}$ of L^1 and assume $y_0 \in (L^1)'$ to vanish on V we see that

$$\langle f, y_0 \rangle = \langle Pf, y_0 \rangle = \langle A_{n_k} f, y_0 \rangle \to \langle f^*, y_0 \rangle.$$

Since otherwise the Hahn–Banach Theorem ([47]) would yield a contradiction, $f - f^*$ is an element of the closure of V. Given $\varepsilon > 0$ we may thus find $g_\varepsilon, h_\varepsilon \in L^1$ with $\|h_\varepsilon\| < \varepsilon$ and $f - f^* = Pg_\varepsilon - g_\varepsilon + h_\varepsilon$. From this representation of $f - f^*$ we get

$$\|A_n f - f^*\| \leq \frac{1}{n} (\|P^n g_\varepsilon\| + \|g_\varepsilon\|) + \varepsilon.$$

Since ε was arbitrary $\lim_{n \to \infty} \|A_n f - f^*\| = 0$. $\square$

Observe that the assumption of P being a Markov operator is not essential for the proof: all we need is that $\|P^n f\|/n \to 0$ as $n \to \infty$ for each $f \in L^1$. On the other hand the usage of the averages $A_n f$ is crucial: the conclusion may fail if $A_n f$ is replaced by $P^n f$. Despite its technical appearance the above theorem has far-reaching implications. In order to prove the existence of an invariant density it suffices to find a weakly convergent subsequence of $(A_n f)_{n \in \mathbb{N}}$ for *some* density f. The corresponding topic of *weak pre-compactness* has been studied extensively in functional analysis ([28, 35]).

In order to deal with some specific applications we shall now follow the second approach mentioned earlier. Instead of analysing abstract Markov operators we shall focus on Frobenius–Perron operators associated with maps on familiar spaces. We call a non-singular map T *statistically stable*, if P_T is asymptotically stable. By imitating the proof of Theorem 4.5 (iv) it is easily shown that a statistically stable map T with $P_T^n f \to f^*$ for all densities is in fact exact with respect to the measure μ^* defined by $\frac{d\mu^*}{d\mu} = f^*$. Intending to present sufficient conditions for statistical stability we shall restrict ourselves to the case $(X, \mathcal{A}, \mu) = ([0, 1], [0, 1] \cap \mathcal{B}^1, \lambda^1|_{[0,1]})$ which is perhaps the most familiar space with rich additional structure. Dealing exclusively with the unit interval is not as special as it might seem: one-dimensional systems from applications can always be transformed to this case if only they take place in a bounded region.

In the sequel we shall exclusively deal with maps $T : [0, 1] \rightarrow [0, 1]$ which are piecewise C^2; this means that for a finite number of points $0 = a_0 < a_1 < \ldots < a_{l-1} < a_l = 1$ the restriction of T to $[a_{i-1}, a_i[$ is C^2 for $i = 1, \ldots, l$ with $T'(a_i)$, $T''(a_i)$ denoting the respective right derivatives. Furthermore we shall impose the condition that $|(\frac{1}{T'})'|$ be bounded. Consider also the properties of *indecomposability*

(i) $T(a_{i-1}) = 0$ for $i = 1, \ldots, l$,

(i') $T(]a_{i-1}, a_i[) =]0, 1[$ for $i = 1, \ldots, l$,

as well as the properties of *uniform expansion*

(ii) $T'(x) \geq \lambda > 1$ for all $x \in [0, 1]$,

(ii') $|T'(x)| \geq \lambda > 1$ for all $x \in [0, 1]$.

By means of these properties we can concisely state the following theorem.

Theorem 4.9 (Lasota, Yorke). *Let $T : [0, 1] \rightarrow [0, 1]$ denote a piecewise C^2-map for which $|(\frac{1}{T'})'|$ is bounded. Then either conditions (i) and (ii) or (i') and (ii') together imply statistical stability of T. Furthermore, condition (ii') alone forces $P_T^n f$ to be asymptotically periodic for every $f \in L^1$ and still guarantees the existence of an invariant density.*

Although we shall not give a proof of this theorem here, let us point out that a more or less explicit expression for the Frobenius–Perron operator P_T may be derived under the above assumptions. Defining the sets $A_i := [a_{i-1}, a_i[$ and $B_i := T(A_i)$ as well as the inverse mappings $\tau_i := (T|_{A_i})^{-1}$ for $i = 1, \ldots, l$ we find

$$\int_A P_T f \, dx = \sum_i \int_{\tau_i(A \cap B_i)} f \, dx = \sum_i \int_{A \cap B_i} f \circ \tau_i |\tau_i'| \, dx = \int_A \sum_i f \circ \tau_i |\tau_i'| \mathbf{1}_{B_i} \, dx$$

for every $A \in [0, 1] \cap \mathcal{B}^1$. Consequently

$$P_T f(x) = \sum_{i=1}^{l} f \circ \tau_i(x) |\tau_i'(x)| \mathbf{1}_{B_i}(x) = \sum_{y:T(y)=x} \frac{f(y)}{|T'(y)|}. \tag{4.6}$$

Formula (4.6) suggests that the expansion properties (ii) and (ii'), respectively, cause a stabilizing effect in the sequence $(P_T^n f)_{n \in \mathbb{N}}$. This effect turns out to be crucial for a rigorous execution of the proof which nevertheless is rather technical (see [13, 35] for the details). Observe that the last assertion in Theorem 4.9 concerning the existence of an invariant density immediately follows from Theorem 4.8: if for some $p \in \mathbb{N}$ the iterates $P_T^{np+i} f$ converge for all $i \in \{0, \ldots, p - 1\}$ then $A_{np} f$ also tends to a limit.

In order to realize the importance of the indecomposability conditions (i) and (i′), respectively, one should have a look at the map

$$T(x) := \begin{cases} 2x & \text{if } x \in \left[0, \frac{1}{4}\right[, \\ 2x - \frac{1}{2} & \text{if } x \in \left[\frac{1}{4}, \frac{3}{4}\right[, \\ 2x - 1 & \text{if } x \in \left[\frac{3}{4}, 1\right], \end{cases}$$

which clearly is expanding; however, conditions (i) and (i′) fail. A straightforward calculation shows that the limit of $(P_T^n f)_{n\in\mathbb{N}_0}$ still exists but depends on f. More specifically,

$$P_T^n f \to 2 \cdot \mathbf{1}_{\left[0,\frac{1}{2}\right[} \int_0^{\frac{1}{2}} f(y)\,dy + 2 \cdot \mathbf{1}_{\left[\frac{1}{2},1\right]} \int_{\frac{1}{2}}^1 f(y)\,dy =: \Phi(f)$$

for all $f \in L^1$ (cf. Figure 4.4). Thus T is not statistically stable – in fact, it is not even ergodic for $T^{-1}\left(\left[0, \frac{1}{2}\right]\right) = \left[0, \frac{1}{2}\right]$.

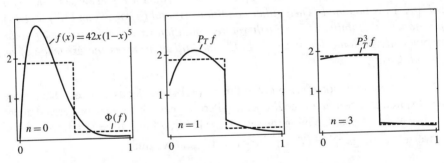

Figure 4.4. The L^1-limit $\Phi(f)$ of $(P_T^n f)_{n\in\mathbb{N}_0}$ obviously depends on f.

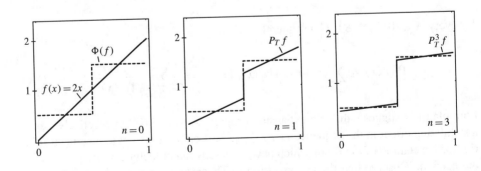

Let us now sketch an interesting application extensively discussed in [13]. Consider the simple model of a rotary drill depicted in Figure 4.5: a toothed cylinder of radius R and mass M is pressed down to a stony base by a force F and moves forward with constant velocity v. Denoting by $(x, p(x))$ the position of the cylinder's centre we have

$$p(x) = \sqrt{R^2 - \left(x - \tfrac{2k+1}{2}T\right)^2}$$

if $kT \leq x \leq (k+1)T$, i.e. the k-th tooth is active. The force K acting on the drill at the contact point follows from Newton's second law, $K = M\frac{d^2p}{dt^2} + F$. Taking into account $\frac{dx}{dt} = v = $ const. we obtain

$$K = M\left(-\frac{R^2v^2}{p^3}\right) + F \approx F\left(1 - \frac{Mv^2}{FR}\right)$$

because $p(x) \approx R$. Introducing the non-dimensional quantity $\Lambda := Mv^2/(FR)$ we are led to distinguish two cases: if $\Lambda < 1$ the contact force will always remain positive, and we expect a rather smooth motion of the drill. If on the other hand $\Lambda > 1$ then a rough motion will take place since the drill will take off over and over again. In the latter case we have to consider two different phases of motion (cf. Figure 4.5): either one tooth is active, and correspondingly

$$\frac{d^2p}{dx^2} = -\frac{R^2}{p^3} \approx -\frac{1}{R},$$

or the drill is flying freely according to

$$\frac{d^2p}{dx^2} = -\frac{F}{Mv^2} = -\frac{1}{\Lambda R}.$$

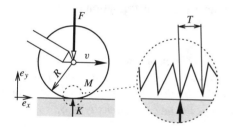

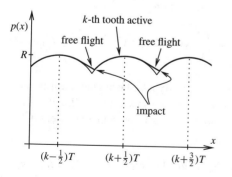

Figure 4.5. The model of the rotary drill and the different phases of motion (below) for $\Lambda > 1$

Geometrically, finding the transition between the two phases corresponds to the intersection of two parabolas. Though elementary in nature, the rigorous treatment of this problem needs some careful analysis which may be found in [13]. After an appropriate normalization one gets as a final result a family $(T_\Lambda)_{\Lambda \geq 1}$ of maps on $[0, 1]$ describing the (normalized) position of the freely flying drill's next impact with the stony base. In Figure 4.6 the graph of T_Λ is shown for various values of the parameter Λ. The family $(T_\Lambda)_{\Lambda \geq 1}$ exhibits a great variety of dynamical effects. Perhaps most interestingly in the present context, T_Λ or some iterate may be shown to be expanding on large parts of the unit interval. This fact implies a rather erratic, unpredictable behaviour of the drill. Finding a sufficiently good approximation of the invariant density clearly is of considerable practical importance: the optimal operating conditions of the drill as well as its expected wear will depend on a reasonable stochastic description.

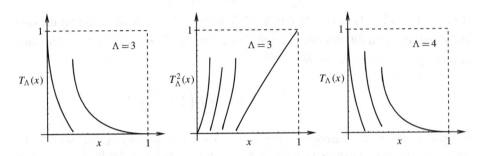

Figure 4.6. Typical maps T_Λ are expanding on (parts of) the unit interval.

No doubt, the assumptions of the Lasota–Yorke Theorem 4.9 are highly restrictive. For example, they are not met by our old friend F_4. A transformation of the problem may nevertheless provide some help.

Lemma 4.10. *Let S be a non-singular map and h a homeomorphism of* $[0, 1]$. *Assume that both h and* h^{-1} *are non-singular (with respect to* $\lambda^1|_{[0,1]}$). *Then the map* $T :=$ $h^{-1} \circ S \circ h$ *is statistically stable if and only if S is. In this case, the unique T-invariant density is given by* $g^* \circ h \cdot |h'|$ *where* g^* *denotes the invariant density of S.*

Proof. Let us first derive an expression for P_T. From

$$\int_A P_T f = \int_{h^{-1}\left(S^{-1}(h(A))\right)} f = \int_{S^{-1}\left(h(A)\right)} f \circ h^{-1} \cdot |(h^{-1})'|$$
$$= \int_{h(A)} P_S\left(f \circ h^{-1} \cdot |(h^{-1})'|\right) = \int_A P_S\left(f \circ h^{-1} \cdot |(h^{-1})'|\right) \circ h \cdot |h'|$$

we deduce that $P_T f = P_S\left(f \circ h^{-1} \cdot |(h^{-1})'|\right) \circ h \cdot |h'|$. By introducing the linear isometry $H : L^1 \to L^1$ with $H(f) := f \circ h \cdot |h'|$ we may neatly express this as $P_T = H \circ P_S \circ H^{-1}$, and thus $P_T^n = H \circ P_S^n \circ H^{-1}$ for all $n \in \mathbb{N}_0$. So T is statistically stable precisely if S is. Furthermore, if $P_S g^* = g^*$ then $P_T\left(H(g^*)\right) = H(g^*)$. $\square$

In order to apply this result to the map F_4 recall the relation $h \circ F_4 \circ h^{-1} = S$ with $h^{-1}(y) := \frac{1}{2}(1 - \cos \pi y)$ and the tent map $S(y) := 2 \min\{y, 1 - y\}$ which we found earlier. Clearly, the Lasota–Yorke Theorem works for the map S, and the unique invariant density g^* for P_S equals $\mathbf{1}_{[0,1]}$. Therefore Lemma 4.10 yields

$$f^*(x) = g^* \circ h(x)|h'(x)| = \frac{1}{|(h^{-1})'(y)|}\Big|_{h^{-1}(y)=x}$$

$$= \frac{2}{\pi \sin \pi y}\Big|_{(\sin \frac{\pi y}{2})^2=x} = \frac{1}{\pi\sqrt{x(1-x)}}$$

as the unique invariant density of F_4. Recall that we have already met f^* several times on a quantitative level (see Figures 1.13 and 4.1); an elementary calculation confirms that $P_{F_4} f^* = f^*$.

A slight modification of Lemma 4.10 also makes possible a concluding analysis of the Newton map N_{q_+} associated with the polynomial $q_+(x) = x^2 + 1$. In Section 2.5 we found $\tilde{N}_{q_+}(x) = 2x \pmod 1$ for the map $\tilde{N}_{q_+}$ induced by N_{q_+}. Making use of the stereographic projection Φ introduced there we are led to study the map $h^{-1} :]0, 1[\to \mathbb{R}$ defined as

$$h^{-1}(y) := \Phi(e^{2\pi i y}) = \cot \pi y,$$

since we have by construction

$$N_{q_+} \circ h^{-1}(y) = \frac{\cot^2 \pi y - 1}{2 \cot \pi y} = \cot 2\pi y = h^{-1} \circ \tilde{N}_{q_+}(y)$$

for $y \neq \frac{1}{2}$. As $\tilde{N}_{q_+}$ is statistically stable with respect to $\lambda^1|_{[0,1]}$, an argument similar to the one in the proof of Lemma 4.10 shows that the original Newton map is statistically stable with respect to $v^* := h^{-1}\lambda^1|_{[0,1]}$. For the density f_+^* of v^* we analogously find

$$f_+^*(x) := \frac{dv^*}{d\lambda^1}(x) = \frac{1}{|(h^{-1})'(y)|}\Big|_{h^{-1}(y)=x} = \frac{\sin^2 \pi y}{\pi}\Big|_{h^{-1}(y)=x} = \frac{1}{\pi(1+x^2)}.$$

Given any density $f \in L^1(\lambda^1)$ we thus have

$$\int_{\mathbb{R}} |P_{N_{q_+}}^n f - f_+^*| d\lambda^1 \to 0 \quad \text{as } n \to \infty.$$

Already in the Introduction we saw that $P_{N_{q_+}} f_+^* = f_+^*$, and we empirically observed that densities rapidly converge towards a limit whereas individual points wander around in a practically unpredictable manner.

The above Lemma 4.10 may also be used to construct a map T on the unit interval which is statistically stable with respect to a given density. To this end assume that S is statistically stable and $g^* = \mathbf{1}_{[0,1]}$. The invariant density of $T = h^{-1} \circ S \circ h$ then is $|h'|$. Given the density f^* (which may be assumed positive almost everywhere), defining h according to $h(x) := \int_0^x f^*(z)\, dz$ thus provides a solution to our problem. As an example, suppose we wished to produce a statistically stable map with an invariant density

$$f_\alpha^*(x) := \alpha x^{\alpha-1} \quad (\alpha > 0).$$

To this end take $S(y) := 2y \pmod 1$ and set $h(x) = \int_0^x f_\alpha^*(z)\, dz = x^\alpha$ from which follows immediately

$$T_\alpha(x) = h^{-1} \circ S \circ h(x) = \begin{cases} 2^{\alpha^{-1}} x & \text{if } 0 \le x < 2^{-\alpha^{-1}}, \\ (2x^\alpha - 1)^{\alpha^{-1}} & \text{if } 2^{-\alpha^{-1}} \le x < 1. \end{cases} \tag{4.7}$$

Some numerical results for the case $\alpha = 2$ are depicted in Figure 4.7.

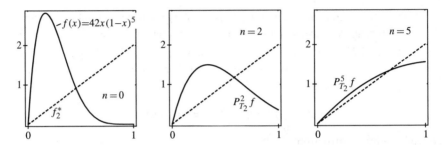

Figure 4.7. The map T_2 as defined in (4.7) is statistically stable, and $f_2^*(x) = 2x$; empirical histograms also exhibit the desired density (below).

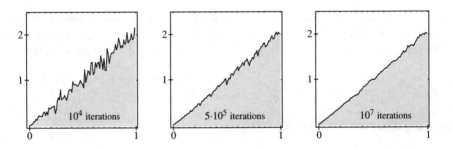

We have already mentioned that the assumptions of the Lasota–Yorke Theorem are highly restrictive. However, there is no straightforward way of weakening them as can be seen from the following example which also is due to Lasota and Yorke ([13, 35]). Consider the map T on the unit interval defined as

$$T(x) := \begin{cases} \frac{x}{1-x} & \text{if } x \in [0, \tfrac{1}{2}[, \\ 2x - 1 & \text{if } x \in [\tfrac{1}{2}, 1], \end{cases} \tag{4.8}$$

the Frobenius–Perron operator of which is given by

$$P_T f(x) = \frac{1}{(1+x)^2} f\left(\frac{x}{1+x}\right) + \frac{1}{2} f\left(\frac{x+1}{2}\right).$$

Observe that T satisfies the conditions (i) and (ii) of Theorem 4.9 except that $T'(0) = 1$. Setting $g_n(x) := x\, P_T^n \mathbf{1}_{[0,1]}(x)$ we find the recursion formula

$$g_{n+1}(x) = \frac{1}{1+x}g_n\left(\frac{x}{1+x}\right) + \frac{x}{1+x}g_n\left(\frac{x+1}{2}\right), \quad g_0(x) = x.$$

By an induction argument $g_n' \geq 0$ for all $n \in \mathbb{N}_0$. Monotonicity implies that the limit $\gamma := \lim_{n\to\infty} g_n(1) \geq 0$ exists because

$$0 \leq g_{n+1}(1) = \frac{1}{2}\left(g_n\left(\frac{1}{2}\right) + g_n(1)\right) \leq g_n(1).$$

Define now $x_k := (1+k)^{-1}$ and observe that

$$g_n(x_{k+1}) = (1+x_k)g_{n+1}(x_k) - x_k g_n\left(\frac{x_k+1}{2}\right),$$

from which we deduce by induction that $\lim_{n\to\infty} g_n(x) = \gamma$ for all $x \in [x_k, 1]$ and all k. Suppose that $\gamma > 0$. Since $x_k \to 0$ as $k \to \infty$ we have for some $\varepsilon > 0$

$$2 < \int_\varepsilon^1 \frac{\gamma}{x}\,dx = \varlimsup_{n\to\infty} \int_\varepsilon^1 P_T^n \mathbf{1}_{[0,1]} \leq 1,$$

an obvious contradiction. Therefore $P_T^n \mathbf{1}_{[0,1]} \to 0$ uniformly on $[\delta, 1]$ for every $\delta > 0$. Concludingly, let us show that densities other than $\mathbf{1}_{[0,1]}$ essentially exhibit the same asymptotic behaviour. Arbitrarily choosing a density $f \in L^1$ and $\beta > 0$ we may find a constant $c_{f,\beta}$ such that $\int_0^1 \max\{f, c_{f,\beta}\} \leq c_{f,\beta} + \beta$. From

$$\int_\delta^1 P_T^n f = \int_\delta^1 P_T^n(f - c_{f,\beta}\mathbf{1}_{[0,1]}) + c_{f,\beta}\int_\delta^1 P_T^n \mathbf{1}_{[0,1]} \leq \beta + c_{f,\beta}\int_\delta^1 P_T^n \mathbf{1}_{[0,1]}$$

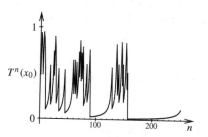

we see that $\int_\delta^1 P_T^n f \to 0$ holds for any density f and any $\delta > 0$. Consequently, the only solution of $P_T f^* = f^*$ in L^1 is $f^* = 0$. Observe that this dramatic conclusion is by no means artificial if one looks at the dynamics of T numerically. As can be seen from Figure 4.8, typical points find themselves in the neighbourhood of the origin quite often – in fact more often than this would be possible if there were a finite, absolutely continuous invariant measure. This observation corresponds to the fact that the origin is a weak (i.e. non-hyperbolic) repelling fixed point for T. In accordance with that, P_T fixes a measurable function which is not an element of L^1; indeed $P_T f^* = f^*$ with

Figure 4.8. A typical orbit of Lasota's map as defined by (4.8)

$f^*(x) := x^{-1}$. The measure μ^* with $\frac{d\mu^*}{d\lambda^1} = f^*$ is not finite but σ-finite (see Appendix A); such σ-finite invariant measures typically occur in case of weakly repelling fixed points ([13, 35]).

Until now our discussion of the Frobenius–Perron operator has focussed on the question of asymptotic stability. Although this type of behaviour will be found in many situations and may also be considered the most desirable one in some applications it is by no means the only possible one. It should come as no surprise that the asymptotic theory of Markov operators offers a great variety of different phenomena, some of which turn out to be physically relevant. Without giving any details we shall close this section by exhibiting one type of asymptotic behaviour quite different from asymptotic stability.

Recall our discussion of the heat equation and let us introduce a family of Markov operators on $L^1(\lambda^1)$ by

$$P_t f(x) := \frac{1}{\sqrt{4\pi t}} \int_{\mathbb{R}} f(y) e^{-\frac{(x+\gamma t - y)^2}{4t}} \, dy \qquad (4.9)$$

for $t > 0$ and $\gamma \in \mathbb{R}$. Again it is easy to show that $(P_t f)_{t>0}$ provides a solution to the initial value problem of a diffusion equation, namely

$$\frac{\partial u}{\partial t} = \frac{\partial^2 u}{\partial x^2} + \gamma \frac{\partial u}{\partial x}, \quad u(0, x) = f(x).$$

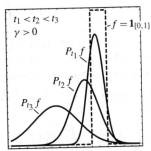

An impression of the dynamics of P_t can be gained from Figure 4.9. What is the ultimate behaviour of $P_t f$ as $t \to \infty$? In order to answer this question, let A denote a bounded interval and observe that

$$\int_A P_t f(x) \, dx \le \frac{\lambda^1(A)}{\sqrt{4\pi t}} \to 0 \quad \text{as } t \to \infty.$$

Figure 4.9. The effect of the Markov operators P_t from (4.9)

This type of behaviour usually is given a name of its own.

Definiton 4.11. A Markov operator P on $L^1(\mu)$ is *sweeping* with respect to the family of sets $\mathcal{A}^* \subseteq \mathcal{A}$ if

$$\lim_{n \to \infty} \int_A P^n f \, d\mu = 0 \quad \text{for all } f \in L^1(\mu), \, A \in \mathcal{A}^*.$$

According to this definition the above operator P_t is sweeping with respect to the family of bounded intervals for any $t > 0$. From this fact it obviously follows that P_t cannot have any invariant density. The physical meaning of the latter observation is quite evident: if we interpret $P_t f$ as the distribution of energy (heat, mass, etc.) at time t then the total amount of this quantity is preserved. In fact, that is just what the Markov property says. On the other hand, due to diffusion the finite amount $\int_{\mathbb{R}} f(x)\,dx$ of energy will be spread across the whole infinite space, thereby leaving arbitrarily little energy in any bounded region. Recalling our microscopic interpretation of this process we may conclude that the wandering particle will leave every bounded set with probability one. If it exists, the mean particle position obeys the deterministic law

$$\int_{\mathbb{R}} x P_t f(x)\,dx = \int_{\mathbb{R}^2} \frac{x}{\sqrt{4\pi t}} f(y) e^{-\frac{(x+\gamma t-y)^2}{4t}}\,dx\,dy = -\gamma t + \int_{\mathbb{R}} y f(y)\,dy,$$

hence moves to the left with velocity γ.

Another sweeping Markov operator is provided by Lasota's example (4.8) from above. Indeed, our findings could be rephrased by saying that in this case P_T is sweeping with respect to the family $\mathcal{A}^* := \{[a, b] \in [0, 1] \cap \mathcal{B}^1 : a > 0\}$. Again an interpretation of the latter fact perfectly agrees with our numerical data: on long time scales typical orbits will heavily concentrate near the origin. Already from this simple example it may also be seen that proving a given Markov operator to be sweeping with respect to a certain non-trivial family $\mathcal{A}^*$ may be rather demanding ([13, 35]).

In general, of course, other types of asymptotic behaviour than asymptotic stability and sweeping may be imagined. For a large and important class of Markov operators, however, it can be shown that these two types are in fact the only possible ones. Provided that a number of technical conditions are met, the so-called *Foguel alternative* states that a Markov operator is either asymptotically stable (or at least asymptotically periodic) or sweeping with respect to a *regular family* $\mathcal{A}^*$, i.e. a family satisfying $\bigcup_{n\in\mathbb{N}} A_n = X$ with $A_n \in \mathcal{A}^*$. (We refer to [35] for details and further references.) Observe that in both examples discussed above the families $\mathcal{A}^*$ are regular.

4.3 Piecewise expanding Markov maps

In this section we shall apply our earlier results to a special class of maps on the unit interval. Despite their simplicity these maps will prove useful in bringing together deterministic and stochastic aspects of dynamics. In the sequel we again assume that T is piecewise C^2 with respect to a finite partition $0 = a_0 < a_1 < \ldots < a_{l-1} < a_l = 1$ of the unit interval; additionally we define $A_i := \,]a_{i-1}, a_i[$ for $i = 1, \ldots, l$.

Definiton 4.12. A map $T : [0, 1] \to [0, 1]$ is called a *piecewise expanding Markov map* if the following two conditions hold:

(i) $|T'(x)| \geq \lambda > 1$ for $0 \leq x \leq 1$;

(ii) for all $i, j \in \{1, \ldots, l\}$ we have $T(A_i) \supseteq A_j$ whenever $T(A_i) \cap A_j \neq \emptyset$ (*"Markov property"*).

Since $T(\{a_0, \ldots, a_l\}) \subseteq \{a_0, \ldots, a_l\}$ by virtue of (ii), the *transition matrix* $A(T) = (a_{ij}(T))_{i,j=1,\ldots,l}$ with

$$a_{ij}(T) := \begin{cases} 1 & \text{if } T(A_i) \supseteq A_j, \\ 0 & \text{if } T(A_i) \cap A_j = \emptyset, \end{cases} \tag{4.10}$$

is well defined. At first sight, dealing with piecewise expanding Markov maps seems to be highly special. Nevertheless these maps are interesting for more than only didactical reasons: there exists a rich theory of approximating more general maps by Markovian ones. (See [13] for an introduction to this interesting topic which is also relevant from a computational point of view.)

One reason for the simplicity of piecewise expanding Markov maps stems from the fact that these maps admit a symbolic description as already suggested by (4.10).

Lemma 4.13. *Let T denote a piecewise expanding Markov map. There exists a continuous map $h : \Sigma_{l,A(T)} \to [0, 1]$ which is onto such that*

$$h \circ \sigma_A\big((x_n)_{n \in \mathbb{N}_0}\big) = T \circ h\big((x_n)_{n \in \mathbb{N}_0}\big)$$

holds unless $h\big((x_n)_{n \in \mathbb{N}_0}\big) \in \{a_0, \ldots, a_l\}$. Moreover, each $x \in [0, 1]$ has at most two pre-images under h.

Proof. Take $(x_n)_{n \in \mathbb{N}_0} \in \Sigma_{l,A(T)}$ and define a decreasing sequence of non-empty nested intervals by $J_n := \bigcap_{k=0}^{n} T^{-k}(A_{x_k})$. Observe that due to the expansivity and the Markov property $\bigcap_{n \in \mathbb{N}_0} J_n$ consists of exactly one point $\bar{x}$. We may thus set $h\big((x_n)_{n \in \mathbb{N}_0}\big) := \bar{x}$. Given $\varepsilon > 0$ choose N_0 such that $\lambda^{-N_0} < \varepsilon$ and set $\delta := 2^{-(N_0+1)}$. As is easily seen $d\big((x_n)_{n \in \mathbb{N}_0}, (y_n)_{n \in \mathbb{N}_0}\big) < \delta$ implies $\big|h\big((x_n)_{n \in \mathbb{N}_0}\big) - h\big((y_n)_{n \in \mathbb{N}_0}\big)\big| < \varepsilon$, hence h is continuous. Since $h(\Sigma_{l,A(T)})$ contains $X := [0, 1] \backslash \bigcup_{n \in \mathbb{N}_0} T^{-n}(\{a_0, \ldots, a_l\})$, a dense set, the map h is onto. As an aside let us mention that h has a continuous inverse on X which once again may be defined via coding ([31, 44]). If $h\big((x_n)_{n \in \mathbb{N}_0}\big) \notin \{a_0, \ldots, a_l\}$ then the $T(J_n) = \overline{T(J_n)}$ for sufficiently large n, hence

$$T \circ h\big((x_n)_{n \in \mathbb{N}_0}\big) \in \bigcap_{n \in \mathbb{N}_0} T(\overline{J_n}) \subseteq \bigcap_{n \in \mathbb{N}_0} \overline{\bigcap_{k=0}^{n} T^{-k}(A_{x_{k+1}})} = \big\{h \circ \sigma_A\big((x_n)_{n \in \mathbb{N}_0}\big)\big\}.$$

It is easy to see that each point in X has exactly one pre-image under h. Finally, if $x \notin X$ then $T^{N_0}(x) \in \{a_0, \ldots, a_l\}$ for some minimal $N_0 \in \mathbb{N}_0$. Let $(x_n)_{n \in \mathbb{N}_0}, (y_n)_{n \in \mathbb{N}_0}$ denote two pre-images of x. Then $x_n = y_n$ for $0 \leq n < N_0$ and either $x_{N_0} = y_{N_0}$ or $|x_{N_0} - y_{N_0}| = 1$. Since the N_0-th entry uniquely determines the remaining entries, there are at most two pre-images for x. $\square$

The following by-product of the proof of Lemma 4.13 is worth mentioning: if we denote by X the set of all points in $[0, 1]$ whose orbits never hit $\{a_0, \ldots, a_l\}$, that is

$$X := \big\{x \in [0, 1] : O^+(x) \cap \{a_0, \ldots, a_l\} = \emptyset\big\} = [0, 1] \backslash \bigcup_{n \in \mathbb{N}_0} T^{-n}(\{a_0, \ldots, a_l\}),$$

then $T(X) \subseteq X$ and $\lambda^1(X) = 1$. Moreover, the set $h^{-1}(X) \subseteq \Sigma_{l,A(T)}$ is shift-invariant, has a countable complement and yields the commutative diagram

$$
\begin{array}{ccc}
h^{-1}(X) & \xrightarrow{\sigma_{A(T)}} & h^{-1}(X) \\
h \downarrow & & \downarrow h \\
X & \xrightarrow{T} & X
\end{array}
\qquad (4.11)
$$

which represents a topological conjugacy. We thus have found an accurate model for the map T on the set X.

Before dealing with the statistical properties of expanding Markov maps we shall take a very short regression to the topological point of view.

Theorem 4.14. *Let the piecewise expanding Markov map T be eventually onto, that is, for each $i \in \{1, \ldots, l\}$ we have $T^{N(i)}(A_i) \supseteq [0, 1]\backslash\{a_0, \ldots, a_l\}$ with an appropriate $N(i) \in \mathbb{N}$. Then T is chaotic on X.*

Proof. Since $T(Y) \supseteq Y$ with $Y := [0, 1]\backslash\{a_0, \ldots, a_l\}$ we have $T^{N_0}(A_i) \supseteq Y$ for all i and some $N_0 \in \mathbb{N}$. Clearly we are done if we can prove that T^{N_0} is chaotic on X. In order to achieve this, take $U = X \cap U'$, $V = X \cap V'$ with non-empty open sets $U', V' \subseteq [0, 1]$. Since obviously $T^{-N_0 m}(U') \cap V' \supseteq W'$ for some non-empty open set W' and $m \in \mathbb{N}$ we see that $T^{-N_0 m}(U) \cap V \supseteq X \cap T^{-N_0 m}(U') \cap V' \supseteq X \cap W'$. Denseness of X therefore implies topological transitivity. On the other hand, given $x \in A_i$ and a small open interval U with $x \in U \subseteq A_i$ we may find an open interval $V \subseteq U$ such that $T^{N_1}(V) = A_i$ for some $N_1 \in \mathbb{N}$ and $T^{N_1}|_V$ is continuous. From this we deduce that there is a periodic point for T in $U \cap X$, hence Per T is dense in X. $\square$

We refrained from calling T *chaotic* on the whole unit interval because we have reserved this notion to continuous maps. Since the set X has full measure the dynamics of T will be pretty complicated anyway.

Let us now turn to the statistical point of view. Clearly, formula (4.6) for the Frobenius–Perron operator P_T remains valid for any expanding Markov map T. As has already been pointed out, the expanding property (i) for T causes a regularizing effect in the sequence $(P_T^n)_{n\in\mathbb{N}}$. Instead of dealing with this phenomenon in a general context we shall exclusively focus on *piecewise linear*, piecewise expanding Markov maps in the sequel. The main reason for this restriction comes from the fact that the analysis of the Frobenius–Perron operator becomes a matter of finite-dimensional linear algebra for this particular class of maps. As will soon turn out, this class is still large enough to cover a lot of interesting applications.

Assume for the rest of this section that T is a piecewise linear, piecewise expanding Markov map on the unit interval. Taking any piecewise constant function $f \in L^1(\lambda^1|_{[0,1]})$, i.e. $f = \sum_{i=1}^{l} f_i \mathbf{1}_{A_i}$ with $f_i \in \mathbb{C}$, we obtain

$$
P_T f(x) = \sum_{i=1}^{l} f_i P_T \mathbf{1}_{A_i}(x) = \sum_{i=1}^{l} f_i \sum_{y:T(y)=x} \frac{\mathbf{1}_{A_i}(y)}{|T'(y)|} = \sum_{i,j=1}^{l} \frac{f_i a_{ij}(T)}{|T'|_{A_i}|} \mathbf{1}_{A_j}(x).
$$

$$(4.12)$$

Since the space of piecewise constant functions may be identified with $\mathbb{C}^l$ via $f \cong \alpha_f := (f_1, \ldots, f_l)$ we can rewrite (4.12) as

$$(P_T f)_j = \sum_{i=1}^{l} f_i \frac{a_{ij}(T)}{\left|T'|_{A_i}\right|} \quad \text{for } j \in \{1, \ldots, l\}.$$

More concisely the latter relation reads

$$\alpha_{P_T f} = \alpha_f P(T) \quad \text{with} \quad P(T) := (p_{ij}(T)) := \left(\frac{a_{ij}(T)}{\left|T'|_{A_i}\right|} \right)_{i,j=1}^{l}. \tag{4.13}$$

If we restrict our analysis to piecewise constant functions the Frobenius–Perron operator may thus be regarded an object of finite-dimensional linear algebra. The question arises whether this point of view is too narrow or not. For example, will it be realistic to hope for a piecewise constant invariant density? In order to deal with such questions we set

$$r := \prod_{i=1}^{l}(a_i - a_{i-1}) > 0 \quad \text{and} \quad R := \text{diag}\left(\frac{a_1 - a_0}{r}, \ldots, \frac{a_l - a_{l-1}}{r} \right). \tag{4.14}$$

Defining $Q(T) := R^{-1} P(T) R$ we deduce from the simple calculation

$$\sum_{j=1}^{l} Q(T)_{ij} = \sum_{j=1}^{l} \frac{a_j - a_{j-1}}{a_i - a_{i-1}} \frac{a_{ij}(T)}{\left|T'|_{A_i}\right|} = \frac{\lambda^1(T(A_i))}{(a_i - a_{i-1})\left|T'|_{A_i}\right|} = 1 \quad (i = 1, \ldots, l)$$

that $P(T)$ is similar to the stochastic matrix $Q(T)$. Observe that $P(T)$ itself will be stochastic if all intervals A_i have the same length $1/l$. Since any piecewise constant invariant density via (4.13) corresponds to an eigenvalue one of $P(T)$, a basic understanding concerning the eigenvalue structure of stochastic matrices will prove helpful. In fact there is a general eigenvalue pattern even for *non-negative* matrices; these are matrices $B = (b_{ij})$ with $b_{ij} \geq 0$ for all i, j. The following theorem due to Perron and Frobenius is a standard result in matrix theory (see [26, 28] for a proof). Recall that we termed a non-negative matrix B irreducible if for any i, j we have $(B^n)_{ij} > 0$ for *some* n whereas B is *aperiodic* if $(B^n)_{ij} > 0$ for *all* sufficiently large n.

Theorem 4.15 (Perron, Frobenius). *Let B denote a non-negative matrix. There exists an eigenvalue $\lambda_{\max} \geq 0$ such that no other eigenvalue has modulus greater than $\lambda_{\max}$; moreover there exists an eigenvector with non-negative coordinates corresponding to $\lambda_{\max}$. If B is irreducible then $\lambda_{\max} > 0$ is simple and all entries of the corresponding eigenvector are positive. If B is aperiodic then there are no other eigenvalues of modulus $\lambda_{\max}$ whatsoever.*

For a *stochastic* matrix B it is easily seen that $\lambda_{\max} = 1$: on the one hand $\lambda_{\max} \geq 1$ because one clearly is an eigenvalue of the transpose B^T. On the other hand, denoting by $x = (x_i)_{i=1}^{l}$ a non-negative eigenvector corresponding to $\lambda_{\max}$ we have $\lambda_{\max} \max_{i=1}^{l} x_i = \max_{i=1}^{l} \sum_{j=1}^{l} b_{ij} x_j \leq \max_{j=1}^{l} x_j$ and hence $\lambda_{\max} \leq 1$.

Since the matrix $P(T)$ from (4.13) is similar to the stochastic matrix $Q(T)$ there is always an invariant density f^* for P_T which is piecewise constant. Moreover, we can explicitly calculate f^* by solving the linear equation $\alpha_{f^*}P(T) = \alpha_{f^*}$ under the additional conditions $f_i^* \geq 0$, $\sum_{i=1}^{l} f_i^* \lambda^1(A_i) = 1$. By imposing a further condition on the Markov map T and then utilizing Theorem 4.15 the density f^* may be shown to be unique and, even better, T turns out to be statistically stable.

Theorem 4.16. *If the piecewise linear, piecewise expanding Markov map T is eventually onto then it is statistically stable.*

Proof. Clearly $P(T)$ is aperiodic; by the Perron–Frobenius Theorem 4.15 there exists a positive invariant density f^* for T which is piecewise constant, that is, $f^* = \sum_{i=1}^{l} f_i^* \mathbf{1}_{A_i}$ with $f_i^* > 0$. We now have to show that the Frobenius–Perron operator as given by (4.12) is asymptotically stable.

Let $\alpha := \{A_1, \ldots, A_l\}$ denote the partition naturally associated with T. From the fact that T is expanding and eventually onto we deduce that α is generating, i.e. $[0,1] \cap \mathcal{B} = A_\sigma(T^{-n}\alpha : n \in \mathbb{N}_0)$. Hence it remains to show that $P_T^n \mathbf{1}_B \to \lambda^1(B) f^*$ for any non-empty set $B := A_{i_0} \cap T^{-1}(A_{i_1}) \cap \ldots \cap T^{-k}(A_{i_k}) \in \bigvee_{i=0}^{k} T^{-i}\alpha$. An explicit calculation yields $\lambda^1(B) = \lambda^1(A_{i_k})|T'|_{A_{i_{k-1}}}|^{-1} \cdot \ldots \cdot |T'|_{A_{i_0}}|^{-1}$ and

$$P_T^k \mathbf{1}_B(x) = \sum_{y:T^k(y)=x} \frac{\mathbf{1}_B(y)}{|T'|_{A_{i_{k-1}}}| \cdot \ldots \cdot |T'|_{A_{i_0}}|} = \frac{\lambda^1(B)}{\lambda^1(A_{i_k})} \mathbf{1}_{A_{i_k}}(x).$$

Therefore we obtain for $n \in \mathbb{N}_0$

$$P_T^{k+n} \mathbf{1}_B = \frac{\lambda^1(B)}{\lambda^1(A_{i_k})} \sum_{j=1}^{l} (P(T)^n)_{i_k j} \mathbf{1}_{A_j}$$

in which expression all quantities are fixed except for $(P(T)^n)_{ij}$.

In order to grasp the asymptotic behaviour of $P(T)^n$ we first tackle the corresponding problem for the aperiodic stochastic matrix $Q(T)$ which by definition is similar to $P(T)$. For $Q(T)$ we have a simple dominant eigenvalue $\lambda_{\max} = 1$ with the corresponding eigenvector $\xi := (f_1^* \lambda^1(A_1), \ldots, f_l^* \lambda^1(A_l))$, and no other eigenvalue has modulus one. Therefore $\lim_{n\to\infty} Q(T)^n$ exists; denote this limit by $\hat{Q} = (\hat{q}_{ij})$, that is, $\lim_{n\to\infty}(Q(T)^n)_{ij} = \hat{q}_{ij}$ for all i, j. Clearly $\hat{Q}$ also is a stochastic matrix, and $\hat{Q}Q(T) = Q(T)\hat{Q} = \hat{Q}$ as well as $\xi\hat{Q} = \xi$. The most important feature of $\hat{Q}$ is that its rows are identical. Indeed, take $\hat{q}_{i_1 j} := \max_i \hat{q}_{ij}$ and assume $\hat{q}_{i_2 j} < \hat{q}_{i_1 j}$ for some i_2; then for sufficiently large n

$$\hat{q}_{i_1 j} = \sum_{k=1}^{l} (Q(T)^n)_{i_1 k} \hat{q}_{kj} < \sum_{k=1}^{l} (Q(T)^n)_{i_1 k} \hat{q}_{i_1 j} = \hat{q}_{i_1 j},$$

an obvious contradiction. Thus $\hat{Q}$ has identical rows which we may calculate as

$$\hat{q}_{ij} = \sum_{k=1}^{l} \xi_k \hat{q}_{kj} = \xi_j = f_j^* \lambda^1(A_j).$$

Returning to the original matrix $P(T)$ we find

$$\lim_{n\to\infty}(P(T)^n)_{ij} = \lim_{n\to\infty}(R\,Q(T)^n\,R^{-1})_{ij} = f_j^*\lambda^1(A_i).$$

If we put all our observations together, we obtain

$$\int_{[0,1]}|P_T^{n+k}\mathbf{1}_B - \lambda^1(B)\,f^*| = \frac{\lambda^1(B)}{\lambda^1(A_{i_k})}\sum_{j=1}^{l}|(P(T)^n)_{i_kj} - \lambda^1(A_{i_k})f_j^*|\lambda^1(A_j) \to 0$$

as $n \to \infty$ for every non-empty $B \in \bigvee_{i=0}^{k}T^{-i}\alpha$. The theorem is thus proved. $\square$

For the special class of maps considered here both interpretations – the determin-istic and the stochastic one – perfectly fit together: given a piecewise linear, piecewise expanding Markov map typical points behave chaotically, and their orbits exhibit sen-sitive dependence on initial conditions. On the other hand, any such map is statistically stable with respect to an absolutely continuous measure with piecewise constant den-sity; the latter may be equally well approximated by iterating the Frobenius–Perron operator and by histograms.

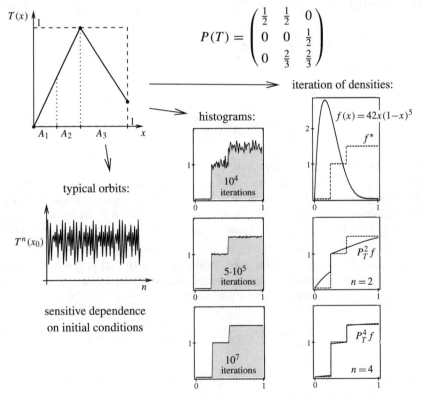

Figure 4.10. Various aspects of the piecewise linear, piecewise expanding Markov map (4.15)

As an example consider the (continuous) map

$$T(x) := \begin{cases} 2x & \text{if } x \in [0, \frac{1}{2}[, \\ \frac{7}{4} - \frac{3}{2}x & \text{if } x \in [\frac{1}{2}, 1], \end{cases} \qquad (4.15)$$

which is Markovian with respect to $(a_0, a_1, a_2, a_3) = (0, \frac{1}{4}, \frac{1}{2}, 1)$. A short calculation yields $\alpha_{f*} = (0, 1, \frac{3}{2})$. (See Figure 4.10; notice also that we have to look at $T|_{[\frac{1}{4}, 1]}$ for directly applying Theorem 4.16.)

Let us now apply the tools developed above to a mechanical system. Suppose that a mathematical pendulum with a small amount of friction is driven according to the following simple rule. Whenever the pendulum's angular velocity $\dot\varphi^-$, which shall denote $\dot\varphi$ in the position $\varphi = 0$, satisfies $0 \leq \dot\varphi^- \leq \omega_0$, a kick instantaneously accelerates the pendulum to $\dot\varphi^+$ (see Figure 4.11). In the sequel we shall assume that $\dot\varphi^+ = f(\dot\varphi^-) \geq \dot\varphi^-$. Restricting ourselves to *long* pendulums, we shall use the linearized equation of motion which reads

$$ml^2\ddot\varphi + c\dot\varphi + mgl\varphi = 0.$$

We shall also assume the damping to be weak, that is, $0 < \rho < 1$ with the non-dimensional damping parameter $\rho := c(2ml^2)^{-1}\sqrt{l/g}$. Denoting by $\dot\varphi_n^-$ the angular velocity immediately before the n-th kick we find

$$\dot\varphi_{n+1}^- = f(\dot\varphi_n^-)e^{-\sigma p(\dot\varphi_n^-)}$$

by virtue of a straightforward calculation; here $p(\dot\varphi_n^-)$ denominates the smallest non-negative integer such that $f(\dot\varphi_n^-)e^{-\sigma p(\dot\varphi_n^-)} \leq \omega_0$, and σ is given by

$$\sigma := \frac{2\pi\rho}{\sqrt{1 - \rho^2}}.$$

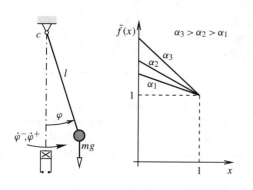

With the abbreviations $x_n := \dot\varphi_n^-/\omega_0$ and $\tilde f(x) := f(x\omega_0)/\omega_0$ we are led to study a map T on the unit interval defined as

$$T(x) := \tilde f(x)e^{-\sigma\left\lceil \frac{\log \tilde f(x)}{\sigma} \right\rceil}. \qquad (4.16)$$

Clearly, the specific form of this map depends on the kick-law $\tilde f$. To keep our analysis as simple as possible, we shall exclusively deal here with the case

Figure 4.11. The model of the pendulum (left) and the kick-law (4.17)

$$\tilde f(x) = 1 + \alpha(1 - x), \qquad \alpha > 0. \qquad (4.17)$$

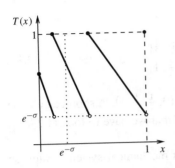

Figure 4.12. The graph of T from (4.16) for $e^{2\sigma} < \alpha + 1 < e^{3\sigma}$

Intuitively speaking, according to this law the kick is the stronger the slower the pendulum and the larger the parameter α. Figure 4.12 shows a typical graph of T. Since $T([0, 1]) \subseteq [e^{-\sigma}, 1]$ we shall restrict our analysis to $T|_{[e^{-\sigma}, 1]}$, nevertheless writing T again for the latter map. By varying the parameters σ and α, quite different dynamical effects may be observed. We shall merely sketch a few results here because the rigorous calculations are quite cumbersome ([9]). For example, it is not difficult to show that T has an attracting fixed point x^* (which corresponds to a periodic motion of the kicked pendulum) if and only if for some $m \in \mathbb{N}$

$$(\sigma, \alpha) \in B_m := \{(\xi, \eta) \in \mathbb{R}_+^2 : b_m^-(\xi) < \eta < b_m^+(\xi)\},$$

with the boundary functions b_m^-, b_m^+ defined as

$$b_m^-(\xi) := \frac{e^{m\xi} - e^{\xi}}{e^{\xi} - 1} \quad \text{and} \quad b_m^+(\xi) := e^{m\xi},$$

respectively. In that case $X_m := \{x \in [e^{-\sigma}, 1] : T^n(x) \not\to x^*\}$ is a set of measure zero. Moreover, $(X_m, T|_{X_m})$ turns out to be topologically conjugate to (Σ_{m-1}, σ). If $m \geq 3$ we therefore observe chaos on a Cantor set while typical points are attracted by x^*. Note how this resembles the situation within the three-periodic window of the logistic family. However, since T is not continuous, changes of the dynamics may take place in a more dramatic way here. If $(\sigma, \alpha) \notin \bigcup_{m \in \mathbb{N}} \overline{B_m}$ then the map T may be seen to be piecewise expanding and eventually onto, and thus is statistically stable, if only the Markov property holds.

But even if T is not Markovian it may be so with respect to a finer partition (cf. Figure 4.13). In fact, parameters (σ, α) giving rise to a Markovian situation are *dense* in $\mathbb{R}^2 \setminus \bigcup_{m \in \mathbb{N}} \overline{B_m}$. Studying some of the latter cases may therefore give an impression of the statistical morphogenesis of our simple pendulum. Figure 4.14 depicts two numerical histograms and also shows the

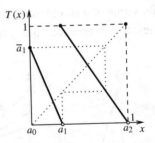

Figure 4.13. The map T is a Markov map with respect to $\{a_0, a_1, \bar{a}_1, a_2\}$ but *not* with respect to $\{a_0, a_1, a_2\}$.

regions B_m in the parameter plane with a few Markovian configurations indicated by dashed lines.

We shall close this section with some theoretical reflections concerning the relation between piecewise linear, piecewise expanding Markov maps and symbolic dynamics. Up to now we have exclusively used the systems $(\Sigma_{l,A}, \sigma_A)$ as topological models. First

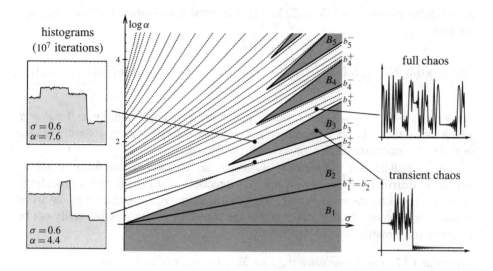

Figure 4.14. The dark regions correspond to transient chaotic behaviour whereas the light regions are fully chaotic. From the family of lines along which T is a Markov map with respect to a refined partition a few representatives are indicated by dashed lines.

of all we are thus asked to define some reasonable probability measures on these spaces. To this end fix a real-valued matrix $P = (p_{ij})_{i,j=1}^{l}$ as well as a vector $\pi = (\pi_i)_{i=1}^{l}$ and define

$$\mu_{\pi,P}([i_0, i_1, \ldots, i_n]) := \pi_{i_0} p_{i_0 i_1} p_{i_1 i_2} \cdots p_{i_{n-1} i_n} \qquad (4.18)$$

for every finite *cylinder*

$$[i_0, i_1, \ldots, i_n] := \{(x_k)_{k \in \mathbb{N}_0} \in \Sigma_{l,A} : x_0 = i_0, \ldots, x_n = i_n\} \subseteq \Sigma_{l,A}.$$

For $\mu_{\pi,P}$ to be a normed measure, P and π clearly have to be non-negative. Moreover, the conditions

$$\sum_{j=1}^{l} \pi_j = 1 \quad \text{and} \quad \pi_i\left(1 - \sum_{j=1}^{l} p_{ij}\right) = 0 \text{ for all } i \qquad (4.19)$$

must hold. Since we intend to make of $\mu_{\pi,P}$ a σ_A-invariant measure, we may assume $\pi_i > 0$ without loss of generality. From (4.19) we then see that P necessarily is a stochastic matrix. If $a_{ij} = 0$ then $[i, j]$ is empty and thus $p_{ij} = 0$. On the other hand it is natural to wish for $\mu_{\pi,P}(C) > 0$ for every non-empty cylinder C since otherwise A could be replaced by a matrix with more zeros. This in turn implies that $p_{ij} > 0$ whenever $a_{ij} = 1$. Via the Extension Theorem A.4 $\mu_{\pi,P}$ uniquely extends to a probability measure on $\left(\Sigma_{l,A}, \mathcal{B}(\Sigma_{l,A})\right)$. Concerning the induced measure $\sigma_A \mu_{\pi,P}$ we find

$$\sigma_A \mu_{\pi,P}([i_0, i_1, \ldots, i_n]) = \sum_{j=1}^{l} \mu_{\pi,P}([j, i_0, i_1, \ldots, i_n]) = \mu_{\pi P,P}([i_0, i_1, \ldots, i_n]),$$

hence $\sigma_A \mu_{\pi,P} = \mu_{\pi P,P}$. Consequently, $\mu_{\pi,P}$ is preserved by the shift map if $\pi P = \pi$ which means that π is a non-negative eigenvector of P corresponding to an eigenvalue one. As P is stochastic such an eigenvector will exist.

The ingredients for the measure $\mu_{\pi,P}$ thus are summarized as follows: take a stochastic matrix $P = (p_{ij})$ with $p_{ij} > 0$ precisely if $a_{ij} = 1$, and a probability vector (π_i) which satisfies $\pi P = \pi$. Then (4.18) uniquely determines a σ_A-invariant probability measure. Notice however that the choice of P and π will usually not be unique given a matrix $A \in \{0, 1\}^{l \times l}$.

Theorem 4.17. *Let the measure $\mu_{\pi,P}$ on $\Sigma_{l,A}$ be σ_A-invariant. Then*

(i) *σ_A is ergodic if and only if A is irreducible;*

(ii) *σ_A is exact if and only if A is aperiodic.*

Proof. Assume that σ_A is ergodic. Given i, j we deduce from $\mu_{\pi,P}([i])\mu_{\pi,P}([j]) > 0$ that $\mu_{\pi,P}(\sigma_A^{-N}([j]) \cap [i]) > 0$ and hence A_{ij}^N for some $N \in \mathbb{N}$. If in turn A is irreducible we have to show that

$$\frac{1}{n} \sum_{k=0}^{n-1} \mu_{\pi,P}(\sigma_A^{-k}(B) \cap C) \to \mu_{\pi,P}(B)\mu_{\pi,P}(C)$$

for all cylinders B, C. The Birkhoff Ergodic Theorem provides us with the almost everywhere relation $\lim_{n\to\infty} \frac{1}{n} \sum_{k=0}^{n-1} \mathbf{1}_B \circ \sigma_A^k = \mathbb{E}(\mathbf{1}_B | \mathcal{A}_{\sigma_A})$. After multiplication by $\mathbf{1}_C$ dominated convergence implies that $\lim_{n\to\infty} \frac{1}{n} \sum_{k=0}^{n-1} \mathbf{1}_B \circ \sigma_A^k \mathbf{1}_C = \mathbb{E}(\mathbf{1}_B | \mathcal{A}_{\sigma_A})\mathbf{1}_C$ in $L^1(\mu_{\pi,P})$. Therefore the limit $\frac{1}{n} \sum_{k=0}^{n-1} \mu_{\pi,P}(\sigma_A^{-k}(B) \cap C)$ exists for all finite cylinders. To be more specific let $B := [i_0, \ldots, i_l]$, $C := [j_0, \ldots, j_m]$ and observe that

$$\mu_{\pi,P}(\sigma_A^{-(k+m+1)}(B) \cap C) = \mu_{\pi,P}(C)\mu_{\pi,P}(B)\frac{(P^{k+1})_{j_m i_0}}{\pi_{i_0}} \quad \text{for all } k \geq 0.$$

The limit $\lim_{n\to\infty} \frac{1}{n} \sum_{k=0}^{n-1}(P^k)_{ij} =: \hat{q}_{ij}$ thus exists for all i, j, and we will be done once we can show that $\hat{q}_{ij} = \pi_j$. Clearly, $\hat{Q} = (\hat{q}_{ij})$ is a stochastic matrix; furthermore $\hat{Q}P = P\hat{Q}$ and $\pi\hat{Q} = \pi$. From the irreducibility of P we deduce that $\hat{Q}$'s rows are identical by exactly the same argument as in the proof of Theorem 4.16. Therefore $\hat{q}_{ij} = \sum_i \pi_i \hat{q}_{ij} = \pi_j$, and σ_A is ergodic.

To prove (ii) first assume that σ_A is exact. Then $\lim_{n\to\infty} \mu_{\pi,P}(\sigma_A^n([i]) \cap [j]) = \pi_j > 0$ for all i, j and thus $A_{ij}^n > 0$ for sufficiently large n. Conversely, we have to verify that for the Frobenius–Perron operator $P_{\sigma_A}^n \mathbf{1}_C \to \mu_{\pi,P}(C)$ strongly, where C is any finite cylinder. If A (and hence P) is aperiodic then the limit $\lim_{n\to\infty} P^n$ is a stochastic matrix with identical rows;

this again is contained in the proof of Theorem 4.16. As before we obtain $\lim_{n\to\infty}(P^n)_{ij} = \pi_j$ for all i, j. By means of the calculations from above we have

$$\int_B P_{\sigma_A}^{k+m+1} \mathbf{1}_C = \int_{\sigma_A^{-(k+m+1)}(B)} \mathbf{1}_C = \mu_{\pi,P}(C)\mu_{\pi,P}(B) \frac{(P^{k+1})_{j_m i_0}}{\pi_{i_0}}$$

from which we deduce the explicit formula

$$P_{\sigma_A}^{k+m+1} \mathbf{1}_C = \mu_{\pi,P}(C) \sum_{i=1}^l \frac{(P^{k+1})_{j_m i}}{\pi_i} \mathbf{1}_{[i]} \quad \text{for } k \in \mathbb{N}_0.$$

But now we are done since

$$\int_{\Sigma_{l,A}} |P_{\sigma_A}^{k+m+1} \mathbf{1}_C - \mu_{\pi,P}(C)| \, d\mu_{\pi,P} = \mu_{\pi,P}(C) \sum_{i=1}^l |(P^{k+1})_{j_m i} - \pi_i| \to 0$$

as $k \to \infty$, and hence σ_A is exact. $\square$

The statistical properties of subshifts thus are easily analysed. In the next section we shall give a stochastic interpretation of the measures $\mu_{\pi,P}$ and also obtain another, purely probabilistic justification of the above theorem. Here we are just going to combine some of our findings.

Let T denote a piecewise linear, piecewise expanding Markov map on the unit interval which is eventually onto. Recall the diagram (4.11) which, as a corollary of Lemma 4.13, provided us with a subspace $\Sigma_T := h^{-1}(X) \subseteq \Sigma_{l,A(T)}$ together with a homeomorphism $h : \Sigma_T \to X$ such that $h \circ \sigma_{A(T)}(x) = T \circ h(x)$ for all $x \in \Sigma_T$. Since $A(T)$ is aperiodic we must have $\mu_{\pi,P}(\Sigma_T) = 1$ no matter which specific measure $\mu_{\pi,P}$ we choose. In any case the probability measure $h\mu_{\pi,P}$ is thus preserved by T. On the other hand, we know by Theorem 4.17 that T admits a unique absolutely continuous invariant measure μ^*. How are all these measures related? Observing that $(\Sigma_T, \sigma_{A(T)})$ and (X, T) both are exact and μ^* is in fact equivalent to $\lambda^1|_{[0,1]}$ we are left with only two possibilities: either $h\mu_{\pi,P}$ is concentrated on a set of Lebesgue measure zero, or $h\mu_{\pi,P} = \mu^*$. In the latter, more interesting case $\mu^*(A_i) = h\mu_{\pi,P}(A_i) = \mu_{\pi,P}([i]) = \pi_i$ for $i = 1, \ldots, l$ and thus the invariance relations

$$\sum_{i=1}^l \pi_i \frac{a_j - a_{j-1}}{a_i - a_{i-1}} \frac{a_{ij}(T)}{|T'|_{A_i}|} = \pi_j \quad \text{and} \quad \sum_{i=1}^l \pi_i p_{ij} = \pi_j \qquad (4.20)$$

must hold simultaneously. In more concise notation (4.20) reads $\pi Q(T) = \pi$ and $\pi P = \pi$, respectively, with the matrix $Q(T)$ introduced after (4.14). Since $Q(T)$ is a stochastic matrix we can simply set $P := Q(T)$. The resulting intimate connection between $(\Sigma_{l,A(T)}, \sigma_{A(T)})$ and $([0,1], T)$ deserves a name of its own.

Definiton 4.18. Let T and S denote two measure-preserving maps on the probability spaces $(X, \mathcal{A}, \mu)$ and $(Y, \mathcal{B}, \nu)$, respectively. If there exist $X_1 \in \mathcal{A}$, $Y_1 \in \mathcal{B}$ with $\mu(X_1) = \nu(Y_1) = 1$ as well as a map $h : X_1 \to Y_1$ such that

(i) $T(X_1) \subseteq X_1$ and $S(Y_1) \subseteq Y_1$,

(ii) h is invertible and measure preserving, i.e. $h\mu|_{X_1} = \nu|_{Y_1}$,

(iii) $h \circ T(x) = S \circ h(x)$ for all $x \in X_1$,

then the systems (X, T) and (Y, S) are called *isomorphic*.

Though it is an immediate consequence of this definition we formulate the following observation as a theorem for its resemblance with the discussion following Definition 2.16.

Theorem 4.19. *Let the systems (X, T) and (Y, S) be isomorphic. If T is ergodic (mixing, exact) then so is S. Moreover, $h_\mu(T) = h_\nu(S)$, i.e., both maps have the same entropy.*

Proof. We shall establish the first assertion(s) by showing that there exists a linear isometric isomorphism $\Phi : L^1(\mu) \to L^1(\nu)$ such that $P_S \circ \Phi = \Phi \circ P_T$. Indeed, it is easily checked that the definition $\Phi(f) := f \circ h^{-1} \mathbf{1}_{Y_1}$ for all $f \in L^1(\mu)$ does the job. Since in $L^1(\nu)$ also $\Phi(\mathbf{1}_X) = \mathbf{1}_{Y_1} = \mathbf{1}_Y$ the claim about ergodicity (mixing, exactness) is proved.

Let β denote a partition of Y. Then $\alpha := h^{-1}(Y_1 \cap \beta)$ is a partition of X_1 with $\bigvee_{i=0}^{n-1} T^{-i}\alpha = h^{-1}(Y_1 \cap \bigvee_{i=0}^{n-1} S^{-i}\beta)$ up to sets of measure zero. From this we deduce

$$\int_X I\left(\bigvee_{i=0}^{n-1} T^{-i}\alpha\right) d\mu = \int_Y I\left(\bigvee_{i=0}^{n-1} S^{-i}\beta\right) d\nu$$

for all $n \in \mathbb{N}$. Therefore $h_\nu(S, \beta) = h_\mu(T, \alpha) \leq h_\mu(T)$, and hence $h_\nu(S) \leq h_\mu(T)$. By interchanging the role of S and T we get $h_\nu(S) \geq h_\mu(T)$. $\square$

Isomorphism between measure-preserving systems is an equivalence relation, and isomorphic systems are as indistinguishable by statistical properties as topologically conjugate systems are by topological ones. Summarizing we may say that any piecewise linear, piecewise expanding, eventually onto Markov map on the unit interval is isomorphic to a subshift of finite type. From the statistical point of view these apparently different objects are but two different manifestations of the same thing.

4.4   A short look at Markov chains

In the course of our statistical analysis of dynamical systems we have encountered similar mathematical phenomena in quite different guises. For example, we have found that in the most elementary setting Markov operators are given by stochastic matrices, but also have observed these objects when studying measures on $\Sigma_{l,A}$ and the Frobenius–Perron operators of very special maps on the unit interval. Analogously, the task of finding a non-negative eigenvector corresponding to an eigenvalue one arose several times. We are now going to show that these analogies are by no means accidental. It is the purely probabilistic concept of *Markov chains* that we shall introduce and discuss in the following and that will lead us to a unified view on much of what has been discussed so far.

As there is a vast literature on this subject we shall content ourselves with an informal discussion of only a few topics. The presentation aims at emphasizing resemblances to and differences from the dynamical systems discussed before. For a more rigorous and comprehensive treatment of Markov chains in their own right we refer to the literature ([14, 41, 55]).

In the sequel we shall deal with a few aspects of Markov chains on a state space I which is at most countable (i.e. finite or countable).

Definiton 4.20. Let P denote a stochastic matrix on I^2, and assume that the nonnegative row-vector $(\lambda_i)_{i \in I}$ satisfies $\sum_{i \in I} \lambda_i = 1$. A sequence $(\xi_n)_{n \in \mathbb{N}_0}$ of random variables on I is called a (homogeneous) (λ, P)-*Markov chain* if

(i) $\mathbb{P}(\xi_0 = i_0) = \lambda_{i_0}$ for all $i_0 \in I$;

(ii) $\mathbb{P}(\xi_{n+1} = i_{n+1} | \xi_n = i_n, \xi_{n-1} = i_{n-1}, \dots, \xi_1 = i_1, \xi_0 = i_0) = p_{i_n i_{n+1}}$ for all $n \in \mathbb{N}_0$ and $i_0, i_1, \dots, i_n, i_{n+1} \in I$.

The possibly infinite matrix $P = (p_{ij})_{i,j \in I}$ is called the *transition matrix* of the chain while λ is referred to as its *initial distribution*. Applications of Markov chains abound in physics, biology, economics and other disciplines. A couple of examples will best illustrate some of the questions one is typically concerned with when dealing with Markov chains. Before sketching these examples let us point out a few basic facts.

After having endowed $(I, \mathcal{P}(I))$ with the counting measure # we have seen earlier that Markov operators on $L^1(\#)$ arise via stochastic matrices. Since

$$\mathbb{P}(\xi_{n+1} = j) = \sum_{i \in I} \mathbb{P}(\xi_{n+1} = j, \, \xi_n = i) = \sum_{i \in I} \mathbb{P}(\xi_n = i) p_{ij}$$

we can now give a stochastic interpretation of these operators: if λ denotes the distribution of ξ_0 then the distribution of ξ_n is λP^n. The long time behaviour of the chain $(\xi_n)_{n \in \mathbb{N}_0}$ thus corresponds to the asymptotic properties of the Markov operator P.

But what, after all, is the intuitive meaning of a stochastic process $(\xi_n)_{n \in \mathbb{N}_0}$ on I being a Markov chain? Assume we know that ξ_m happened to be i_m and set $\eta_n := \xi_{n+m}$ $(n \in \mathbb{N}_0)$. As an immediate consequence of property (ii) in Definition 4.20 we then have $\mathbb{P}(\eta_{n+1} = j_{n+1} | \eta_n = j_n) = p_{j_n j_{n+1}}$ and $\mathbb{P}(\eta_0 = j_0) = \delta_{i_m j_0}$. Here and throughout, δ_{ij} equals one if $i = j$, and zero otherwise; moreover, $\delta_k := (\delta_{ki})_{i \in I}$ denotes the distribution corresponding to a unit mass at k. The new process $(\eta_n)_{n \in \mathbb{N}_0}$ is a (δ_{i_m}, P)-Markov chain not depending on ξ_k for $0 \le k < m$. This fact is often expressed by saying that a Markov chain "has no memory": the probability of jumping from state i to state j does not depend on how the process managed to reach i; at every step the process completely forgets its history. By introducing additional states a chain with bounded memory can be reduced to the case discussed here, just as a higher order differential equation can be transformed to a system of first order equations.

Another, more symmetric way of interpreting the Markov property (ii) is as follows. Assuming $\mathbb{P}(\xi_n = i_n) > 0$ we have

$$\mathbb{P}(\xi_{n+1} = i_{n+1}, \xi_{n-1} = i_{n-1}|\xi_n = i_n) = \frac{\mathbb{P}(\xi_{n+1} = i_{n+1}, \xi_n = i_n, \xi_{n-1} = i_{n-1})}{\mathbb{P}(\xi_n = i_n)}$$

$$= \mathbb{P}(\xi_{n+1} = i_{n+1}|\xi_n = i_n)\mathbb{P}(\xi_{n-1} = i_{n-1}|\xi_n = i_n)$$

and therefore ξ_{n+1} (the "future") and ξ_{n-1} (the "past") are independent given the outcome of ξ_n (the "presence"). Due to this observation, which is symmetric in time, one could try to run a Markov chain backwards. Unfortunately, if $\mathbb{P}(\xi_n = i) > 0$ then the transition probabilities

$$\mathbb{P}(\xi_{n-1} = j|\xi_n = i) = p_{ji}\frac{\mathbb{P}(\xi_{n-1} = j)}{\mathbb{P}(\xi_n = i)}$$

will in general depend on n and therefore give rise to a non-homogeneous chain. However, homogeneity is guaranteed if $\mathbb{P}(\xi_n = i) = \pi_i$ for all n which means that the chain $(\xi_n)_{n\in\mathbb{N}_0}$ is *stationary* (or *in equilibrium*).

As a first example consider the following crude Markovian model of a single game of tennis.

We naively assume that the player A wins each individual point with probability $p \in [0, 1]$ while the player B wins with probability $q = 1 - p$. More realistic effects, e.g. varying fighting spirit, are not taken into account here. Lumping together equivalent states (e.g. "30:30" and "Deuce") we obtain a Markov chain with 17 states as depicted in Figure 4.15.

Probably the most interesting quantities for this simple model are

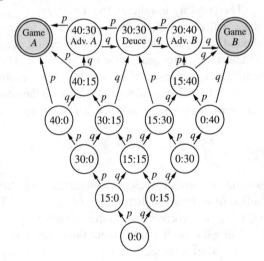

Figure 4.15. Let's play a game of tennis!

$$\Phi(p) := \mathbb{P}(A \text{ wins the game}) \quad \text{and} \quad \Psi(p) := \mathbb{E}\,T$$

where $T := \min\{n : \xi_n \in \{\text{"Game } A\text{"}, \text{"Game } B\text{"}\}, \xi_0 = \text{"0:0"}\}$ denotes the random duration of the game. By means of techniques discussed later one finds

$$\Phi(p) = \frac{p^4(3 - 2p)(5 - 8p + 4p^2)}{1 - 2p + 2p^2}, \quad \Psi(p) = 4 + 4p(1-p)\frac{1 + 6p^2 - 12p^3 + 6p^4}{1 - 2p + 2p^2}.$$

$$(4.21)$$

The graphs of Φ and Ψ are shown in Figure 4.16. By symmetry $\Phi(0.5) = 0.5$; however, since $\Phi'(0.5) = 2.5$ small differences between the two players are magnified. Not surprisingly, the maximal mean length occurs in the case of equal players, $\Psi(0.5) = 6.75$. Despite its simplicity our model motivates some questions of general interest: Starting at i, what is the probability of the process to ever hit the set $J \subseteq I$? How long will that take (in the mean)? Will the state i be met again? We shall discuss these questions below.

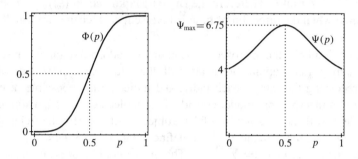

Figure 4.16. Player A wins with probability $\Phi(p)$; the game's mean duration is $\Psi(p)$.

An interesting class of Markov chains arises from *random walks*. As a simple example imagine a particle which moves on the set of integers by jumping one unit to the right or to the left with probabilities depending only on its actual position. Clearly, this process can be modelled as a Markov chain on $\mathbb{Z}$ with

$$\mathbb{P}(\xi_{n+1} = j | \xi_n = i) = \begin{cases} p_{ij} & \text{if } |i - j| = 1, \\ 0 & \text{otherwise}, \end{cases}$$

subject to $p_{i,i+1} + p_{i,i-1} = 1$ for all $i \in \mathbb{Z}$; in the simplest case these probabilities do not depend on i, that is $p_{i,i+1} = 1 - p_{i,i-1} = p$. In an analogous manner random walks on the lattice $\mathbb{Z}^d$ and other spaces may be defined. One of the most important topics concerning random walks is *recurrence*. Given a state i, what is the probability of returning to i sometime in the future? If this probability equals one, will the mean return time be finite? For example, in case of the simplest random walk on $\mathbb{Z}$ mentioned above we shall see later that for all $i \in \mathbb{Z}$

$$\mathbb{P}(\xi_n = i \text{ for some } n \in \mathbb{N} | \xi_0 = i) = 2 \min\{p, 1 - p\}$$

which is smaller than one unless the transition matrix is symmetric $(p = \frac{1}{2})$.

As a last example we are going to discuss a simple model of diffusion due to Ehrenfest ([21, 33]). Consider a certain amount of gas contained in an adiabatic chamber which is divided into two equal parts by a permeable membrane. We assume that the total number of gas molecules equals $2M$. Furthermore, at each instant of time one randomly chosen molecule moves from one half of the chamber to the other. Assigning to each molecule the number zero or one depending on whether at the moment it is in the left or the right half-chamber respectively, we are led to study a Markov chain $(\xi_n)_{n \in \mathbb{N}_0}$ on $I := \{0, 1\}^{2M}$, i.e. the set of all zero-one-sequences of length $2M$. The latter set may be interpreted as the set of vertices of the unit cube in $\mathbb{R}^{2M}$. Moreover, the possible transitions which all have equal probability $(2M)^{-1}$ uniquely correspond to the edges of that cube. It is easy to see that the transition matrix P of this Markov chain is symmetric and that $\pi = \pi P$ where π denotes the uniform distribution on I, i.e. $\pi_i = 2^{-2M}$ for all i. Therefore the (π, P)-Markov chain $(\xi_n)_{n \in \mathbb{N}_0}$ completely looks the same when the direction of time is reversed; such chains are usually called *reversible*.

Despite its simplicity the microscopic point of view adopted so far may be of minor interest in many physical applications: after all, the actual value of M is usually very large, and recording the fate of each individual molecule is impossible in practice. Consequently, a macroscopic model considers the molecules to be indistinguishable and merely deals with the total number of these objects within the two half-chambers. In order to put this more mathematically, let us define a map $\rho : I \to \tilde{I} := \{0, 1, \ldots, 2M\}$ according to $\rho(x_1, \ldots, x_{2M}) := \sum_{i=1}^{2M} x_i$. The macroscopic process $(\tilde{\xi}_n)_{n \in \mathbb{N}_0}$ on $\tilde{I}$ is given by $\tilde{\xi}_n := \rho(\xi_n)$ for all $n \in \mathbb{N}_0$. Is $(\tilde{\xi}_n)_{n \in \mathbb{N}_0}$ a Markov chain? And if so, what does its transition matrix $\tilde{P}$ look like?

A simple calculation shows that

$$\tilde{p}_{ij} := \mathbb{P}(\tilde{\xi}_{n+1} = j | \tilde{\xi}_n = i) = \begin{cases} 1 - \frac{i}{2M} & \text{if } j = i+1, \\ \frac{i}{2M} & \text{if } j = i-1, \\ 0 & \text{otherwise}, \end{cases}$$

not depending on n. Therefore $(\tilde{\xi}_n)_{n \in \mathbb{N}_0}$ is a $(\tilde{\pi}, \tilde{P})$-Markov chain with

$$\tilde{\pi}_i := \sum_{j:\rho(j)=i} \pi_j = 2^{-2M} \binom{2M}{i} \quad \text{for all } i \in \{0, 1, \ldots, 2M\}.$$

Although this chain is reversible too, there should be less symmetry than in the microscopic case. If the macroscopic model bears any resemblance to reality we would expect the time $T_{0 \to M}$ of equalization (starting with the completely asymmetric configuration of one empty and one full half-chamber) to be significantly smaller than the time $T_{M \to \{0, 2M\}}$ of total separation (starting with the equalized configuration of two equally filled half-chambers; cf. Figure 4.17). In addition, we intuitively feel that equalized situations ($\tilde{\xi}_n \approx M$) should occur much more often than extreme ones ($|\tilde{\xi}_n - M| \approx M$). Later we shall see that the macroscopic chain definitely exhibits the behaviour we expect.

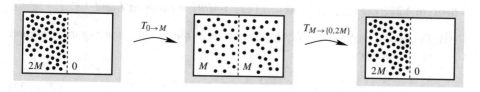

Figure 4.17. We expect $T_{0 \to M} \ll T_{M \to \{0, 2M\}}$ for large M (see also Figure 3.6).

4.4.1 Class structure, absorption probabilities, and hitting times

Let $(\xi_n)_{n \in \mathbb{N}_0}$ denote a (λ, P)-Markov chain on I and take $i, j \in I$. We say that i *leads to* j (symbolized by $i \to j$) if $(P^n)_{ij} > 0$ for some $n \in \mathbb{N}_0$. We call i and j *communicating* (writing $i \leftrightarrow j$) if both $i \to j$ and $j \to i$. The relation $\leftrightarrow$ is easily seen to be an equivalence relation and thus partitions I into *classes of communicating states*. A chain $(\xi_n)_{n \in \mathbb{N}_0}$ is called *irreducible* if there is only one such class. Clearly $(\xi_n)_{n \in \mathbb{N}_0}$ is irreducible if and only if the corresponding transition matrix P is irreducible. Irreducibility may be considered a stochastic property of non-decomposability. Recall that we have got to know such properties in other areas before, e.g. *ergodicity* and *topological transitivity*. Theorem 4.17 suggests that these notions may somehow be related. We shall discuss this topic at the end of the present chapter.

The individual communicating classes may be ordered according to the following rule: for two classes $C, D \subseteq I$ define

$$C \succ D \quad \text{if and only if } i \to j \text{ for some } i \in C, j \in D.$$

The minimal elements of this partial ordering are called *closed* (or *absorbing*) classes: once entered, such a class will never be left again. Observe that in the case of a finite state space I there is at least one absorbing class. For example, our simple model of tennis is made up of 15 classes, two of which ("Game A" and "Game B") are closed (cf. Figure 4.15). The only class consisting of more than one state is {"40:30", "30:30", "30:40"}.

If there are several closed classes, what is the probability of eventually reaching one particular class? More generally, given a set $J \subseteq I$, what is the probability of eventually hitting J? And how long will it take us to get to J? In order to answer these questions consider the hitting time of J,

$$H_J := \begin{cases} \min\{n \in \mathbb{N}_0 : \xi_n \in J\} & \text{if } \{n : \xi_n \in J\} \neq \emptyset, \\ \infty & \text{otherwise}, \end{cases}$$

which is a random variable with values in $\mathbb{N}_0 \cup \{\infty\}$. We are interested in the quantities

$$x_i^J := \mathbb{P}(H_J < \infty \,|\, \xi_0 = i) \quad \text{and} \quad y_i^J := \mathbb{E}(H_J \,|\, \xi_0 = i),$$

which are the probability of ever hitting J and the mean hitting time of J, respectively, when starting at i.

Theorem 4.21. *Let* $(\xi_n)_{n\in\mathbb{N}_0}$ *denote a* (λ, P)-*Markov chain on* I *and* $J \subseteq I$.

(i) *The hitting probabilities* $(x_i^J)_{i\in I}$ *are given as the minimal non-negative solution of the linear equations*

$$
\begin{aligned}
x_i &= 1 && \text{if } i \in J\,, \\
x_i &= \textstyle\sum_{j\in I} p_{ij} x_j && \text{if } i \notin J\,.
\end{aligned}
\qquad (4.22)
$$

(ii) *The mean hitting times* $(y_i^J)_{i\in I}$ *provide the minimal non-negative solution to the system*

$$
\begin{aligned}
y_i &= 0 && \text{if } i \in J\,, \\
y_i &= 1 + \textstyle\sum_{j\in I} p_{ij} y_j && \text{if } i \notin J\,.
\end{aligned}
\qquad (4.23)
$$

Proof. In order to prove (i) we first show that $(x_i^J)_{i\in I}$ satisfies the given system of equations. If $i \in J$ clearly $x_i^J = 1$, and if $i \notin J$ we have

$$
x_i^J = \sum_{n\in\mathbb{N}} \mathbb{P}(H_J = n | \xi_0 = i) = \sum_{j\in J} p_{ij} + \sum_{n\geq 2} \sum_{j\in I\setminus J} \mathbb{P}(H_J = n,\, \xi_1 = j | \xi_0 = i)
$$

$$
= \sum_{j\in J} p_{ij} + \sum_{j\in I\setminus J} p_{ij} x_j^J = \sum_{j\in I} p_{ij} x_j^J\,.
$$

Assume in turn that the non-negative vector $(x_i)_{i\in I}$ satisfies (4.22). Then for $i \notin J$

$$
x_i = \sum_{j\in J} p_{ij} + \sum_{j\notin J, k\in J} p_{ij} p_{jk} + \sum_{j\notin J, k\notin J} p_{ij} p_{jk} x_k \geq \mathbb{P}(H_J = 1 | \xi_0 = i) + \mathbb{P}(H_J = 2 | \xi_0 = i)\,.
$$

By induction $x_i \geq \mathbb{P}(H_J \leq n | \xi_0 = i)$ for any $n \in \mathbb{N}$; hence $x_i \geq x_i^J$ which proves (i).

The proof of (ii) proceeds along the same lines. Since $x_i^J < 1$ implies $y_i^J = \infty$ we may assume $x_i^J = 1$ for $i \in I\setminus J$ without loss of generality. Then $y_i^J = 0$ if $i \in J$ and

$$
y_i^J = \sum_{n\in\mathbb{N}} n\mathbb{P}(H_J = n | \xi_0 = i) = \sum_{j\in J} p_{ij} + \sum_{n\geq 2} \sum_{j\in I\setminus J} n\mathbb{P}(H_J = n,\, \xi_1 = j | \xi_0 = i)
$$

$$
= \sum_{j\in J} p_{ij} + \sum_{n\geq 1} \sum_{j\in I\setminus J} (n+1)\mathbb{P}(H_J = n | \xi_0 = j) p_{ij} = 1 + \sum_{j\notin J} p_{ij} y_j^J
$$

if $i \notin J$. Assume in turn that the non-negative row $(y_i)_{i\in I}$ satisfies (4.23). Then for $i \notin J$

$$
y_i = 1 + \sum_{j\notin J} p_{ij} + \sum_{j\notin J, k\notin J} p_{ij} p_{jk} y_k \geq \mathbb{P}(H_J \geq 1 | \xi_0 = i) + \mathbb{P}(H_J \geq 2 | \xi_0 = i)\,.
$$

Again, an induction argument yields $y_i \geq \sum_{l=1}^{n} \mathbb{P}(H_J \geq l | \xi_0 = i)$ for all $n \in \mathbb{N}$ and therefore $y_i \geq \sum_{l=1}^{\infty} \mathbb{P}(H_J \geq l | \xi_0 = i) = \mathbb{E}(H_J | \xi_0 = i)$; in particular, $y_i = \infty = y_i^J$ if $x_i^J < 1$. $\square$

The above theorem makes possible not only a verification of the results (4.21) concerning our simple Markovian model of tennis, but also an analysis of the game

Snakes and Ladders mentioned in the Introduction. As another simple application of Theorem 4.22 let us consider the following problem: a gambler enters a casino with a fortune of $c \, €$. Gambling $1 \, €$ at a time he (or she) either doubles or loses his (her) stake with probabilities p and q, respectively. Assuming the reserves of the casino to be infinite, what is the probability $\Phi(c)$ that the gambler will end up broke? If we describe the situation by means of a Markov chain $(\xi_n)_{n \in \mathbb{N}_0}$ on $\mathbb{N}_0$ with ξ_n denoting the gambler's fortune after the n-th game, we have

$$p_{ij} := \begin{cases} p & \text{if } j = i + 1 \text{ and } i \geq 1, \\ q & \text{if } j = i - 1 \text{ and } i \geq 1, \\ 1 & \text{if } i = j = 0, \\ 0 & \text{otherwise}, \end{cases}$$

and $\Phi(c) = \mathbb{P}(H_{\{0\}} < \infty \mid \xi_0 = c)$. Setting $x_i := \mathbb{P}(H_{\{0\}} < \infty \mid \xi_0 = i)$ we have to solve the system $x_0 = 1$, $x_i = q x_{i-1} + p x_{i+1}$ for $i \geq 1$, the general solution of which is given by

$$x_i = \begin{cases} 1 + \gamma \left(\left(\frac{q}{p} \right)^i - 1 \right) & \text{if } p \neq q, \\ 1 + \gamma i & \text{if } p = q, \end{cases}$$

with an arbitrary constant γ. For $p \leq q$ (which is the case in most thriving casinos) the uniform restriction $0 \leq x_i \leq 1$ implies $\gamma = 0$; if $p > q$ we find $\gamma = 1$ by the minimality condition. Summarizing we have

$$\Phi(c) = x_c = \begin{cases} 1 & \text{if } p \leq q, \\ \left(\frac{q}{p} \right)^c & \text{if } p > q. \end{cases}$$

So even if the casino is fair (i.e. $p = q = \frac{1}{2}$) the gambler will surely end up broke. This rather surprising result is often referred to as the *gambler's ruin paradox*. To tell the truth, this effect is not altogether paradoxical: as our calculations show it is merely due to the infinite reserves of the casino. In addition, there is a cold comfort. Setting $\Psi(c) := \mathbb{E}(H_{\{0\}} \mid \xi_0 = c)$ we obtain from Theorem 4.22 $\Psi(c) = c(q - p)^{-1}$ if $p < q$. Despite the certainty of its end the game is therefore likely to last quite a long time if the casino is nearly fair.

Another example of hitting times is provided by the macroscopic Ehrenfest model of diffusion. Recall that the latter gave rise to a Markov chain $(\xi_n)_{n \in \mathbb{N}_0}$ on $\{0, 1, \ldots, 2M\}$ with

$$p_{ij} = \begin{cases} 1 - \frac{i}{2M} & \text{if } j = i + 1, \\ \frac{i}{2M} & \text{if } j = i - 1, \\ 0 & \text{otherwise}. \end{cases}$$

By means of Theorem 4.21 we can now estimate the behaviour of the expected times of equalization and separation $T_{0 \to M}$ and $T_{M \to \{0,2M\}}$, respectively. The mean hitting times $r_i := \mathbb{E}(H_{\{0,2M\}} \mid \xi_0 = i)$ satisfy

$$r_0 = r_{2M} = 0, \quad r_i = 1 + p_{i,i-1} r_{i-1} + p_{i,i+1} r_{i+1} \quad \text{for } i = 1, \ldots, 2M - 1.$$

The solution of this recursion may be explicitly written down: evidently we have $T_{M \to \{0,2M\}} = r_M$ and therefore

$$T_{M \to \{0,2M\}} = \sum_{i=1}^{M} \left(\frac{(M-i)!\,(M+i-1)!}{M!\,(M-1)!} + 2M \sum_{j=1}^{i-1} \frac{(M-i)!\,(M+i-1)!}{(M+j)!\,(M-j)!} \right).$$

Having at one's disposal this formula it is a cumbersome but elementary task to show ([33]) that

$$\lim_{M \to \infty} \frac{T_{M \to \{0,2M\}}}{2^{2M-1}} = 1.$$

Hence the mean time of total separation grows exponentially with M. Taking into account typical values of M (e.g. $M \approx 2.430 \cdot 10^{22}$ for one cm^3 of CO_2 at 298 K temperature and 100 kPa pressure) we conclude that such a total separation is *extremely* unlikely on short time scales. (Recall our discussion of this topic within the scope of the Poincaré Recurrence Theorem.)

The mean equalization time $T_{0 \to M}$ can be calculated analogously. Setting $s_i :=$ $\mathbb{E}(H_{\{M\}} | \xi_0 = i)$ we have

$$s_0 = 1 + s_1, \quad s_M = 0, \quad s_i = 1 + p_{i,i-1}s_{i-1} + p_{i,i+1}s_{i+1} \text{ for } i = 1, \ldots, M-1$$

which by virtue of a few straightforward manipulations yields

$$T_{0 \to M} = s_0 = 2M \sum_{i=1}^{M} \sum_{j=0}^{i-1} \frac{(i-1)!\,(2M-i)!}{j!\,(2M-j)!}.$$

Again, a certain endurance is required in order to deduce from the latter formula the asymptotic estimates

$$\underline{\lim}_{M \to \infty} \frac{T_{0 \to M}}{M} \geq 2, \quad \overline{\lim}_{M \to \infty} \frac{T_{0 \to M}}{M \log M} \leq 1.$$

The mean equalization time $T_{0 \to M}$ therefore does not grow faster than $M \log M$ and thus is *much* smaller than the mean separation time $T_{M \to \{0,2M\}}$ already for moderate numbers of molecules.

4.4.2   Recurrence and transience: dynamical classification of states

We are now going to investigate the dynamical properties of individual states. Since the state space I is assumed to be discrete, these properties have to be formulated in terms of the stochastic structure alone. For every state i we introduce the *first passage time* to i according to

$$T_i := \begin{cases} \min\{n \geq 1 : \xi_n = i\} & \text{if } \{n \geq 1 : \xi_n = i\} \neq \emptyset, \\ \infty & \text{otherwise}. \end{cases}$$

Definiton 4.22. A state i is called *recurrent* for the Markov chain $(\xi_n)_{n \in \mathbb{N}_0}$ if the chain is sure to come back to i when started at i, i.e. $\mathbb{P}(T_i < \infty | \xi_0 = i) = 1$; otherwise the state is called *transient*.

Recall that we have repeatedly observed phenomena of recurrence when analysing dynamical systems (e.g. Theorem 3.2 and Exercise 2.7). In the completely stochastic setting considered here, the notion of recurrence looks different though we shall soon find some striking resemblances.

It may easily be seen that a state is recurrent if and only if it is visited (with probability one) infinitely often by the chain. The question arises how this property can be detected in practice.

Theorem 4.23. *Let $(\xi_n)_{n \in \mathbb{N}_0}$ denote a (λ, P)-Markov chain.*

(i) *A state i is recurrent if and only if $\sum_{n=1}^{\infty} (P^n)_{ii}$ diverges;*

(ii) *If i is recurrent and $i \leftrightarrow j$ then j also is recurrent (recurrence thus is a class property);*

(iii) *Every recurrent class is closed; for a finite class the converse is also true.*

Proof. By introducing the quantities $x_i^{(n)} := \mathbb{P}(T_i = n | \xi_0 = i)$ we obtain the relation

$$(P^n)_{ii} = \mathbb{P}(\xi_n = i | \xi_0 = i) = \sum_{k=1}^{n} \mathbb{P}(\xi_n = i, \, T_i = k | \xi_0 = i) = \sum_{k=1}^{n} (P^{n-k})_{ii} x_i^{(k)}.$$

As an auxiliary device let us define the power series (usually referred to as *generating functions*)

$$X_i(z) := \sum_{n=1}^{\infty} x_i^{(n)} z^n, \quad \Pi_i(z) := \sum_{n=0}^{\infty} (P^n)_{ii} z^n \quad (i \in I),$$

the radii of convergence of which are at least one. The above relation then reads $\Pi_i(z) = 1 + X_i(z)\Pi_i(z)$ from which (i) follows. In order to prove (ii) assume that i is recurrent and $i \leftrightarrow j$. Then for some $n_1, n_2 \in \mathbb{N}$ we have $(P^{n_1})_{ji} > 0$ and $(P^{n_2})_{ij} > 0$. Consequently $(P^{n_1 + n + n_2})_{jj} \geq (P^{n_1})_{ji} (P^n)_{ii} (P^{n_2})_{ij}$ for all $n \in \mathbb{N}$ implying that j is recurrent. If the class containing the recurrent state i were not closed and $i \to k$ but $k \nrightarrow i$ then $(P^{n_3})_{ik} > 0$ for some $n_3 \in \mathbb{N}$ and $\mathbb{P}(T_i < \infty | \xi_0 = i) \leq \mathbb{P}(\xi_{n_3} \neq k | \xi_0 = i) = 1 - (P^{n_3})_{ik} < 1$ contradicting the recurrence of i. Finally every finite closed class is recurrent because at least one (and hence any) state is sure to be visited infinitely often. $\square$

Deciding on recurrence and transience thus is a non-trivial task for infinite closed classes only. As an example thereof we shall consider the simple random walk on $\mathbb{Z}$ according to

$$p_{ij} = \begin{cases} p & \text{if } j = i+1, \\ q & \text{if } j = i-1, \\ 0 & \text{otherwise,} \end{cases} \quad \text{with } 0 < p, q, < 1 \text{ and } p + q = 1.$$

Since the resulting Markov chain is irreducible it suffices to find out whether the state 0 is recurrent. In order to travel from 0 back to 0 in $2n$ steps we have to perform exactly n steps to the left and n steps to the right. Therefore

$$(P^{2n})_{00} = \binom{2n}{n} p^n q^n, \quad (P^{2n+1})_{00} = 0 \quad \text{for all } n \in \mathbb{N}_0,$$

and the corresponding generating function Π_0 may be calculated explicitly,

$$\Pi_0(z) = \sum_{n=0}^{\infty} \binom{2n}{n} p^n q^n z^{2n} = \frac{1}{\sqrt{1 - 4pqz^2}}.$$

From this we deduce $X_0(z) = 1 - \sqrt{1 - 4pqz^2}$ and thus for the probability of ever returning $\mathbb{P}(T_0 < \infty | \xi_0 = 0) = X_0(1) = 2 \min\{p, 1 - p\}$ which equals one only in the symmetric case ($p = q = \frac{1}{2}$). But even in this case the expected first passage time from 0 back to 0 is infinite:

$$\mathbb{E}(T_0 | \xi_0 = 0) = \sum_{n=1}^{\infty} 2n \mathbb{P}(T_0 = 2n | \xi_0 = 0) = \lim_{z \to 1} X_0'(z) = \lim_{z \to 1} \frac{z}{\sqrt{1 - z^2}} = \infty.$$

This weak type of recurrence is termed *null recurrence*. On the other hand, a state i is called *positive recurrent* if the stronger relation $\mathbb{E}(T_i | \xi_0 = i) < \infty$ holds. Summarizing our observations we may say that the states of a simple random walk on $\mathbb{Z}$ are transient if $p \neq q$ and null-recurrent if $p = q = \frac{1}{2}$. (Null and positive recurrence turn out to be class properties also, see [41].)

Let us finally generalize the above result to simple random walks on the lattice $\mathbb{Z}^d$. We assume that at each step the chain $(\xi_n)_{n \in \mathbb{N}_0}$ is allowed to move from $s \in \mathbb{Z}^d$ to $t \in \mathbb{Z}^d$ only if $\sum_{i=1}^{d} |s_i - t_i| = 1$. In the latter case we have

$$p_{s,t} = \begin{cases} p_i & \text{if } t_i = s_i + 1, \\ q_i & \text{if } t_i = s_i - 1, \end{cases}$$

with $\sum_{i=1}^{d} (p_i + q_i) = 1$; otherwise $p_{s,t} = 0$. By considering the one-dimensional chain which only moves when the projection of $(\xi_n)_{n \in \mathbb{N}_0}$ onto the i-th coordinate moves we see that in case of recurrence necessarily $p_i = q_i$ for all $i \in \{1, \ldots, d\}$. Clearly $(P^{2n+1})_{0,0} = 0$ and

$$(P^{2n})_{0,0} = \sum_{\substack{i_1 + \ldots + i_d = n \\ i_1, \ldots, i_d \geq 0}} \frac{(2n)!}{(i_1! \ldots i_d!)^2} p_1^{2i_1} \ldots p_d^{2i_d}$$

$$= 2^{-2n} \binom{2n}{n} \sum_{\substack{i_1 + \ldots + i_d = n \\ i_1, \ldots, i_d \geq 0}} \left(\frac{n!}{i_1! \ldots i_d!} \right)^2 (2p_1)^{2i_1} \ldots (2p_d)^{2i_d}.$$

By virtue of the elementary inequality ([55])

$$\max_{\substack{i_1+\ldots+i_d=n \\ i_1,\ldots,i_d\geq 0}} \frac{n!}{i_1!\ldots i_d!}(2p_1)^{i_1}\ldots(2p_d)^{i_d} \leq \frac{c_d(p_1,\ldots,p_d)}{(\pi n)^{\frac{d-1}{2}}}$$

with the quantity $c_d(p_1,\ldots,p_d)$ not depending on n, we can estimate the asymptotic behaviour of $(P^{2n})_{0,0}$ as

$$(P^{2n})_{0,0} \leq 2^{-2n}\binom{2n}{n}\frac{c_d(p_1,\ldots,p_d)}{(\pi n)^{\frac{d-1}{2}}} = \frac{c_d(p_1,\ldots,p_d)}{(\pi n)^{\frac{d}{2}}}(1+\varepsilon_n)$$

with $\lim_{n\to\infty}\varepsilon_n = 0$. If $d \geq 3$ we have $\sum_{n=0}^{\infty}(P^n)_{0,0} < \infty$ and therefore the origin (as any other state) is transient. The remaining case $d = 2$ can be dealt with similarly ([55]), but there is a simple alternative argument: the two one-dimensional chains $\eta_n^+ := (\xi_n)_1 + (\xi_n)_2$ and $\eta_n^- := (\xi_n)_1 - (\xi_n)_2$ are independent, and

$$\mathbb{P}(\xi_{2n} = 0|\xi_0 = 0) = \mathbb{P}(\eta_{2n}^+ = 0|\eta_0^+ = 0)\mathbb{P}(\eta_{2n}^- = 0|\eta_0^- = 0)$$

$$= \binom{2n}{n}^2(p_1+p_2)^{4n} = \frac{1+\varepsilon_n}{\pi n}$$

with $\lim_{n\to\infty}\varepsilon_n = 0$ provided that $p_1 = q_1$, $p_2 = q_2$. The simple symmetric random walk on $\mathbb{Z}^2$ therefore is recurrent ("all roads lead to Rome").

4.4.3 The long-time behaviour of Markov chains

The concept of invariant probability measures which has been an important element of our earlier studies has a stochastic counterpart in the Markovian context.

Definiton 4.24. Let $(\xi_n)_{n\in\mathbb{N}_0}$ be a (λ, P)-Markov chain. The distribution $(\pi_i)_{i\in I}$ is said to be *invariant* if $\pi P = \pi$, which means that $\sum_{j\in J}\pi_j p_{ji} = \pi_i$ for all $i \in I$. More generally, any non-negative row vector $(\mu_i)_{i\in I}$ with $\mu P = \mu$ is called an *invariant measure*.

For finite I a non-zero invariant measure can always be normalized to give an invariant distribution. For infinite I it might happen that $\sum_{i\in I}\mu_i$ diverges for any non-trivial invariant measure μ in which case things may become more complicated. However, we again expect invariant measures and distributions to tell us something about the chain's behaviour in the long run. Before confirming this expectation, we should therefore ask for conditions guaranteeing the existence of such precious objects. Since invariant distributions are necessarily supported on closed classes and any two of them may be combined in a convex sum we have to focus on irreducible chains.

Lemma 4.25. *Let $(\xi_n)_{n\in\mathbb{N}_0}$ be an irreducible Markov chain and μ, ν two invariant measures. If one (and hence any) state is recurrent then there exists $c \geq 0$ such that $\nu_i = c\mu_i$ for all $i \in I$.*

Proof. Fix a recurrent state k and denote by γ_i^k the mean number of visits to i between two successive visits to k, that is

$$\gamma_i^k := \mathbb{E}\Big(\sum_{n=0}^{T_k-1} \mathbf{1}_{\{i\}}(\xi_n)|\xi_0 = k\Big).$$

An elementary calculation confirms that $\gamma^k = (\gamma_i^k)_{i\in I}$ constitutes an invariant measure. Indeed, since $\mathbb{P}(T_k < \infty|\xi_0 = k) = 1$ we find

$$\gamma_i^k = \mathbb{E}\Big(\sum_{n=1}^{T_k} \mathbf{1}_{\{i\}}(\xi_n)|\xi_0 = k\Big) = \sum_{n=1}^{\infty} \mathbb{P}(\xi_n = i, T_k \geq n|\xi_0 = k)$$

$$= \sum_{j\in I}\sum_{n=1}^{\infty} \mathbb{P}(\xi_n = i, \xi_{n-1} = j, T_k \geq n|\xi_0 = k)$$

$$= \sum_{j\in I} p_{ji} \sum_{n=1}^{\infty} \mathbb{P}(\xi_{n-1} = j, T_k \geq n|\xi_0 = k)$$

$$= \sum_{j\in I} p_{ji}\mathbb{E}\Big(\sum_{n=0}^{T_k-1} \mathbf{1}_{\{j\}}(\xi_n)|\xi_0 = k\Big) = \sum_{j\in I} \gamma_j^k p_{ji},$$

where we have also used the fact that the event $\{T_k \geq n\}$ solely depends on the outcomes of $\xi_1, \ldots, \xi_{n-1}$. Furthermore $\gamma_k^k = 1$ and, by irreducibility $\gamma_i^k \geq (P^{n_1})_{ki} > 0$ as well as $1 \geq \gamma_i^k(P^{n_2})_{ik} > 0$ for some $n_1, n_2 \in \mathbb{N}$. Hence $0 < \gamma_i^k < \infty$ for all $i \in I$.

Assume now that $\mu = (\mu_i)_{i\in I}$ is invariant and $\mu_k > 0$. Then

$$\mu_i = \sum_{j\neq k} \mu_j p_{ji} + \mu_k p_{ki} = \sum_{j\neq k, l\neq k} \mu_l p_{lj} p_{ji} + \mu_k \sum_{j\neq k} p_{kj} p_{ji} + \mu_k p_{ki}$$

$$\geq \mu_k\big(\mathbb{P}(\xi_2 = i, T_k \geq 2|\xi_0 = k) + \mathbb{P}(\xi_1 = i, T_k \geq 1|\xi_0 = k)\big)$$

and thus, by virtue of an induction argument

$$\mu_i \geq \mu_k \sum_{n=1}^{\infty} \mathbb{P}(\xi_n = i, T_k \geq n|\xi_0 = k) = \mu_k \gamma_i^k$$

for all $i \in I$. Therefore $\mu \geq \mu_k \gamma^k$, and $\tilde{\mu} := \mu - \mu_k \gamma^k \geq 0$ is also invariant. Given $i \in I$ we observe

$$0 = \tilde{\mu}_k = \sum_{j\in I} \tilde{\mu}_j(P^{n_2})_{jk} \geq \tilde{\mu}_i(P^{n_2})_{ik} \geq 0.$$

Consequently $\tilde{\mu} \equiv 0$, and the invariant measure μ is a multiple of γ^k. This clearly proves the lemma. $\square$

Lemma 4.25 may be rephrased by saying that any two invariant measures for a recurrent irreducible Markov chain are multiples of each other. Moreover, the proof shows that there always exists an invariant measure in this case. On the other hand it is easily seen that the assumptions must not be weakened in order to achieve the desired uniqueness result (see Exercise 4.14).

We are now in a position to precisely say under which conditions an invariant *distribution* exists.

Theorem 4.26. *Let $(\xi_n)_{n\in\mathbb{N}_0}$ denote an irreducible Markov chain on I. The following statements are equivalent:*

(i) *There exists an invariant distribution $(\pi_i)_{i\in I}$;*

(ii) *Every state is positive recurrent.*

Moreover, if (i) *or* (ii) *holds then $\pi_i = \mathbb{E}(T_i|\xi_0 = i)^{-1}$ for all $i \in I$.*

Proof. As a preliminary observation we note that with the notion from the preceding proof

$$\sum_{i\in I} \gamma_i^k = \sum_{i\in I}\sum_{n\in\mathbb{N}} \mathbb{P}(\xi_n = i, T_k \geq n|\xi_0 = k) = \mathbb{E}(T_k|\xi_0 = k).$$

Let now $(\pi_i)_{i\in I}$ denote an invariant distribution. Clearly $\pi_k > 0$ for some k, and $\pi_i \geq \pi_k\gamma_i^k$ for all $i \in I$ as in the proof of Lemma 4.25. But then

$$\mathbb{E}(T_k|\xi_0 = k) = \sum_{i\in I} \gamma_i^k \leq \sum_{i\in I} \frac{\pi_i}{\pi_k} = \frac{1}{\pi_k} < \infty,$$

so k (as any other state) is positive recurrent. Assume in turn that our chain is positive recurrent. Then $(\gamma_i^k)_{i\in I}$ is invariant and can be normalized to give a distribution

$$\pi_i := \frac{\gamma_i^k}{\sum_{j\in I} \gamma_j^k} = \frac{\gamma_i^k}{\mathbb{E}(T_k|\xi_0 = k)} \quad (i \in I),$$

where evidently $\pi_k = \mathbb{E}(T_k|\xi_0 = k)^{-1}$. Since k has been arbitrary the proof is complete. $\square$

As a corollary we note that the invariant distribution of an irreducible Markov chain is uniquely determined if it exists at all. Furthermore one should notice the striking resemblance of Theorem 4.26 to Kac's lemma. Although we are now in a completely stochastic setting the familiar relation between invariant distributions and mean return times persists.

To better put these analogies in perspective let us assume for the moment that $I = \{1, \ldots, l\}$ and that $(\xi_n)_{n\in\mathbb{N}_0}$ is a (λ, P)-Markov chain on I with invariant distribution $(\pi_i)_{i\in I}$. Now recall the σ_A-invariant measures $\mu_{\pi,P}$ on $\Sigma_{l,A}$ constructed earlier. If we define $A := (a_{ij}) \in \{0, 1\}^{l\times l}$ with $a_{ij} = 1$ precisely if $p_{ij} > 0$ then $\Sigma_{l,A}$ has an obvious interpretation: it is just the set of all possible stochastic histories which the chain $(\xi_n)_{n\in\mathbb{N}_0}$ may undergo. If $(\xi_n)_{n\in\mathbb{N}_0}$ is irreducible then so is the matrix A, and hence σ_A is ergodic by virtue of Theorem 4.17. Let us fix a state $i \in I$ and take a specific history $x = (\xi_k)_{k\in\mathbb{N}_0} \in \Sigma_{l,A}$. Then

$$t_{[i]}(x) = \min\{n : \sigma_A^n(x) \in [i]\} = \min\{n \in \mathbb{N} : \xi_n = i\} = T_i,$$

and Kac's Lemma 3.10 immediately yields

$$1 = \int_{[i]} t_{[i]} \, d\mu_{\pi,P} = \pi_i \mathbb{E}(T_i | \xi_0 = i).$$

When we look at it in this way the content of Theorem 4.26 is hardly surprising!

For another utilization of our earlier tools we again fix a state $i \in I$ together with the corresponding density $f_i := \pi_i^{-1} \mathbf{1}_{[i]} \in L^1(\mu_{\pi,P})$. A short calculation supplies us with a neat expression for the Frobenius–Perron operator,

$$P_{\sigma_A}^n(f_i) = \sum_{i_1,\ldots,i_n \in I} p_{ii_1} \cdots p_{i_{n-1}i_n} f_{i_n} = \sum_{j \in I} \frac{\mathbb{P}(\xi_n = j | \xi_0 = i)}{\pi_j} \mathbf{1}_{[j]}.$$

If the transition matrix P is aperiodic then so is A, and by Theorem 4.17 σ_A is exact. This in turn implies that $P_{\sigma_A}^n(f_i) \to \mathbf{1}$ in $L^1(\mu_{\pi,P})$ from which it is straightforward to deduce that

$$\lim_{n\to\infty} \mathbb{P}(\xi_n = j | \xi_0 = i) = \lim_{n\to\infty} \mathbb{P}(\xi_n = j) = \pi_j \qquad (4.24)$$

for all $i, j \in I$. Again, a purely stochastic fact has been established without invoking probabilistic arguments.

Some care has to be taken in proving (4.24) for *infinite* I. To this purpose let us call a state i *aperiodic* for an irreducible Markov chain if $(P^n)_{ii} > 0$ for all sufficiently large n. The careful reader may wish to check that aperiodicity is a class property which, in the finite case, coincides with the aperiodicity of the transition matrix P. In the light of the above discussion the following theorem which establishes (4.24) for arbitrary I is barely surprising. However, the argument now is completely probabilistic in nature (and could in turn give a new, independent proof of Theorem 4.17).

Theorem 4.27. *Let $(\xi_n)_{n\in\mathbb{N}_0}$ denote an irreducible (λ, P)-Markov chain with invariant distribution $(\pi_i)_{i\in I}$. If one (and hence any) state is aperiodic then for all $i, j \in I$*

$$\lim_{n\to\infty} \mathbb{P}(\xi_n = j | \xi_0 = i) = \lim_{n\to\infty} \mathbb{P}(\xi_n = j) = \pi_j.$$

Proof. We apply a so-called *coupling* argument. Let $(\eta_n)_{n\in\mathbb{N}_0}$ denote a (π, P)-Markov chain on I which is independent of $(\xi_n)_{n\in\mathbb{N}_0}$. The compound $(\xi_n, \eta_n)_{n\in\mathbb{N}_0}$ is easily seen to be an irreducible $(\lambda \otimes \pi, \tilde{P})$-Markov chain on $I \times I$ with $\tilde{P}_{(i,j),(k,l)} = p_{ik}p_{jl}$ for all pairs $(i, j), (k, l) \in I \times I$. Furthermore, it has $\pi \otimes \pi$ as an invariant distribution and hence is (positive) recurrent. Fix a state $i_0 \in I$ and define the coupling time T as

$$T := \min\{n \in \mathbb{N} : (\xi_n, \eta_n) = (i_0, i_0)\}$$

with the usual convention that $T = \infty$ if the set on the right is empty. Due to recurrence we have $\mathbb{P}(T < \infty) = 1$. Since $\xi_T = \eta_T = i_0$ and the further evolution does not depend on the past once the two chains have met in i_0 we have

$$\mathbb{P}(\xi_n = j, T \le n | \xi_0 = i) = \mathbb{P}(\eta_n = j, T \le n | \xi_0 = i) \qquad \text{for all } i, j \in I, n \in \mathbb{N}.$$

But then

$$\left|\mathbb{P}(\xi_n = j|\xi_0 = i) - \pi_j\right| = \left|\mathbb{P}(\xi_n = j|\xi_0 = i) - \mathbb{P}(\eta_n = j|\xi_0 = i)\right|$$

$$= \left|\mathbb{P}(\xi_n = j, \ T > n|\xi_0 = i) - \mathbb{P}(\eta_n = j, \ T > n|\xi_0 = i)\right| \le \mathbb{P}(T > n|\xi_0 = i) \to 0$$

as $n \to \infty$. For all $i, j \in I$ therefore $\lim_{n\to\infty} \mathbb{P}(\xi_n = j|\xi_0 = i) = \pi_j$ and, by dominated convergence, also $\lim_{n\to\infty} \mathbb{P}(\xi_n = j) = \pi_j$. $\square$

Before further discussing the resemblances between deterministic and stochastic notions let us illustrate the above theorem. The following urn model may be considered a slight modification of the Ehrenfest model of diffusion discussed earlier. Assume that M white and M black balls are distributed to two urns, each containing M balls of random colour. At any step one ball is independently chosen at random from each urn, and the chosen balls are interchanged. (One could think of two species of flea which live on a pair of dogs sleeping side by side.) Denoting by ξ_n the number of black balls in one fixed urn after the n-th interchange we have

$$p_{i,i-1} = \left(\frac{i}{M}\right)^2, \quad p_{i,i} = 2\frac{i}{M}\left(1 - \frac{i}{M}\right), \quad p_{i,i+1} = \left(1 - \frac{i}{M}\right)^2, \qquad (4.25)$$

as well as $p_{ij} = 0$ if $|i - j| \ge 2$. Clearly the chain $(\xi_n)_{n\in\mathbb{N}_0}$ is aperiodic with the unique invariant distribution (calculated from $\pi P = \pi$)

$$\pi_i = \frac{(M!)^2}{(2M)!}\left(\binom{M}{i}\right)^2 \quad (i = 0, \dots, M).$$

(It is worth noting that π is in fact a hypergeometric distribution which – in a certain sense – maximizes the uncertainty in the present model. The fascinating topic of characterizing equilibrium states by means of variational principles is thoroughly treated in [32].) No matter which initial distribution λ is chosen, λP^n always converges to π rather quickly (in fact exponentially, see Figure 4.18). Intuitively speaking, π resembles the symmetric binomial distribution $B_{M,\frac{1}{2}}$ with parameters M and $\frac{1}{2}$. Indeed,

$$\mathbb{E}\pi = \mathbb{E}B_{M,\frac{1}{2}} = \frac{M}{2} \quad \text{but} \quad \text{Var}\,\pi = \frac{M^2}{4(2M-1)} \approx \frac{M}{8} = \frac{1}{2}\,\text{Var}\,B_{M,\frac{1}{2}}.$$

The invariant distribution π thus concentrates around $M/2$ more heavily than $B_{M,\frac{1}{2}}$ does. For the rest of this section we shall concentrate on unifying our view on deterministic and stochastic dynamics. To this end let us first briefly discuss an ergodic property of Markov chains. Recall that studying the asymptotic behaviour of the relative frequencies $h_A(x, n)$ and the slightly more general averages $S_n f(x)$ was one of our main motivations for entering ergodic theory. Given a Markov chain $(\xi_n)_{n\in\mathbb{N}_0}$ on I and a function $f : I \to \mathbb{C}$ we can equally well assign an average to f,

$$S_n f := \frac{1}{n}\sum_{k=0}^{n-1} f(\xi_k),$$

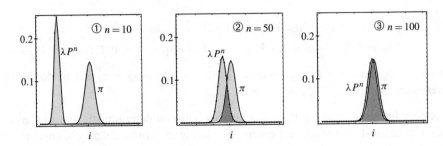

Figure 4.18. Visualizing and quantifying (below) the convergence to equilibrium for the Ehren-fest model (4.25) with $M = 60$, $\lambda = 0.2 \cdot \mathbf{1}_{\{0,1,2,3,4\}}$

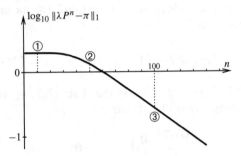

which in contrast to (3.9) is a random quantity. Despite its random nature, is there any reasonable long-time behaviour of $S_n f$? Given our discussion earlier in this section, it is not difficult to provide an affirmative answer to this question in case of a finite irre-ducible Markov chain with invariant distribution $(\pi_i)_{i \in I}$. By applying Theorem 4.17 and the Birkhoff Ergodic Theorem one finds that for any initial distribution

$$S_n f \to \sum_{i \in I} f(i) \pi_i = \int_I f \, d\pi \quad \text{with probability one}. \tag{4.26}$$

Fortunately, (4.26) holds in general, as the following theorem shows.

Theorem 4.28. *Let $(\xi_n)_{n \in \mathbb{N}_0}$ denote an irreducible Markov chain on I with invariant distribution $(\pi_i)_{i \in I}$. For any bounded function $f : I \to \mathbb{C}$*

$$\lim_{n \to \infty} S_n f = \sum_{i \in I} f(i) \pi_i$$

holds with probability one.

Proof. Again we argue in a probabilistic manner to first fix the asymptotic behaviour of $h_{\{i\}}(n) :=$ $S_n 1_{\{i\}}$. Define the r-th passage time to i, in symbols $T_i^{(r)}$, according to

$$T_i^{(0)} := 0, \quad T_i^{(1)} := T_i, \quad T_i^{(r+1)} := \min\{n \geq T_i^{(r)} + 1 : \xi_n = i\}.$$

Since $(\xi_n)_{n \in \mathbb{N}_0}$ is recurrent $\mathbb{P}(T_i^{(r)} < \infty) = 1$ for all $r \in \mathbb{N}_0$. By the Markov property the excursion times $R_i^{(r)} := T_i^{(r)} - T_i^{(r-1)}$ ($r \in \mathbb{N}_0$) form a sequence of independent, identically distributed random variables with $\mathbb{E}(R_i^{(r)}) = \mathbb{E}(T_i | \xi_0 = i)$. By virtue of the Strong Law of Large Numbers

$$\lim_{r \to \infty} \frac{R_i^{(1)} + \ldots + R_i^{(r)}}{r} = \lim_{r \to \infty} \frac{T_i^{(r)}}{r} = \mathbb{E}(T_i | \xi_0 = i)$$

for all $i \in I$ with probability one. Set $g_i(n) := n h_{\{i\}}(n)$ for the sake of brevity and observe that from the definitions $T_i^{(g_i(n))} \leq n - 1$ and $T_i^{(g_i(n)+1)} \geq n$. Taking into account that by recurrence $g_i(n) \to \infty$ with probability one we deduce that

$$\mathbb{P}\left(\lim_{n \to \infty} h_{\{i\}}(n)^{-1} = \mathbb{E}(T_i | \xi_0 = i)\right) = 1.$$

Since $\mathbb{E}(T_i | \xi_0 = i) = \pi_i^{-1} > 0$ we obtain $\lim_{n \to \infty} h_{\{i\}}(n) = \pi_i$ with probability one.

Given a bounded function $f : I \to \mathbb{C}$ it is evident that $S_n f = \sum_{i \in I} f(i) h_{\{i\}}(n)$. Let $\|f\|_\infty := \sup_{i \in I} |f(i)|$ and take $\varepsilon > 0$. Furthermore choose a finite set J_ε with $\sum_{i \notin J_\varepsilon} \pi_i < \varepsilon$. Then

$$\left| S_n f - \sum_{i \in I} f(i) \pi_i \right| \leq \|f\|_\infty \sum_{i \in J_\varepsilon} |h_{\{i\}}(n) - \pi_i| + \|f\|_\infty \sum_{i \notin J_\varepsilon} |h_{\{i\}}(n) - \pi_i|$$

$$\leq \|f\|_\infty \sum_{i \in J_\varepsilon} |h_{\{i\}}(n) - \pi_i| + \|f\|_\infty \sum_{i \notin J_\varepsilon} (h_{\{i\}}(n) + \pi_i)$$

$$\leq 2\|f\|_\infty \sum_{i \in J_\varepsilon} |h_{\{i\}}(n) - \pi_i| + 2\|f\|_\infty \sum_{i \notin J_\varepsilon} \pi_i .$$

Therefore with probability one $\overline{\lim}_{n \to \infty} |S_n f - \sum_{i \in I} f(i) \pi_i| \leq 2\|f\|_\infty \varepsilon$. Since ε was arbitrary, the theorem is proved. $\square$

Let us now summarize our occupation with Markov chains. Constituting the most basic and elementary class of stochastic processes, these models deserve detailed study in their own right. However, some facts about the long-time behaviour (Theorems 4.26 and 4.28 for instance) suggest that these processes may also be studied under a dynamical systems perspective. Concepts of ergodic theory prove useful in anticipating several results of purely stochastic nature. In particular, we experienced that a simple modification of one's point of view suffices to pass from Markov chains to shift spaces: the latter are just the path spaces of the former! If we look at all possible complete histories rather than at individual states, the Markov chain becomes completely deterministic with the influence of chance being replaced by the action of σ_A, i.e. by shifting through history. In turn, the simple discrete space on which the Markov chain lives is replaced by a space of sequences which is rather complicated from a geometric point of view.

On the other hand, there are many ways of arriving at Markov chains when starting in a deterministic context. Perhaps the simplest of such possibilities is provided by Markov maps on the unit interval. As a specific example consider the map

$$T(x) := \begin{cases} 3x & \text{if } x \in [0, \frac{1}{4}[, \\ \frac{5}{4} - 2x & \text{if } x \in [\frac{1}{4}, \frac{1}{2}[, \\ 3x - \frac{5}{4} & \text{if } x \in [\frac{1}{2}, \frac{3}{4}[, \\ \frac{5}{2} - 2x & \text{if } x \in [\frac{3}{4}, 1], \end{cases}$$

the graph of which is depicted in Figure 4.19. Assume for the moment that someone wanted us to calculate $T(x)$ but instead of providing x (s)he just told us in which of the sets A_i the point x is located. What can we say about $T(x)$? Without any additional information available we shall certainly assume all points in A_i to be equally likely. The best we can obtain in this situation is a probability distribution on $\{1, \ldots, 4\}$ which describes where $T(x)$ might be. Continuing in this way we get a sequence of predictions for $T^n(x)$ in the form of a stochastic process which turns out to be a $(\delta_i, Q(T))$-Markov chain since T is a piecewise linear Markov map. When investigating the map

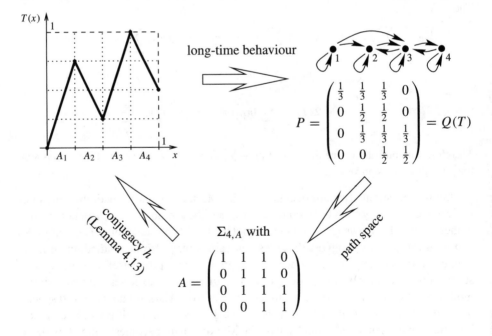

Figure 4.19. Depending on one's point of view similar objects may look quite different.

T numerically we have at our disposal much more information than just "$x \in A_i$" for some $i \in \{1, \ldots, 4\}$. However, the sets of points sharing the same fate with respect to the partition $(A_i)_{i=1}^4$ for at least n iterations shrink exponentially as n increases. (Compare this assertion with our discussion of entropy in Chapter Three.) For some not-too-large number N (in case of double precision floating point arithmetic one finds $N \approx 53$) therefore T^N is a Markov map on a refined partition to which the above considerations about the emergence of a Markov *chain* apply. Calculating $T^n(x)$ for $n \geq N$ thus may be regarded as simulating a Markov chain. If we chose at random an initial point x, and if our calculations yield the result that $T^n(x) \in A_i$ for some large n then this statement will be false with approximate probability $1 - \pi_i$ where π_i denotes the invariant distribution at i of the corresponding Markov chain. (In the above example $1 - \pi_i \geq \frac{4}{7}$ for all i !) Explicit calculations of $T^n(x)$ should thus be looked at with a more than sceptical eye.

By the above discussion piecewise expanding Markov maps which earlier we found to be chaotic incorporate both deterministic *and* stochastic features: while purely deterministic on a short time-scale their long-time behaviour is nevertheless best understood and described under a statistical perspective. Though the specific situation may be considerably more intricate this observation seems to persist for many other chaotic systems. After all, a notable portion of chaotic systems' fascination comes from the fact that they are associated with two seemingly different worlds.

Exercises

(4.1) Verify the assertions made at the end of Section 4.1. More precisely, prove that the diameter of the lattice $A^{-n}\mathbb{Z}^d$ tends to zero as $n \to \infty$ (and hence T_A is exact on $\mathbb{T}^d$) for the matrices

$$A = \begin{pmatrix} 3 & -1 \\ -1 & 1 \end{pmatrix} \quad \text{and} \quad A = \begin{pmatrix} 2 & 0 & 0 \\ 0 & 7 & 1 \\ 0 & 3 & 1 \end{pmatrix},$$

while this is not true for

$$A = \begin{pmatrix} 2 & 0 & 0 \\ 0 & 7 & 2 \\ 0 & 3 & 1 \end{pmatrix}.$$

(4.2) We have seen that there exists a unique piecewise constant invariant density for a piecewise expanding map on the unit interval if this map is piecewise linear, eventually onto, and has the Markov property. Give examples showing that this conclusion may fail if any of the three assumptions is dropped.

(4.3) Consider the so-called *β-transformation*

$$T_\beta : x \mapsto \beta x \quad (\text{mod } 1)$$

on the unit interval with $\beta > 1$. Clearly, T_β is piecewise linear and expanding, but under which conditions does it exhibit the Markov property? Determine the invariant density for T_β with β satisfying $\beta^3 - \beta^2 - \beta - 1 = 0$. (The specific value

$$\beta = \frac{1 + \sqrt[3]{19 - 3\sqrt{33}} + \sqrt[3]{19 + 3\sqrt{33}}}{3} \approx 1.839$$

is of minor importance for the calculation.)

(4.4) A continuous map $T : [0, 1] \to [0, 1]$ is called a *generalized tent transformation* if T is monotonically increasing on $[0, \frac{1}{2}]$ and $T(1-x) = T(x)$ for all $x \in [0, 1]$. Given a positive density $f \in L^1(\lambda^1|_{[0,1]})$ prove that there is a unique generalized tent transformation T_f which has f as an invariant density. Provide conditions for T_f to be statistically stable. As an example, calculate T_f with $f(x) := 1/\sqrt{4x}$.

(4.5) Let $(\xi_n)_{n \in \mathbb{N}_0}$ be the Markov chain corresponding to our simple model of one game of tennis (cf. Figure 4.15) and denote by T the (random) duration of the game, that is

$$T := \min \left\{ n : \xi_n \in \{\text{``Game } A\text{''}, \text{``Game } B\text{''}\}, \; \xi_0 = \text{``0:0''} \right\}.$$

For $k \in \mathbb{N}$ explicitly calculate the probability $\mathbb{P}(T = k)$ and verify the relation

$$\mathbb{E}\,T = 4 + 4p(1 - p)\frac{1 + 6p^2 - 12p^3 + 6p^4}{1 - 2p + 2p^2}.$$

Draw a sample of T from a match of your favourite tennis player(s). Does our simple model fit the data? Why not?

(4.6) Consider the following stochastic motion on the vertex set of a finite connected graph: from a given vertex the process can move with equal probability to any of the adjacent vertices. Describe this motion in terms of a reversible Markov chain and explicitly calculate the (unique) invariant distribution. As an application, calculate the mean return time of a knight starting in the corner of an empty chessboard and always choosing with equal probability among the admissible moves. (Compare the elegant calculation to the task of solving 64 linear equations according to Theorem 4.21!)

(4.7) For the subshift of finite type $(\Sigma_{l,A}, \sigma_A)$ calculate the entropy $h_{\mu_{\pi,P}}(\sigma_A)$ where $\mu_{\pi,P}$ denotes the σ_A-invariant measure defined by (4.18). Characterize the case that $h_{\mu_{\pi,P}}(\sigma_A) = 0$ for *all* admissible P and π. The following problems address the question how $h_{\mu_{\pi,P}}(\sigma_A)$ depends on the specific choice of P given $A \in \{0, 1\}^{l \times l}$.

  (i) Consider the full shift on l symbols, that is $a_{ij} = 1$ for all $1 \le i, j \le l$. Show that $h_{\mu_{\pi,P}}(\sigma_A)$ is maximal precisely if all p_{ij} and π_i are equal.

  (ii) For the subshift $(\Sigma_{3,A}, \sigma_A)$ with

$$A := \begin{pmatrix} 0 & 1 & 1 \\ 1 & 1 & 1 \\ 1 & 0 & 0 \end{pmatrix}$$

encountered earlier in the context of horseshoes (Figure 2.29) explicitly find P^* and π^* such that $h_{\mu_{\pi,P}}(\sigma_A)$ is maximized by μ_{π^*,P^*}. Confirm that $h_{\mu_{\pi^*,P^*}}(\sigma_A) = \log \lambda$ where λ denotes the eigenvalue of A with maximal modulus.

For aperiodic A a general theorem asserts that the maximum of $h_\mu(\sigma_A)$ over *all* σ_A-invariant probability measures on $(\Sigma_{l,A}, \mathcal{B}(\Sigma_{l,A}))$ is attained at a unique measure μ_{π^*,P^*} commonly referred to as the *Parry measure*. Moreover, as in the special cases above $h_{\mu_{\pi^*,P^*}}(\sigma_A) = \log(\max_{\lambda \in \sigma(A)} |\lambda|)$ holds ([31]).

(4.8) Demonstrate that $h_{\mu^*}(F_4) = \log 2$ where μ^* with

$$\frac{d\mu^*}{d\lambda^1} = \frac{1}{\pi\sqrt{x(1-x)}} 1_{]0,1[}(x)$$

denotes the unique F_4-invariant probability measure which is absolutely continuous with respect to λ^1. (Again, a further analysis reveals that in fact $h_\mu(F_4) < h_{\mu^*}(F_4)$ for *any* other F_4-invariant probability measure on $([0, 1], [0, 1] \cap \mathcal{B}^1)$.)

(4.9) Deduce from Exercise 4.7 and Theorem 4.19 that the Bernoulli shifts over $(X, \mathcal{P}(X), \mu)$ with $X := \{1, \ldots, m\}$ and $\mu = m^{-1}\#$ cannot be isomorphic for different values of $m \in \mathbb{N}$. Show that nevertheless any two such shifts are spectrally isomorphic (cf. Exercise 3.13).

(4.10) Let $(\xi_n)_{n \in \mathbb{N}_0}$ denote a (λ, P)-Markov chain on I. For $J \subseteq I$ define a new stochastic process $(\eta_n)_{n \in \mathbb{N}_0}$ in J by observing $(\xi_n)_{n \in \mathbb{N}_0}$ only when it is in J. More formally, set

$$T_0 := \inf\{k \ge 0 : \xi_k \in J\}$$

and, for $n \in \mathbb{N}_0$,

$$T_{n+1} := \inf\{k > T_n : \xi_k \in J\}.$$

Assuming that $\mathbb{P}(T_n < \infty) = 1$ for all $n \in \mathbb{N}_0$ we may define $\eta_n := \xi_{T_n}$. Show that $(\eta_n)_{n \in \mathbb{N}_0}$ is a Markov chain and determine its transition matrix.

Any map $\varphi : I \to I$ straightforwardly yields another stochastic process $(\zeta_n)_{n \in \mathbb{N}_0}$ via the simple definition $\zeta_n := \varphi(\xi_n)$. Give an example showing that $(\zeta_n)_{n \in \mathbb{N}_0}$ need not be a Markov chain.

(4.11) In [35] an entropy-like quantity for Markov operators is introduced, which allows to quantify convergence to equilibrium and also resembles the use of the term *entropy* in thermodynamics. Let $(\xi_n)_{n \in \mathbb{N}_0}$ denote an aperiodic Markov chain on I with invariant distribution $(\pi_i)_{i \in I}$. Define

$$H(\xi_n) := -\sum_{i \in I} \mathbb{P}(\xi_n = i) \log \frac{\mathbb{P}(\xi_n = i)}{\pi_i}.$$

Prove that $H(\xi_n)$ monotonically increases towards a finite limit. (The quantity H thus allows to define a "direction of time" for non-equilibrium Markov chains [21].)

(4.12) Fix $k \in \mathbb{N} \setminus \{1\}$ and consider the $\lambda_{\mathbb{T}^1}$-preserving map $T_{(k)} : x \mapsto kx$. Does the associated Frobenius–Perron operator $P_{T_{(k)}}$ posses any eigenfunctions, i.e., are there non-zero elements $f \in L^1(\lambda_{\mathbb{T}^1})$ such that $P_{T_{(k)}} f = \alpha f$ for an appropriate $\alpha \in \mathbb{C}$? Show that for any $|\alpha| < 1$ the eigenspace $E_\alpha := \{f \in L^1 : P_{T_{(k)}} f = \alpha f\}$ has in fact infinite dimension by verifying that $f_{\alpha,m} \in E_\alpha$ with

$$f_{\alpha,m}(x) := \sum_{n=0}^{\infty} \alpha^n e^{2\pi i m k^n x} \quad \text{for } |\alpha| < 1, m \in \mathbb{Z} \setminus k\mathbb{Z}.$$

Are there any polynomials in E_α? As a first step towards an answer prove that for every $l \in \mathbb{N}$ there exists a polynomial B_l of degree l in $E_{k^{-l}}$. (The polynomial B_l is unique up to multiplication by a constant and is traditionally referred to as the l-th *Bernoulli polynomial*.) Relate your findings to the statistical stability of $T_{(k)}$.

(4.13) For measure-preserving maps on compact metric spaces (endowed with the corresponding Borel σ-algebra) both notions of equivalence, *isomorphism* and *topological conjugacy*, equally apply. Is there any implication between these notions, i.e., are isomorphic systems topologically conjugate and/or vice versa?

(4.14) By Lemma 4.25 any recurrent irreducible Markov chain has an invariant measure which is unique up to multiplication by a constant. Show that without the recurrence assumption an invariant measure need not be unique, or may fail to exist at all. Also give an example showing that recurrence is generally not implied by the existence of an invariant measure which is unique up to multiplication by a constant.

(4.15) For $n \in \mathbb{N}$ determine c_n such that $T_n : x \mapsto c_n x (1 - x^n)$ maps the unit interval onto itself; evidently $T_1 = F_4$, an old friend of ours. Show by means of a coordinate transformation that T_n is statistically stable in this case. Can you give an approximation for the invariant density of P_{T_n}?

Chapter 5
The dynamical evolution of measures

Up to now our statistical analysis of dynamical systems has made use of a special class of measures – those represented by densities. As we have seen, powerful theorems and interesting applications come with them. However, not least in the light of several problems from applications, the techniques presented so far may be restrictive. For example, the Dirac measure at an attracting fixed point should be stable in some sense although it will typically not have a density, so that the Frobenius–Perron terminology of Chapter Four does not apply. Also we would like to consider point-like heat distributions in the realm of the heat equation, but how to deal with them? Analogously, the effect of sweeping proved physically relevant but could not be treated satisfyingly by means of densities. Observations like these suggest that we study the dynamical evolution of (probability) measures in a more general context. No doubt, such a generalization could make our analysis more ambitious and technically demanding. The present chapter, however, merely aims at a gentle and rather informal introduction to the subject, guided by examples and applications. The discussion thus mainly intends to motivate further studies but will nevertheless deepen our understanding of dynamics. For example, fractal sets, a few aspects of which we are going to discuss, are commonly regarded to be intimately related to chaos in nature, or even to be the geometric manifestation thereof.

5.1   Basic examples and concepts

To get an idea why densities may fail to yield a proper description of dynamical systems let us first look at some simple examples. The dynamics of the map $T_\alpha : \mathbb{R} \to \mathbb{R}$ with $x \mapsto \alpha x$ and $0 < |\alpha| < 1$ is easily analysed: the hyperbolic fixed point at the origin is globally attracting. However, since for any density $f \in L^1(\lambda^1)$ and any $\varepsilon > 0$

$$\int_{\mathbb{R}\setminus[-\varepsilon,\varepsilon]} P_{T_\alpha}^n f \, d\lambda^1 = 1 - \int_{[-\varepsilon,\varepsilon]} |\alpha|^{-n} f\left(\frac{x}{\alpha^n}\right) dx = 1 - \int_{\mathbb{R}} f \mathbf{1}_{[-\frac{\varepsilon}{|\alpha|^n}, \frac{\varepsilon}{|\alpha|^n}]} d\lambda^1 \to 0$$

the Frobenius–Perron operator only reflects this elementary fact in a rather diffuse way: P_{T_α} is sweeping with respect to the family of closed intervals not containing the origin. Consequently there is neither an $L^1(\lambda^1)$-limit for $P_{T_\alpha}^n f$ nor any invariant density at all. If, however, we fix a density f and define a sequence of probability measures $(\mu_n)_{n\in\mathbb{N}_0}$ according to $\frac{d\mu_n}{d\lambda^1} := P_{T_\alpha}^n f$, then we observe for any bounded continuous function $h : \mathbb{R} \to \mathbb{R}$ that

$$\int_{\mathbb{R}} h \, d\mu_n = \int_{\mathbb{R}} h(\alpha^n x) f(x) \, dx \to h(0) = \int_{\mathbb{R}} h \, d\delta_0 \quad \text{as } n \to \infty. \tag{5.1}$$

Therefore, the Dirac measure δ_0 corresponding to the attracting fixed point is itself attracting in a certain sense.

A second example refers to our earlier treatment of the heat equation by means of Markov operators. Let μ denote a probability measure on $(\mathbb{R}, \mathcal{B}^1)$ and define for $t > 0$ an $L^1(\lambda^1)$-density $u_\mu(t, \cdot)$ according to

$$u_\mu(t, x) := \frac{1}{\sqrt{4\pi t}} \int_{\mathbb{R}} e^{-\frac{(x-y)^2}{4t}} \, d\mu(y), \tag{5.2}$$

which generalizes (4.3). A straightforward calculation shows that u_μ still satisfies the heat equation $\frac{\partial u}{\partial t} = \frac{\partial^2 u}{\partial x^2}$ for $t > 0$. But how about the limit $t \to 0$? No doubt, the latter will not exist as an L^1-limit in general. Nevertheless, by dominated convergence we see that again

$$\lim_{t \to 0} \int_{\mathbb{R}} u_\mu(t, x) h(x) \, d\lambda^1(x) = \int_{\mathbb{R}} h(x) \, d\mu(x) \tag{5.3}$$

for every bounded continuous function $h :$ $\mathbb{R} \to \mathbb{R}$. Without being too formal here we may thus consider $u_\mu(t, x)$ as a solution of the heat equation corresponding to the initial (heat) distribution μ. For example,

$$u_{\delta_{x_0}}(t, x) = \frac{1}{\sqrt{4\pi t}} e^{-\frac{(x-x_0)^2}{4t}}$$

may consequently be thought of as a description of the process of heat diffusion which starts at $t = 0$ when all the heat is concentrated at the point x_0 (cf. Figure 5.1).

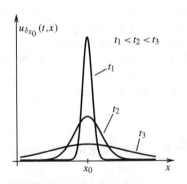

Figure 5.1. The evolution of $u_{\delta_{x_0}}$

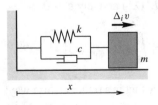

Figure 5.2. A periodically excited mechanical system

As a final example let us consider the following stochastic mechanical system. Assume that the velocity $\dot{x}$ of a simple linear visco-elastic oscillator governed by the equation

$$m\ddot{x} + c\dot{x} + kx = 0 \tag{5.4}$$

(see Figure 5.2) is instantaneously changed by an amount Δv all τ seconds. Between two kicks the motion of the oscillator obeys (5.4). If we denote by $(x_n, \dot{x}_n)$ the state of the system immediately after the

n-th kick then by a straightforward calculation we end up with an affine map $\Phi_{\Delta v}$: $(x_n, \dot{x}_n) \mapsto (x_{n+1}, \dot{x}_{n+1})$ of the $(x, \dot{x})$ phase plane. More precisely, $\Phi_{\Delta v} = C + T$ where C is a linear contraction which corresponds to (5.4) and T is a translation along the $\dot{x}$-axis which corresponds to the instantaneous velocity change. The ultimate behaviour of this system is easily analysed: there exists a globally attracting fixed point corresponding to a periodic motion of the oscillator (cf. Figure 5.3). This easy-to-survey situation may, however, be changed drastically by a slight modification of the system. Assume that we have at our disposal a finite number of different kick

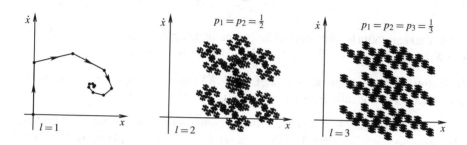

Figure 5.3. In the deterministic case ($l = 1$, left) there is a globally attracting fixed point; in the stochastic case ($l > 1$) a complicated attractor may be observed.

intensities, say $\Delta_1 v, \ldots, \Delta_l v$ from which at each kick $\Delta_i v$ is chosen with probability $p_i > 0$. Equivalently, we may choose and apply with probability p_i the affine map $\Phi_{\Delta_i v} = C + T_i$ from a family of such maps. The resulting stochastic orbits exhibit a remarkable long-time behaviour: they all tend towards a rather complicated attractor (see Figure 5.3 for the cases $l = 2, 3$). How can this long-term behaviour be explained? And what can be said about the geometric structure of the complicated attractors shown in Figure 5.3? After all, our human brain tends to interpret pictures like Figure 5.3 as planar histograms ([19]). Having developed the appropriate mathematical tools we shall face these questions later in this chapter.

First of all we are going to introduce a few basic concepts concerning the dynamical evolution of measures. In the sequel we shall always assume that (X, d) is a complete separable metric space and endow it with the Borel σ-algebra $\mathcal{B}(X)$. Closed (but also open [6]) subsets of $\mathbb{R}^d$ may serve as examples of such spaces. We denote by $\mathcal{M}(X)$ the set of all finite and by $\mathcal{M}_1(X)$ the set of all probability measures on $(X, \mathcal{B}(X))$, respectively.

Definiton 5.1.   A sequence $(\mu_n)_{n \in \mathbb{N}}$ in $\mathcal{M}(X)$ converges *weakly* to $\mu \in \mathcal{M}(X)$, symbolically $\mu_n \overset{w}{\to} \mu$, if

$$\int_X h \, d\mu_n \to \int_X h \, d\mu \quad \text{as } n \to \infty$$

for all bounded continuous functions $h : X \to \mathbb{R}$.

Evidently, we have met this type of convergence in our above examples: in the case of T_α we observed $\mu_n \overset{w}{\to} \delta_0$ as $n \to \infty$, while (5.3) may equivalently be rephrased as $\mu_t \overset{w}{\to} \mu$ for $t \to 0$ with $\frac{d\mu_t}{d\lambda^1} := u_\mu(t, \cdot)$.

Although the notion of weak convergence turns out to be natural in many circumstances we are going to introduce a second important concept. For any two measures $\mu, \nu \in \mathcal{M}(X)$ we define

$$d(\mu, \nu) := \sup_n \sum_{i=1}^{n} |\mu(X_i) - \nu(X_i)|$$

where the supremum is taken over all finite measurable partitions $(X_i)_{i=1}^n$, which means that $X_i \in \mathcal{B}(X)$ with $X = \bigcup_{i=1}^n X_i$ and $X_i \cap X_j = \emptyset$ if $i \neq j$.

Definiton 5.2.   A sequence $(\mu_n)_{n \in \mathbb{N}}$ in $\mathcal{M}(X)$ converges *strongly* to the measure $\mu \in \mathcal{M}(X)$, symbolically $\mu_n \to \mu$, if $d(\mu_n, \mu) \to 0$ as $n \to \infty$.

As the reader may wish to check, strong convergence implies weak convergence. The converse is easily seen to be false in general. Take for example $\mu_n \in \mathcal{M}_1(\mathbb{R})$ with $\frac{d\mu_n}{d\lambda^1} = n\mathbf{1}_{[0,n^{-1}]}$. Then $\mu_n \overset{w}{\to} \delta_0$ but $d(\mu_n, \delta_0) = 2$ for all $n \in \mathbb{N}$. More generally, if μ_n and μ are absolutely continuous with respect to some $\nu \in \mathcal{M}_1(X)$, having densities f_n and f, respectively, then

$$d(\mu_n, \mu) = \int_X |f_n - f| \, d\nu = \|f_n - f\|_{L^1(\nu)}.$$

Since $L^1(\nu)$ is complete, the strong limit of a sequence of absolutely continuous measures is itself absolutely continuous if it exists at all. For example, in the case of the heat equation we have $\mu_t \to \mu$ as $t \to 0$ if and only if μ is absolutely continuous with respect to λ^1.

Markov operators acting on densities (or L^1-functions) have played a prominent role in the previous chapter. We shall now carry over this concept to $\mathcal{M}(X)$.

Definiton 5.3. A map $P : \mathcal{M}(X) \to \mathcal{M}(X)$ is called a *Markov operator on measures* if

   (i) $P(r\mu + s\nu) = rP(\mu) + sP(\nu)$ for all $r, s \geq 0$ and $\mu, \nu \in \mathcal{M}(X)$;

   (ii) $P(\mu)(X) = \mu(X)$ for all $\mu \in \mathcal{M}(X)$.

Obviously, $P(\mathcal{M}_1) \subseteq \mathcal{M}_1$ for any Markov operator. Observe that – contrary to the case of densities – a Markov operator on measures need not be continuous in any reasonable sense: take as an example

$$P : \mu \mapsto \mu([0, 1[) \delta_0 + \mu|_{[1,\infty[} \quad \text{on } \mathcal{M}([0, +\infty[) \, ;$$

clearly $\delta_{1-n^{-1}} \overset{w}{\to} \delta_1$ but $P(\delta_{1-n^{-1}}) = \delta_0$ for all $n \in \mathbb{N}$ whereas $P(\delta_1) = \delta_1$. In the sequel we shall, however, not be concerned with such subtleties; moreover, we shall write $P\mu$ instead of $P(\mu)$ for the sake of lucidity.

If P preserves absolute continuity with respect to some fixed $\nu \in \mathcal{M}_1$, then there exists a unique Markov operator $\hat{P}$ on $L^1(\nu)$ such that the diagram

$$
\begin{array}{ccc}
\mathcal{M}_{ac}(X) & \overset{P}{\longrightarrow} & \mathcal{M}_{ac}(X) \\
{\scriptstyle \frac{d}{d\nu}} \downarrow & & \downarrow {\scriptstyle \frac{d}{d\nu}} \\
L^1(\nu) & \overset{\hat{P}}{\longrightarrow} & L^1(\nu)
\end{array}
$$

commutes; here $\mathcal{M}_{ac}(X) \subseteq \mathcal{M}(X)$ denotes the class of ν-absolutely continuous finite measures on $(X, \mathcal{B}(X))$. The concept of Markov operators on measures may therefore be regarded a generalization of our earlier Definition 4.4; for a countable space endowed with the counting measure both notions coincide.

Recall that for us the most important Markov operator on densities has been the Frobenius–Perron operator. In fact, this tool can equally be defined in the more general setting of measures: given a $\mathcal{B}(X)$-measurable map $T : X \to X$ we call

$$
P_T : \begin{cases} \mathcal{M}(X) & \to & \mathcal{M}(X), \\ \mu & \mapsto & T\mu, \end{cases}
$$

the *Frobenius–Perron operator* on $\mathcal{M}(X)$ associated with T. Clearly, P_T is a Markov operator on measures which is weakly continuous if T is continuous. On the other hand, the space X may be embedded into $\mathcal{M}_1(X)$ via $x \mapsto \delta_x$, and $P_T \delta_x = \delta_{T(x)}$. Thus the general notions presented here in some sense comprise and unify the statistical and the geometrical point of view which we discussed in earlier chapters.

Definiton 5.4.   A measure $\mu \in \mathcal{M}_1(X)$ is called *invariant* (or *stationary*) for the Markov operator P if $P\mu = \mu$.

Proving the existence of an invariant measure for a given Markov operator necessitates a certain amount of structure for the space under consideration. In case of familiar spaces like $\mathbb{R}^d$ and $\mathbb{T}^d$ there are powerful tools for dealing with this question. Since we are mainly interested in dynamical systems we shall sketch just one aspect.

Lemma 5.5. *Let P denote a weakly continuous Markov operator on $\mathcal{M}(X)$. Assume that some $\mu \in \mathcal{M}_1(X)$ has the following property: for any $\varepsilon > 0$ there exists a compact set $K_\varepsilon \subseteq X$ such that $P^n \mu(K_\varepsilon) \geq 1 - \varepsilon$ for all sufficiently large n. Then there exists an invariant measure $\mu^* \in \mathcal{M}_1(X)$ for P.*

Proof. Defining a sequence of probability measures by $\mu_n := \frac{1}{n} \sum_{i=0}^{n-1} P^i \mu$ we see that $\mu_n(K_\varepsilon) \geq 1 - 2\varepsilon$ with at most a finite number of exceptions. By a theorem due to Prohorov ([10, 55]) there is a subsequence $(\mu_{n_k})_{k \in \mathbb{N}}$ which converges weakly (to $\mu^* \in \mathcal{M}_1(X)$, say). But then

$$
\left| \int_X h \, dP\mu_{n_k} - \int_X h \, d\mu_{n_k} \right| = \frac{1}{n_k} \left| \int_X h \, dP^{n_k}\mu - \int_X h \, d\mu \right| \leq \frac{2 \sup_{x \in X} |h(x)|}{n_k} \to 0
$$

and hence $\int_X h \, dP\mu^* = \int_X h \, d\mu^*$ by continuity. Since two measures $\mu, \nu \in \mathcal{M}(X)$ are equal precisely if $\int_X h \, d\mu = \int_X h \, d\nu$ for all bounded continuous functions h, we have $P\mu^* = \mu^*$. $\square$

As a corollary we notice that a continuous map on a compact metric space always admits at least one invariant probability measure (Theorem of Krylov–Bogolyubov).

In order to understand the meaning of the seemingly artificial condition in Lemma 5.5 let us take a look at the familiar case $X = \mathbb{R}$. Every probability measure μ on the real line can be identified with its *distribution function* $F_\mu(x) := \mu(]-\infty, x])$ which is a right-continuous and non-decreasing function showing the additional property that

$$\lim_{x \to -\infty} F_\mu(x) = 0 \quad \text{and} \quad \lim_{x \to +\infty} F_\mu(x) = 1. \tag{5.5}$$

Given a sequence $(\mu_n)_{n \in \mathbb{N}}$ in $\mathcal{M}_1(\mathbb{R})$ we may study F_{μ_n}. By Helly's Theorem ([6, 55]) there exists a subsequence $(n_k)_{k \in \mathbb{N}}$ of natural numbers and a right-continuous non-decreasing function F such that $\lim_{k \to \infty} F_{\mu_{n_k}}(x) = F(x)$ at each point of continuity x of F. Is there a probability measure μ^* on $\mathbb{R}$ such that $F = F_{\mu^*}$? (Observe that this would imply $\mu_{n_k} \overset{w}{\to} \mu^*$, see Exercise 5.8.) In general, the answer is negative: mass may escape to infinity as the following examples show.

Denoting by P_T the Frobenius–Perron operator associated with the translation map $T(x) := x + 1$ we find

$$F_{P_T^n \mu}(x) = P_T^n \mu(]-\infty, x]) = \mu(]-\infty, x - n]) = F_\mu(x - n)$$

for every $\mu \in \mathcal{M}_1(\mathbb{R})$. Therefore $F(x) := \lim_{n \to \infty} F_{P_T^n \mu}(x) \equiv 0$ which implies that there is no T-invariant probability measure at all. If we think of μ as a mass distribution on $\mathbb{R}$ then the last conclusion is fairly obvious: eventually the whole mass will wander to infinity at the right.

As a second example consider the family of probability measures $(\mu_t)_{t>0}$ with densities according to (5.2) which describe the evolution of the initial heat distribution μ. A straightforward calculation yields

$$F_{\mu_t}(x) = \int_\mathbb{R} \Phi\left(\frac{x - y}{\sqrt{2t}}\right) d\mu(y)$$

where Φ denotes the distribution function of the standard normal distribution ([55]). By dominated convergence we have

$$F(x) := \lim_{t \to \infty} F_{\mu_t}(x) = \Phi(0) = \frac{1}{2} \quad \text{for all } x \in \mathbb{R}.$$

Again the non-existence of an invariant distribution for the evolution operator $P_t : \mu \mapsto \mu_t$ is easily understood: eventually the finite amount of heat spreads over the whole line and thus disappears. In order to guarantee the existence of an invariant distribution we therefore have to rule out such a loss. This goal is achieved only if

$$\lim_{x \to -\infty} F(x) = 0 \quad \text{and} \quad \lim_{x \to +\infty} F(x) = 1 \tag{5.6}$$

because otherwise some mass will have disappeared. Condition (5.6) is equivalent to the following statement: given $\varepsilon > 0$ there exist real numbers $a_\varepsilon, b_\varepsilon$ such that $F(a_\varepsilon) < \varepsilon$ and $F(b_\varepsilon) > 1 - \varepsilon$. Since F has at most a countable number of jumps this latter statement may be rewritten precisely in the form of Lemma 5.5.

A convenient tool for analysing probability measures on $\mathbb{R}^d$ is provided by Fourier analysis. Given $\mu \in \mathcal{M}_1(\mathbb{R}^d)$ we call the complex-valued function

$$\varphi_\mu(x) := \int_{\mathbb{R}^d} e^{i\langle x, y \rangle} \, d\mu(y) \tag{5.7}$$

the *Fourier transform* of μ; it is easy to see that φ_μ is uniformly continuous and $|\varphi_\mu(x)| \leq 1 = \varphi_\mu(0)$. Since we merely intend to apply this tool to the above examples we shall not enter into the details of the theory (see for instance [47]) and merely quote without proof that for $\mu, \nu, \mu_n \in \mathcal{M}_1(\mathbb{R}^d)$

$$\varphi_\mu = \varphi_\nu \quad \text{if and only if} \quad \mu = \nu$$

as well as

$$\mu_n \xrightarrow{w} \mu \quad \text{if and only if} \quad \varphi_{\mu_n}(x) \to \varphi_\mu(x) \quad \text{for every } x \in \mathbb{R}^d.$$

As far as our examples are concerned, we find

$$\varphi_{P_T^n \mu}(x) = e^{ixn} \varphi_\mu(x) \quad \text{and} \quad \varphi_{\mu_t}(x) = e^{-tx^2} \varphi_\mu(x),$$

the former for the Frobenius–Perron operator of the translation and the latter for the heat equation. In both cases the limit either fails to exist or is not the Fourier transform of any probability measure. Observe how Fourier theory converts a difficult question about measures into a simple matter of calculus.

We close this section by sketching how to apply Lemma 5.5 to an important class of systems. Let $(Y, \mathcal{C}, \nu)$ denote an arbitrary probability space and assume that the map $T : X \times Y \to X$ is continuous on X for every fixed $y \in Y$ as well as $\mathcal{C}$-$\mathcal{B}(X)$-measurable for every fixed $x \in X$. A convenient way of interpreting such a map consists in regarding it as a family $(T_y)_{y \in Y}$ of maps on X with the choice of the specific family members $T_y := T(\cdot, y)$ being governed by ν. Equivalently, we may study the *random dynamical system*

$$x_{n+1} := T(x_n, \xi_{n+1}) \tag{5.8}$$

on X, where $(\xi_n)_{n \in \mathbb{N}}$ denotes a sequence of independent random variables in Y all distributed according to ν. Since by (5.8) ξ_{n+1} does not depend on the earlier outcomes $x_0, \ldots, x_n$, we have

$$\mu_{n+1}(A) := \mathbb{P}(x_{n+1} \in A) = \int_Y \mathbb{P}(T(x_n, \xi_{n+1}) \in A | \xi_{n+1} = y) \, d\nu(y)$$

$$= \int_X \int_Y \mathbf{1}_A \circ T(x, y) \, d\nu(y) \, d\mu_n(x)$$

and thus naturally obtain an operator P on $\mathcal{M}_1(X)$ as

$$P\mu(A) := \int_X \int_Y \mathbf{1}_A \circ T(x, y) \, d\nu(y) \, d\mu(x) \quad \text{for all } A \in \mathcal{B}(X). \quad (5.9)$$

In more concise notation (see Appendix A) this reads $P\mu := T(\mu \otimes \nu)$. It is easy to see that for fixed ν the operator P is in fact a continuous Markov operator on measures. Specifically, $P^n \delta_{x_0}$ describes the distribution after n steps of the random dynamical system (5.8) if we start at the point x_0. If X is compact then there exists an invariant measure for P by Lemma 5.5. In more general situations the concept of a Lyapunov function may be helpful. We call a measurable function $L : X \to [0, +\infty[$ a *Lyapunov function* if $L^{-1}([0, a])$ is compact for all a. (In an analogous manner as in the following theorem Lyapunov functions are used in stability theory to rule out an escape to infinity [61].)

Theorem 5.6. *Let the Markov operator be defined by* (5.9). *Assume that there exists a Lyapunov function L and real constants α, β with $0 \leq \alpha < 1$ such that*

$$\int_Y L \circ T(x, y) \, d\nu(y) \leq \alpha L(x) + \beta \quad \text{for all } x \in X.$$

Then there exists an invariant measure for P.

Proof. Set $K_a := L^{-1}([0, a])$ and $\mu_n := P^n \mu_0$ where $a \in \mathbb{R}$ and $\mu_0 \in \mathcal{M}_1(X)$ will be specified later. By virtue of the Chebyshev inequality A.12(vi) we have

$$\nu(\{y : T(x, y) \in K_a\}) \geq 1 - \frac{1}{a} \int_Y L \circ T(x, y) \, d\nu(y)$$

for all $x \in X$. Therefore

$$\mu_{n+1}(K_a) = \int_X \nu(\{y : T(x, y) \in K_a\}) \, d\mu_n(x) \geq 1 - \frac{\beta}{a} - \frac{\alpha}{a} \int_X L(x) \, d\mu_n(x).$$

Assume for the moment that $\sup_{n \in \mathbb{N}_0} \int_X L(x) \, d\mu_n(x) < \infty$ for some $\mu_0 \in \mathcal{M}_1(X)$. In that case we are done because the compact set K_a will satisfy the assumptions of Lemma 5.5 if we choose a sufficiently large. From the present assumptions on L we get

$$\int_X L(x) \, d\mu_{n+1}(x) = \int_X \int_Y L \circ T(x, y) \, d\nu(y) d\mu_n(x) \leq \alpha \int_X L(x) \, d\mu_n(x) + \beta$$

and thus by induction

$$\int_X L \, d\mu_{n+1} \leq \alpha^{n+1} \int_X L \, d\mu_0 + \frac{\beta}{1 - \alpha}.$$

Therefore $\int_X L \, d\mu_n$ is bounded from above for $\mu_0 := \delta_{x_0}$ for any $x_0 \in X$. $\square$

Let us apply this theorem to a specific example. Assume that we intended to calculate $\sqrt{a}$ for some $a > 0$ by means of Newton's method (2.8) for the polynomial $p(x) := x^2 - a$, i.e. by iterating

$$x \mapsto N_p(x) = \frac{1}{2}\left(x + \frac{a}{x}\right) \quad (5.10)$$

on $X := \mathbb{R}^+$. As we know from Section 2.3 $\lim_{n \to \infty} N_p^n(x) = \sqrt{a}$ for all $x \in X$. However, assume that due to internal errors our computer does not give the exact value $N_p(x)$ but rather provides $T(x, \xi) := N_p(x) + \xi$ where ξ denotes a small non-negative stochastic perturbation. In order to prove the existence of an invariant distribution for the Markov operator arising from this random dynamical system we take $L(x) := N_p(x)$ as a Lyapunov function. Since

$$\mathbb{E}L \circ T(x, \xi) = \int_{[0,+\infty[} N_p(N_p(x) + y)\, d\nu(y) \le \frac{1}{2}\left(L(x) + \mathbb{E}\xi + \sqrt{a}\right),$$

Theorem 5.6 applies provided that $\mathbb{E}\xi < \infty$, that is, if the stochastic perturbation ξ has finite mean. Figure 5.4 shows some histograms from a numerical simulation with the stochastic perturbations ξ_n obeying an exponential distribution.

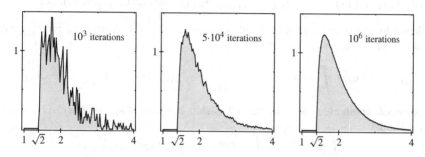

Figure 5.4. Empirical histograms for the stochastically perturbed Newton scheme (5.10) with $a = 2$ and $\xi \sim E_2$, i.e. $\nu([0, y]) = 1 - e^{-2y}$

5.2 Asymptotic stability

Since we have introduced two different notions of convergence for sequences of measures (*weak* and *strong*, respectively) one may guess that there also are two notions of asymptotic stability. In this short section we shall briefly discuss both concepts by focusing on special cases (and referring to [35] for further details).

Definiton 5.7. A Markov operator P on $\mathcal{M}(X)$ is said to be *weakly (strongly) asymptotically stable* if there exists a unique invariant distribution $\mu^* \in \mathcal{M}_1(X)$ for P and

$$P^n \mu \xrightarrow{w} \mu^* \quad (P^n \mu \to \mu^*) \qquad \text{for all } \mu \in \mathcal{M}_1(X).$$

In the sequel we shall exclusively deal with Markov operators of the form (5.9) arising from the random dynamical system (5.8). A simple sufficient condition for weak asymptotic stability of these Markov operators is provided by the following result.

Theorem 5.8. *Let P denote a Markov operator of the form (5.9). Assume that for some $0 \leq \alpha < 1$*

$$\int_Y d\big(T(x_1, y), T(x_2, y)\big) \, d\nu(y) \leq \alpha d(x_1, x_2) \quad \text{for all } x_1, x_2 \in X; \qquad (5.11)$$

in addition assume that there exists an invariant measure $\mu^* \in \mathcal{M}_1(X)$ for P. Then P is weakly asymptotically stable.

Proof. In a first step we are going to show that $|\int_X h \, dP^n \mu_1 - \int_X h \, dP^n \mu_2| \to 0$ for all $\mu_1, \mu_2 \in \mathcal{M}_1(X)$ with bounded support and all bounded continuous functions $h : X \to \mathbb{R}$. Without loss of generality we may assume that h satisfies a Lipschitz condition (cf.[10]). Define inductively

$$T_1(x, y_1) := T(x, y_1), \quad T_{n+1}(x, y_1, \ldots, y_{n+1}) := T(T_n(x, y_1, \ldots, y_n), y_{n+1}),$$

and denote by F a bounded closed set with $\mu_1(F) = \mu_2(F) = 1$. Also set $\nu_n := \otimes_{k=1}^n \nu$ for the sake of brevity. Given $\varepsilon > 0$ we find

$$\left| \int_X h \, dP^n \mu_1 - \int_X h \, dP^n \mu_2 \right| \leq \varepsilon + \int_{Y^n} \left| h \circ T_n(z_1, y) - h \circ T_n(z_2, y) \right| d\nu_n(y)$$

for appropriate points $z_1, z_2 \in F$. Using the Lipschitz condition of h as well as (5.11) the latter relation implies

$$\left| \int_X h \, dP^n \mu_1 - \int_X h \, dP^n \mu_2 \right| \leq \varepsilon + L_h \alpha^n \, \text{diam } F;$$

since $\varepsilon > 0$ was arbitrary and $\alpha < 1$ we are done with the first step.

Let us now show that $|\int_X h \, dP^n \mu_1 - \int_X h \, dP^n \mu_2| \to 0$ holds for arbitrary measures $\mu_1, \mu_2 \in \mathcal{M}_1(X)$ which in turn will prove the theorem. Given $\varepsilon > 0$ and $\mu_i \in \mathcal{M}_1(X)$ there exists a bounded closed set F_ε satisfying $\mu_i(X \backslash F_\varepsilon) \leq \varepsilon$ $(i = 1, 2)$. Define measures $\overline{\mu}_i$ supported on F_ε according to

$$\overline{\mu}_i(A) := \frac{\mu_i(A \cap F_\varepsilon)}{\mu_i(F_\varepsilon)} = \mu_i(A | F_\varepsilon) \quad (A \in \mathcal{B}(X)).$$

Given a bounded continuous function $h : X \to \mathbb{R}$ we observe

$$\left| \int_X h \, dP^n \mu_i - \int_X h \, dP^n \overline{\mu}_i \right| \leq \left| \mu_i(F_\varepsilon)^{-1} - 1 \right| \left| \int_{F_\varepsilon \times Y^n} h \circ T_n(x, y) \, d\mu_i(x) \, d\nu_n(y) \right|$$

$$+ \left| \int_{X \backslash F_\varepsilon \times Y^n} h \circ T_n(x, y) \, d\mu_i(x) \, d\nu_n(y) \right| \leq 3\varepsilon \|h\|,$$

where $\|h\| := \sup_{x \in X} |h(x)|$ and, without loss of generality, $\varepsilon < \frac{1}{2}$. The proof is thus complete. $\square$

If we disregard the technical details, then Theorem 5.8 just says that P is asymptotically stable if the family $(T_y)_{y \in Y}$ is contracting *in the mean*: in probabilistic terms (5.11) reads

$$\mathbb{E} d\big(T(x_1, \xi), T(x_2, \xi)\big) \leq \alpha d(x_1, x_2).$$

Suppose for example that Y is finite (corresponding to the random choice among a *finite* number of maps) and that each T_y has a Lipschitz constant L_y. If furthermore $X = \mathbb{R}^d$ and $p_y := \nu(\{y\})$ then the associated Markov operator is weakly asymptotically stable provided that $\sum_{y \in Y} p_y L_y < 1$. In the next section we shall deal with this specific situation in more detail.

Let us apply Theorem 5.8 to the stochastically perturbed Newton scheme (5.10). Since

$$|T(x_1, y) - T(x_2, y)| = \frac{1}{2}|x_1 - x_2| \cdot \left|1 - \frac{a}{x_1 x_2}\right|$$

does not depend upon y, the contraction condition (5.11) is met on $[\sqrt{a/2}, +\infty[$. Any invariant measure of the associated Markov operator P is supported on $[\sqrt{a}, +\infty[$. Therefore P is weakly asymptotically stable if only $\mathbb{E}\xi < \infty$ (cf. Figure 5.4).

For the rest of this section we shall focus on strong asymptotic stability. In order to avoid excessive technicalities we shall still restrict ourselves to Markov operators given by (5.9). Additionally we assume that $X, Y \subseteq \mathbb{R}^d$ and also that T is of the special form

$$T(x, y) = T_0(x) + y$$

where $T_0 : X \to X$ and $x + y \in X$ whenever $x \in X$ and $y \in Y$. Though seemingly rather special, systems of this type have found a great variety of applications ([35]).

Lemma 5.9. *Let P denote the Markov operator on measures associated with*

$$T : \begin{cases} X \times Y & \to & X \\ (x, y) & \mapsto & T_0(x) + y \end{cases} \tag{5.12}$$

according to (5.9). Assume that T_0 is non-singular with respect to Lebesgue measure on X. Then

(i) *absolute continuity is preserved by P, i.e. $P(\mathcal{M}_{ac}(X)) \subseteq \mathcal{M}_{ac}(X)$;*

(ii) *if ν has a non-trivial absolutely continuous part, that is $\nu_{ac}(Y) \neq 0$, then $\lim_{n \to \infty} (P^n \mu)_{ac}(X) = 1$ for all $\mu \in \mathcal{M}_1(X)$.*

Proof. Denoting by P_{T_0} the Frobenius–Perron operator (on densities) associated with T_0 and writing f for the λ^d-density of μ we see that

$$\begin{aligned} P\mu(A) &= \int_X \int_Y \mathbf{1}_A(T_0(x) + y) \, d\nu(y) \, f(x) \, dx \\ &= \int_Y \int_X \mathbf{1}_A(x - y) P_{T_0} f(x) \, dx \, d\nu(y) \\ &= \int_A \int_Y \mathbf{1}_X(x - y) P_{T_0} f(x - y) \, d\nu(y) \, dx, \end{aligned}$$

hence $P\mu$ is absolutely continuous with density $\int_Y \mathbf{1}_X(x - y) P_{T_0} f(x - y) \, d\nu(y)$.

In order to prove (ii) denote by g the Radon–Nikodym derivative of ν_{ac} and observe that

$$P\mu_s(A) \geq \int_X \int_Y \mathbf{1}_A(T_0(x) + y)g(y)\,dy\,d\mu_s(x)$$
$$= \int_A \int_X \mathbf{1}_Y(y - T_0(x))g(y - T_0(x))\,d\mu_s(x)\,dy$$
$$= T(\mu_s \otimes \nu_{ac})(A) =: \rho(A);$$

hence the measure ρ is absolutely continuous, and $P\mu \geq P\mu_{ac} + \rho$. Since the right-hand side is itself absolutely continuous we find

$$(P\mu)_{ac}(X) \geq P\mu_{ac}(X) + \rho(X) = \mu_{ac}(X) + \rho(X).$$

On the other hand $\rho(X) = \mu_s(X) \cdot \nu_{ac}(Y)$ and consequently

$$(P\mu)_{ac}(X) \geq \mu_{ac}(X) + \big(1 - \mu_{ac}(X)\big)\nu_{ac}(Y)$$

for every $\mu \in \mathcal{M}_1(X)$. Setting $\alpha_n := (P^n\mu)_{ac}(X)$ we thus obtain the inequality

$$\alpha_{n+1} \geq \alpha_n + (1 - \alpha_n)\nu_{ac}(Y)$$

from which we conclude that $\alpha_n \geq 1 - (1 - \alpha_0)\big(1 - \nu_{ac}(Y)\big)^n$. Since $\nu_{ac}(Y)$ has been assumed positive, assertion (ii) of the lemma follows. $\square$

The special systems considered here thus are regularizing in a strong sense: no matter how irregular the initial measure μ is, the singular part of $P^n\mu$ will eventually vanish if only the stochastic perturbations ξ_n obey a non-singular law, which means that $\nu_{ac} \neq 0$. In other words, even the smallest non-trivial absolutely continuous part of the perturbations ξ_n forces the dynamical system (5.8) to evolve according to an absolutely continuous measure in the long run. As can be seen from the proof of Lemma 5.9 this effect is mainly due to the special structure (5.12) of the maps considered here. In this context also the following theorem should not come as a surprise. Although its assertions are quite easy to grasp, the proof actually necessitates some technical devices from the spectral theory of Markov operators and thus is omitted here (see [35] for the details).

Theorem 5.10. *Assume that both the map T_0 and the distribution ν are non-singular (with respect to Lebesgue measure). If the Markov operator associated with (5.9) is weakly asymptotically stable then it is also strongly asymptotically stable. Moreover, the unique invariant measure is absolutely continuous.*

Returning to our Newtonian example (5.10) we see that this system is in fact strongly asymptotically stable if only the perturbations ξ_n are non-singular. Thus no matter what ugly initial measure we choose, we will always end up with the same absolutely continuous invariant distribution. The density of the latter may be approximated by means of histograms as we have already seen in Figure 5.4. However, observe that the conclusion of Theorem 5.10 need not hold if the perturbations are singular. As a simple illustration of this fact take $\nu = \delta_0$ (no perturbation at all). Then for $x \neq \sqrt{a}$ neither does $P^n\delta_x = \delta_{T_0^n(x)}$ converge strongly to the unique invariant distribution $\mu^* = \delta_{\sqrt{a}}$ nor is the latter absolutely continuous.

5.3   Back to geometry: fractal sets and measures

For the random dynamical systems which we considered in the last section we found a strongly regularizing behaviour: any probability measure is attracted by a unique absolutely continuous invariant distribution. Now we are going to discuss a class of Markov operators which behave just the other way round in that even the nicest measure becomes increasingly singular under iteration of these operators. By analysing some of the bizarre features of such systems we are led back to our studies of chaotic dynamics as well as to a reinforced geometric point of view. Though invitingly appearing at the horizon, *fractal geometry* is a topic definitely beyond the scope of this book. We shall therefore restrict ourselves to the most basic definitions. Nevertheless our simple observations may provide a strong motivation for the curious to go further in this direction.

As always in this chapter let (X, d) denote a complete separable metric space. Assume that we are given a finite family of continuous maps $(T_i)_{i=1}^{l}$ on X among which we choose and apply T_i with probability $p_i > 0$. We then are in the situation of (5.8) with $Y = \{1, \dots, l\}$ and $\nu(\{i\}) = p_i$. In the sequel we shall furthermore assume that $d\big(T_i(x), T_i(y)\big) \leq L_i d(x, y)$ for all $x, y \in X$, i.e., L_i is a Lipschitz constant for T_i. We call $(T_i)_{i=1}^{l}$ an *iterated function system* with ratio list $(L_i)_{i=1}^{l}$ and probabilities $(p_i)_{i=1}^{l}$.

As an immediate consequence of Theorem 5.8 we have

Proposition 5.11. *The Markov operator* (5.9) *arising from the iterated function system* $(T_i)_{i=1}^{l}$ *is given by*

$$P\mu = \sum_{i=1}^{l} p_i P_{T_i} \mu = \sum_{i=1}^{l} p_i T_i \mu .$$

If $\sum_{i=1}^{l} p_i L_i < 1$ *this operator is weakly asymptotically stable.*

In the literature an iterated function system for which the stronger contraction condition $\max_{i=1}^{l} L_i < 1$ holds is called *hyperbolic*. Specifically, for the Markov operator P arising from such a system we have $P^n \mu \xrightarrow{w} \mu^*$ for every $\mu \in \mathcal{M}_1(X)$. What can be said about the limiting measure μ^*? Before dealing with this question in general we take a look at a specific example. To this end let $X := \mathbb{R}^2$ and

$$T_1(x) := \frac{1}{2}x, \quad T_2(x) := \frac{1}{2}x + \left(\frac{1}{2}, 0\right), \quad T_3(x) := \frac{1}{2}x + \left(\frac{1}{4}, \frac{\sqrt{3}}{4}\right);$$

to be specific also choose $p_1 = p_2 = p_3 = \frac{1}{3}$. In order to get an impression how the corresponding Markov operator works, let μ be the uniform distribution on the equilateral triangle

$$\Delta_0 := \{(x_1, x_2) : 0 \leq x_2 \leq \sqrt{3}x_1, \ \sqrt{3}x_1 + x_2 \leq \sqrt{3}\} \subseteq \mathbb{R}^2,$$

which means that $\frac{d\mu}{d\lambda^2} = \lambda^2(\Delta_0)^{-1}\mathbf{1}_{\Delta_0}$. Evidently, $P^n\mu$ is the uniform distribution on $\Delta_n := \bigcup_{i_1,\ldots,i_n=1}^3 T_{i_1} \circ \ldots \circ T_{i_n}(\Delta)$. As can be seen from Figure 5.5 the sets Δ_n become increasingly complicated from a geometric point of view. For example we find

$$\lambda^2(\Delta_n) = \lambda^2\Big(\bigcup_{i_1,\ldots,i_n=1}^3 T_{i_1} \circ \ldots \circ T_{i_n}(\Delta_0)\Big) = \frac{1}{2} \cdot \Big(\frac{3}{4}\Big)^{n+\frac{1}{2}},$$

$$\lambda^1(\partial\Delta_n) = \lambda^1\Big(\partial\bigcup_{i_1,\ldots,i_n=1}^3 T_{i_1} \circ \ldots \circ T_{i_n}(\Delta_0)\Big) = 2 \cdot \Big(\frac{3}{2}\Big)^{n+1},$$

so that the circumference of Δ_n grows exponentially while the corresponding area tends to zero. Observe now that for all $k \in \mathbb{N}_0$

$$P^n\mu(\Delta_k) = 1 \quad \text{if } n \geq k.$$

Denoting by μ^* the unique weak limit of $(P^n\mu)_{n\in\mathbb{N}_0}$ we thus find $\mu^*(\Delta_k) = 1$ for all k (Exercise 5.10). Consequently

$$\mu^*(\Delta_\infty) = 1 \quad \text{with} \quad \Delta_\infty := \bigcap_{k\geq 0}\Delta_k,$$

hence μ^* is supported on Δ_∞, a set which traditionally is referred to as the *Sierpinski triangle*. Intuitively, Δ_∞ is a compact set in the plane with infinite circumference but zero area and thus exhibits a rather complex geometric structure. How can one describe such sets and quantify their geometric complexity? And how about the measure μ^* itself? For example, we expect μ^* to be something like the uniform distribution on Δ_∞ if $p_1 = p_2 = p_3 = \frac{1}{3}$; but if not all the probabilities are equal then the structure of μ^* will be more complicated (see Figure 5.6). Below we shall discuss these questions by introducing a concept of *dimension* which e.g. assigns a value between one and two to the Sierpinski triangle Δ_∞ – just as we would expect in the light of our above observations.

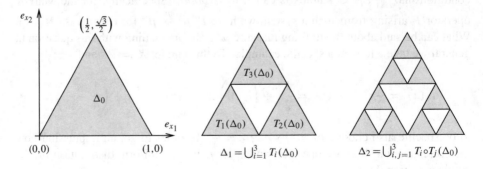

Figure 5.5. The first steps towards the Sierpinski triangle Δ_∞

First of all we have to provide a formal setting for thoroughly dealing with such sets as the Sierpinski triangle. To this purpose we denote by $\mathcal{K}(X)$ the family of non-empty compact subsets of X. Given $K \in \mathcal{K}(X)$ and $\varepsilon > 0$ we denominate an open set K_ε containing K as

$$K_\varepsilon := \{x \in X : d(x,k) < \varepsilon \text{ for some } k \in K\}.$$

In addition, for any two sets $K_1, K_2 \in \mathcal{K}(X)$ we define

$$d_H(K_1, K_2) := \inf\{\varepsilon : (K_1)_\varepsilon \supseteq K_2 \text{ and } (K_2)_\varepsilon \supseteq K_1\}.$$

Since compact sets are bounded, $d_H(K_1, K_2)$ always is finite; it is called the *Hausdorff distance* between K_1 and K_2.

Lemma 5.12. $(\mathcal{K}(X), d_H)$ *is a complete separable metric space if* (X, d) *is. Moreover, according to*

$$\tau : \begin{cases} \mathcal{K}(X) & \to & \mathcal{K}(X) \\ K & \mapsto & \bigcup_{i=1}^{l} T_i(K) \end{cases}$$

every hyperbolic iterated function system $(T_i)_{i=1}^{l}$ *on* X *induces a contraction map* τ *on* $(\mathcal{K}(X), d_H)$.

Proof. A straightforward argument shows that d_H is indeed a metric on $\mathcal{K}(X)$. If $\{x_n : n \in \mathbb{N}\}$ is dense in X then $\{\{x_{n_1}, \ldots, x_{n_l}\} : l \in \mathbb{N}\}$ is dense in $\mathcal{K}(X)$. It remains to show that $(\mathcal{K}(X), d_H)$ is complete. Suppose that $(K_n)_{n \in \mathbb{N}}$ is a Cauchy sequence in $\mathcal{K}(X)$, i.e. $d_H(K_m, K_n) < \varepsilon$ for all $m, n \geq N_0(\varepsilon)$. We aim at showing that $d_H(K_n, K^*) \to 0$ with

$$K^* := \{x : \text{there exists a sequence } (x_n)_{n \in \mathbb{N}} \text{ with } x_n \in K_n \text{ and } x_n \to x\}.$$

Given $\varepsilon > 0$ and $x \in K^*$ we have $d_H(K_n, K_m) < \frac{1}{2}\varepsilon$ for $m, n \geq N_0(\frac{\varepsilon}{2})$ and $d(x_n, x) < \frac{1}{2}\varepsilon$ for $n \geq N_1(\varepsilon, x)$. But then for $m \geq N_0(\frac{\varepsilon}{2})$ and n sufficiently large $d(x, y_m) \leq d(x, x_n) + d(x_n, y_m) < \varepsilon$ with an appropriate $y_m \in K_m$; hence $K^* \subseteq (K_m)_\varepsilon$ provided that $m \geq N_0(\frac{\varepsilon}{2})$. In order to prove the reverse inclusion $K_m \subseteq K_\varepsilon^*$ for sufficiently large m we fix m such that $d_H(K_m, K_n) < \frac{1}{2}\varepsilon$ for all $n \geq m$. Furthermore we may find $n_1 < n_2 < \ldots < n_k < \ldots$ with $n_1 = m$ such that $d_H(K_{n_j}, K_n) < 2^{-j}\varepsilon$ for all $n \geq n_j$. Given an arbitrary $x \in K_m$ we can inductively construct a sequence $(x_n)_{n \geq m}$ with $x_m = x$ and $x_n \in K_n$ such that $d(x_{n_j}, x_n) < 2^{-j}\varepsilon$ for all n with $n_j + 1 \leq n \leq n_{j+1}$. Since $(x_n)_{n \geq m}$ is a Cauchy sequence it converges to some $\bar{x} \in K^*$. (This also implies that $K^* \neq \emptyset$.) Due to $d(x, \bar{x}) \leq \sum_{j=1}^{\infty} d(x_{n_j}, x_{n_{j+1}}) \leq \varepsilon$ we have $K_m \subseteq K_{2\varepsilon}^*$ and consequently $d_H(K_m, K^*) \to 0$ as $n \to \infty$.

We still have to show that K^* is compact. Since any finite ε-net for K_m naturally yields a finite 2ε-net for K^* if $d_H(K_m, K^*) < \varepsilon$ the set K^* is totally bounded. Furthermore, if $x_n \in K^*$ and $x_n \to x$ then $d(x, y_n) \leq d(x, x_n) + d_H(K^*, K_n)$ for some $y_n \in K_n$. Hence K^* is closed and thus compact. The assertion concerning τ follows directly from a simple observation: if $K_1, K_2, M_1, M_2 \in \mathcal{K}(X)$ then we find for the Hausdorff distances $d_H(K_1 \cup M_1, K_2 \cup M_2) \leq \max\{d_H(K_1, K_2), d_H(M_1, M_2)\}$ and $d_H\big(T_i(K_1), T_i(K_2)\big) \leq L_i d_H(K_1, K_2)$. Consequently τ is a contraction map with contraction rate not larger than $\max_{i=1}^{l} L_i$. $\square$

Although it takes some time to get acquainted with the Hausdorff distance of compact sets even in familiar spaces, a small value of d_H is – after some contemplation – seen to reasonably formalize the property of sets being near to each other. Lemma 5.12 together with the following result gives a sound basis to the existence of sets like the Sierpinski triangle. Moreover, a method for approximating such bizarre sets is provided.

Theorem 5.13. *Let $(T_i)_{i=1}^l$ denote a hyperbolic iterated function system on X with μ^* being its unique invariant probability measure. Then*

$$d_H(\tau^n(K), \operatorname{supp} \mu^*) \to 0 \quad as \; n \to \infty$$

for every $K \in \mathcal{K}(X)$.

Proof. We shall first show that $\operatorname{supp} P\mu = \tau(\operatorname{supp} \mu)$ for any $\mu \in \mathcal{M}_1(X)$ whose support is an element of $\mathcal{K}(X)$. Since $P\mu(\tau(\operatorname{supp} \mu)) = \sum_i p_i \mu(T_i^{-1}(\tau(\operatorname{supp} \mu))) \geq \sum_i p_i = 1$ clearly $\operatorname{supp} P\mu \subseteq \tau(\operatorname{supp} \mu)$. On the other hand, if $x \notin \operatorname{supp} P\mu$ then $P\mu(U) = 0$ for some open set U containing x. Consequently $T_i \mu(U) = 0$ for all i and $\tau(\operatorname{supp} \mu) \subseteq U^c$; hence $x \notin \tau(\operatorname{supp} \mu)$. The theorem will follow if we are able to show that $\operatorname{supp} \mu^*$ is compact and fixed by τ. In order to achieve this we shall briefly develop a tool useful also for our further investigations.

Fix $y \in X$ and define a map $h : \Sigma_l \to X$ according to

$$h\big((x_n)_{n\in\mathbb{N}_0}\big) := \lim_{n\to\infty} T_{x_0} \circ \ldots \circ T_{x_n}(y).$$

It is easy to see that h is well defined and does not depend on the choice of y. Let us endow Σ_l with the metric

$$d_\tau\big((x_n)_{n\in\mathbb{N}_0}, (y_n)_{n\in\mathbb{N}_0}\big) := \begin{cases} 1 & \text{if } x_0 \neq y_0, \\ L_{x_0}L_{x_1}\ldots L_{x_{m-1}} & \text{if } m = \min\{n : x_n \neq y_n\} \geq 1, \\ 0 & \text{otherwise}, \end{cases} \quad (5.13)$$

where we tacitly and without loss of generality assume all Lipschitz constants to be positive. Definition (5.13) is a generalization of the metric (2.3) which we exclusively dealt with in earlier chapters – there we had $L_i = \frac{1}{2}$ for all i. Observe that the metrics arising from different contraction ratio lists are generally not equivalent although they induce the same topology on Σ_l. However, we notice that $d\big(h((x_n)_{n\in\mathbb{N}_0}), h((y_n)_{n\in\mathbb{N}_0})\big) \leq \kappa d_\tau\big((x_n)_{n\in\mathbb{N}_0}, (y_n)_{n\in\mathbb{N}_0}\big)$ with some constant κ. Thus h is Lipschitz continuous, and $K^* := h(\Sigma_l)$ is compact.

Clearly $\tau(K^*) = K^*$, hence $d_H(\operatorname{supp} P^n \mu, K^*) \to 0$. Our final task is to prove that $\operatorname{supp} \mu^* = K^*$. To this end take $x \in X \backslash K^*$. For some U containing x and all sufficiently large n we have $P^n \mu(U) = 0$. By $P^n \mu \overset{w}{\to} \mu^*$ and the so-called *Portmanteau Theorem* (Exercise 5.10) $\mu^*(U) = 0$ which shows that $\operatorname{supp} \mu^* \subseteq K^*$. Therefore $\operatorname{supp} \mu^*$ is compact, and we are done because K^* is the only non-empty compact set fixed by τ. $\square$

So far the probabilities $(p_i)_{i=1}^l$ have not played any role. While μ^* is expected to depend on these probabilities we have just seen that $\operatorname{supp} \mu^*$ does not. The following theorem offers a stochastic possibility to construct $\operatorname{supp} \mu^*$. Recall that according to (5.8) we may interpret an iterated function system as a stochastic iteration scheme $x_{n+1} := T_{\xi_{n+1}}(x_n)$ where the ξ_n are independent random variables on $\{1, \ldots, l\}$ with $\mathbb{P}(\xi_n = i) = p_i$. We assume that $p_i > 0$ for all i and denote by $\{x_n : n \in \mathbb{N}_0\} \subseteq X$ the stochastic orbit of the initial point x_0.

Theorem 5.14. *Let $(T_i)_{i=1}^{l}$ denote a hyperbolic iterated function system with invariant measure μ^*. Then with probability one*

$$d_H\left(\overline{\{x_n, x_{n+1}, \ldots\}}, \operatorname{supp}\mu^*\right) \to 0 \quad \text{as } n \to \infty.$$

Proof. Since $d_H(\tau^n(\{x_0\}), \operatorname{supp}\mu^*) \to 0$ the stochastic orbit of x_0 has compact closure. Given $\varepsilon > 0$ we shall show that $\overline{\lim}_{n\to\infty} d_H\left(\overline{\{x_n, x_{n+1}, \ldots\}}, \operatorname{supp}\mu^*\right) < 3\varepsilon$ with probability one. Since $\operatorname{supp}\mu^*$ is compact there exists a finite ε-net $\{y_1, \ldots, y_r\}$ for that set. Choose m sufficiently large so that $d_H(\tau^m(\{x_0\}), \operatorname{supp}\mu^*) < \varepsilon$ as well as $(\max_i L_i)^m \operatorname{diam} \overline{\{x_0, x_1, \ldots\}} < \varepsilon$. For appropriate numbers $\alpha(1, i), \ldots, \alpha(m, i)$ from $\{1, \ldots, l\}$ we then find

$$d(T_{\alpha(1,i)} \circ T_{\alpha(2,i)} \circ \ldots \circ T_{\alpha(m,i)}(z), y_i) < 2\varepsilon \quad \text{for all } z \in \overline{\{x_0, x_1, \ldots\}} \text{ and } i \in \{i, \ldots, r\}.$$

The probability that each of the r finite sequences of symbols $(\alpha(m, i), \ldots, \alpha(1, i))$ appears infinitely often within a sequence from Σ_l equals one. Therefore with probability one for any n

$$d\left(y_i, \overline{\{x_n, x_{n+1}, \ldots\}}\right) < 2\varepsilon \quad \text{for all } i \in \{1, \ldots, r\}$$

implying that $\operatorname{supp}\mu^* \subseteq \overline{\{x_n, x_{n+1}, \ldots\}}_{3\varepsilon}$ with probability one, too. Conversely, since $d_H(\tau^n(\overline{\{x_0, x_1, \ldots\}}), \operatorname{supp}\mu^*) \to 0$ and hence $\overline{\{x_n, x_{n+1}, \ldots\}} \subseteq (\operatorname{supp}\mu^*)_{3\varepsilon}$ for sufficiently large n we finally get with probability one

$$\overline{\lim}_{n\to\infty} d_H\left(\overline{\{x_n, x_{n+1}, \ldots\}}, \operatorname{supp}\mu^*\right) < 3\varepsilon$$

from which the theorem follows. $\square$

According to Theorem 5.14 the possibly complicated set $\operatorname{supp}\mu^*$ may be approximated by simulating a stochastic orbit and omitting a sufficiently large initial part thereof. This construction elucidates our earlier observation concerning the dynamics of a stochastically excited mechanical system (cf. Figure 5.3). Although $\operatorname{supp}\mu^*$ does not depend on the probabilities $(p_i)_{i=1}^{l}$ the rate of convergence can be rather poor if these probabilities are not chosen appropriately. In Figure 5.6 two stochastic approximations of the Sierpinski triangle corresponding to different choices of (p_1, p_2, p_3) are depicted.

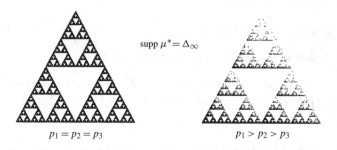

$$\operatorname{supp}\mu^* = \Delta_\infty$$

$$p_1 = p_2 = p_3 \qquad\qquad\qquad p_1 > p_2 > p_3$$

Figure 5.6. The measure μ^* depends on the choice of $(p_i)_{i=1}^{l}$ but $\operatorname{supp}\mu^*$ does not.

Having ensured a mathematically sound existence and sketched a method of generation for such complicated objects as the Sierpinski triangle we shall now have a closer look at the geometry of these sets. Our intention is to describe the elements of $\mathcal{K}(X)$, or even $\mathcal{B}(X)$, by a quantity which we call *dimension*. As one might imagine, there are many ways to define dimensions of sets, the utility of each such concept depending decisively on the questions one wants to answer. As a substantial discussion of the latter statement is far beyond the scope of this book, we restrict ourselves to an informal treatment of one special type of dimension usually referred to as *Hausdorff* (sometimes also *Hausdorff–Besicovitch*) *dimension*. This special dimension is quite hard to determine even for relatively simple sets. Nevertheless it is probably the most important type of dimension for its mathematical versatility. In the sequel we are going to briefly introduce Hausdorff measures and Hausdorff dimension and also sketch the latter's application to iterated function systems. The interested reader is encouraged to consult [24] for further details.

Fix a subset A of the complete separable metric space X. A sequence $(A_n)_{n\in\mathbb{N}}$ of sets is called a δ-*cover* of A if $\bigcup_{n\in\mathbb{N}} A_n \supseteq A$ and diam $A_n \leq \delta$ for all n. Given real numbers $s \geq 0$, $\delta > 0$ let

$$H_\delta^s(A) := \inf\left\{\sum_{n\in\mathbb{N}}(\text{diam } A_n)^s : (A_n)_{n\in\mathbb{N}} \text{ is a } \delta\text{-cover of } A\right\},$$

where $H_\delta^s(A)$ is understood to be infinite if the set on the right-hand side happens to be empty. Since $H_\delta^s(A)$ is non-decreasing for decreasing δ the (possibly infinite) limit

$$H^s(A) := \lim_{\delta\to 0} H_\delta^s(A) \geq 0$$

exists for every set $A \subseteq X$. The following theorem summarizes some of the properties of the quantities H^s. Since parts of the proof necessitate specific technical tools not available here we completely omit it. The assertions are, however, easy to grasp.

Theorem 5.15. *For the set functions H^s defined above the following properties hold:*

(i) *The restriction of H^s to $\mathcal{B}(X)$ defines a measure, called the s-dimensional Hausdorff measure on X;*

(ii) *H^0 is just the counting measure #;*

(iii) *If $s > 0$ then $H^s(A) = 0$ for any countable set A;*

(iv) *Let Y denote another metric space. If $T : X \to Y$ is a Hölder map of exponent $\alpha \in]0, 1]$, which means that $d(T(x), T(y)) \leq C d(x, y)^\alpha$ with some constant $C > 0$, then $H^{\frac{s}{\alpha}}(T(A)) \leq C^{\frac{s}{\alpha}} H^s(A)$;*

(v) *For $X = \mathbb{R}^d$ with the Euclidean metric the measure H^d is a multiple of λ^d; more precisely, $H^d = c_d^{-1}\lambda^d$ where $c_d = 2^{-d}\pi^{\frac{d}{2}}/\Gamma\left(\frac{d}{2}+1\right)$ denotes the volume of the d-dimensional ball with diameter one.*

We deduce from (iv) that $H^s(T(A)) = L^s H^s(A)$ if T is a similarity, i.e. $d(T(x), T(y)) = Ld(x, y)$ for all x, y. As a consequence of property (v) we have $H^d(U) > 0$ for any non-empty open subset $U \subseteq \mathbb{R}^d$; the quantities H^s thus may be thought of as a generalization of λ^d to non-integer dimensions.

For us the measures H^s themselves are of little interest since for a given set A the value of $H^s(A)$ turns out to be zero or infinite for most s. Indeed, if $H^{s_0}(A) < \infty$ then $H^s(A) = 0$ for all $s > s_0$. If on the other hand $H^{s_0}(A) > 0$ then necessarily $H^s(A) = \infty$ for all $s < s_0$.

Definiton 5.16. For any set $A \subseteq X$ the uniquely determined number

$$\dim_H A := \sup\{s \geq 0 : H^s(A) = \infty\} = \inf\{s \geq 0 : H^s(A) = 0\} \in [0, +\infty]$$

is called the *Hausdorff dimension* of A. (In order to avoid ambiguities we set $\sup \emptyset := 0$ and $\inf \emptyset := +\infty$.)

The versatility of Hausdorff dimension mainly comes from the fact that its definition rests on the quantities H^s which are measures and thus mathematically well-understood. However, given a set $A \subseteq X$ it is generally impossible to calculate $\dim_H A$ right from the definition. In the sequel we shall therefore estimate Hausdorff dimensions by means of ad-hoc techniques. A more systematic treatment (if there is one at all) definitely lies outside the scope of this text (see for instance [24]).

Theorem 5.17. *The following properties hold for the Hausdorff dimension:*

(i) *If $A \subseteq B$ then $\dim_H A \leq \dim_H B$;*

(ii) *For every sequence $(A_n)_{n \in \mathbb{N}}$ of sets $\dim_H \bigcup_{n \in \mathbb{N}} A_n = \sup_{n \in \mathbb{N}} \dim_H A_n$;*

(iii) *If T is a Hölder map of exponent α then $\dim_H T(A) \leq \alpha^{-1} \dim_H A$ for all $A \subseteq X$;*

(iv) *If $\dim_H X < 1$ then X is totally disconnected (and hence zero-dimensional as a topological space);*

(v) *For every open subset $U \subseteq \mathbb{R}^d$ $\dim_H U = d$.*

Proof. Properties (i) and (ii) immediately follow from $H^s_\delta(A) \leq H^s_\delta(B)$ for $A \subseteq B$ and the countable sub-additivity of H^s, respectively, while (iii) is implied by (iv) of Theorem 5.15. In order to prove (iv) take two different points $x_1, x_2 \in X$; we are going to show that these two points belong to different open components. To this end define a Lipschitz continuous map $T : X \to \mathbb{R}$ according to $T(x) := d(x, x_1)$. Consequently $\dim_H T(X) < 1$ by (iii), and $\mathbb{R} \backslash X$ is dense in $\mathbb{R}$. Take $\rho \in \mathbb{R} \backslash X$ with $0 < \rho < d(x_1, x_2)$. The disjoint sets $T^{-1}([0, \rho[)$ and $T^{-1}(]\rho, +\infty[)$ both are open, furthermore their union is X. Therefore X is totally disconnected. Finally, every ball in $\mathbb{R}^d$ with non-empty interior has finite positive H^d-measure by property (v) of Theorem 5.15; this clearly implies (v). $\square$

The above theorem may be summarized by saying that Hausdorff dimension exhibits several properties one is likely to expect for any notion of dimension: it is *monotonic* (property (i)), *countably stable* (property (ii)), gives dimension zero to single points, and $\dim_H \mathbb{R}^d = d$. Additionally, due to (iii) Hausdorff dimension is a *metric invariant*, which means that two spaces have the same Hausdorff dimension if there exists a bijective isometry between them.

What we are most interested in here is $\dim_H K^*$ where K^* denotes the attractor of a hyperbolic iterated function system $(T_i)_{i=1}^l$ with ratio list $(L_i)_{i=1}^l$ and $0 < L_i < 1$. To determine $\dim_H K^*$ we shall first calculate $\dim_H \Sigma_l$, where we endow Σ_l with the metric d_τ according to

$$
d_\tau\left((x_n)_{n\in\mathbb{N}_0}, (y_n)_{n\in\mathbb{N}_0}\right) := \begin{cases} 1 & \text{if } x_0 \neq y_0, \\ L_{x_0} L_{x_1} \ldots L_{x_{m-1}} & \text{if } m = \min\{n : x_n \neq y_n\} \geq 1, \\ 0 & \text{otherwise}; \end{cases}
$$

this metric is a generalization of the metric on Σ_l introduced in Chapter Two and has already been used in the proof of Theorem 5.13. Let s^* denote the unique real number which satisfies

$$
L_1^{s^*} + \ldots + L_l^{s^*} = 1. \tag{5.14}
$$

Our aim is to show that $\dim_H \Sigma_l = s^*$. Indeed, setting $\delta_n := (\max_{i=1}^l L_i)^n$ we notice that $\{[i_0, \ldots, i_n] : i_0, \ldots, i_n \in \{1 \ldots, l\}\}$ is a δ_n-cover for Σ_l. Therefore $H_{\delta_n}^s(\Sigma_l) \leq \sum_{i_0,\ldots,i_n=1}^l (L_{i_0} \ldots L_{i_n})^s = \left(\sum_{i=1}^l L_i^s\right)^n$, and $H^s(\Sigma_l) = 0$ for any $s > s^*$ which in turn implies that $\dim_H \Sigma_l \leq s^*$. In order to obtain the reverse inequality we first observe that any set $A \subseteq \Sigma_l$ containing at least two elements is itself contained in a finite cylinder C with $\operatorname{diam} A = \operatorname{diam} C$. We may thus restrict ourselves to coverings by finite cylinders. But then, given $\varepsilon > 0$, by virtue of (5.14)

$$
H_{\delta_n}^{s^*}(\Sigma_l) + \varepsilon \geq \sum_{i_0,\ldots,i_n=1}^l (\operatorname{diam}[i_0,\ldots,i_n])^{s^*} = 1
$$

and thus $\dim_H \Sigma_l \geq s^*$. Note that a variation of the quantities L_i will typically change s^* while the *topology* of Σ_l, i.e. the family of its open sets, remains unchanged. Hausdorff dimension therefore is not a topological invariant: homeomorphic spaces may exhibit different Hausdorff dimensions.

Piecing our considerations together we obtain the following result about the objects we are mainly interested in.

Theorem 5.18. *Let* $(T_i)_{i=1}^l$ *be a hyperbolic iterated function system with Lipschitz constants* L_i *and denote by* $K^* \in \mathcal{K}(X)$ *the unique attractor of this system. Then the relation* $\dim_H K^* \leq s^*$ *holds with* s^* *calculated from (5.14).*

Proof. We just have to combine our earlier findings. Recall that for any fixed $y \in X$ the map $h : \Sigma_l \to X$ defined as $h\left((x_n)_{n\in\mathbb{N}_0}\right) := \lim_{n\to\infty} T_{x_0} \circ \ldots \circ T_{x_n}(y)$ turns out to be a Lipschitz map with respect to the metric d_τ. According to the proof of Theorem 5.13 we have $h(\Sigma_l) = K^*$, and thus the result follows. $\square$

In the case of the Sierpinski triangle Δ_∞ for example we have $\frac{\log 3}{\log 2} \approx 1.585$ as an upper bound for the Hausdorff dimension. In general the estimate provided by Theorem 5.18 will not be accurate. As a simple illustration take $X := \mathbb{R}$ and

$$T_1(x) := \frac{2}{3}x, \quad T_2(x) := \frac{2}{3}x + \frac{1}{3}. \tag{5.15}$$

Here $s^* = \frac{\log 2}{\log 3 - \log 2} \approx 1.710$ but $K^* = [0, 1]$, and so $\dim_H K^* = 1 < s^*$. However, if we confine ourselves to a special class of iterated function systems the inequality in Theorem 5.18 is in fact an equality.

Theorem 5.19. *Assume that the hyperbolic iterated function system $(T_i)_{i=1}^l$ exclusively consists of similarities on $\mathbb{R}^d$. If for some non-empty bounded open set U*

$$U \supseteq \bigcup_{i=1}^l T_i(U) \quad \text{and} \quad T_i(U) \cap T_j(U) = \emptyset \ \text{if } i \neq j$$

then, with the notation of Theorem 5.18, $\dim_H K^* = s^*$.

Proof. We include the proof for the sake of completeness. In its course it will become evident that the actual computation of Hausdorff dimensions may necessitate a certain amount of non-trivial calculations; our proof closely follows [23]. In a first step we are going to show that for any set $A \subseteq K^*$ the cardinality of

$$\Xi_A := \{(i_0, \ldots, i_n) \in \bigcup_{k \in \mathbb{N}} \{1, \ldots, l\}^k : T_{i_0} \circ \ldots \circ T_{i_n}(U) \cap A \neq \emptyset,$$
$$\text{diam } T_{i_0} \circ \ldots \circ T_{i_n}(U) < \text{diam } A \leq \text{diam } T_{i_0} \circ \ldots \circ T_{i_{n-1}}(U)\}$$

is bounded by a constant not depending on A. Setting $L := \min_{i=1}^l L_i > 0$ we have the estimate $\text{diam } T_{i_0} \circ \ldots \circ T_{i_n}(U) \geq L \, \text{diam } A$ for all $(i_0, \ldots, i_n) \in \Xi_A$ because the T_i are similarities. But then

$$\lambda^d \left(T_{i_0} \circ \ldots \circ T_{i_n}(U)\right) = \lambda^d(U)(L_{i_0} \ldots L_{i_n})^d \geq \lambda^d(U)L^d \left(\frac{\text{diam } A}{\text{diam } U}\right)^d.$$

Given $x \in A$ and $(i_0, \ldots, i_n) \in \Xi_A$ every point in $T_{i_0} \circ \ldots \circ T_{i_n}(U)$ is at most 2 diam A apart from x. Consequently $\bigcup_{(i_0, \ldots, i_n) \in \Xi_A} T_{i_0} \circ \ldots \circ T_{i_n}(U)$ is contained in a ball of diameter 2 diam A, and thus

$$(2 \, \text{diam } A)^d \lambda^d(\text{unit ball}) \geq \#\Xi_A \lambda^d(U)L^d \left(\frac{\text{diam } A}{\text{diam } U}\right)^d$$

from which we deduce that $\#\Xi_A$ is bounded by a quantity ξ which only depends on U and d.

By the properties of U it is easy to see that $h^{-1}(A) \subseteq \bigcup_{(i_0, \ldots, i_n) \in \Xi_A} [i_0, \ldots, i_n]$ for any $A \subseteq K^*$ open in K^*; here h denotes the map $h : \Sigma_l \to K^*$ introduced in the proof of Theorem 5.13. But then

$$H^{s^*}\left(h^{-1}(A)\right) \leq \sum_{(i_0, \ldots, i_n) \in \Xi_A} H^{s^*}([i_0, \ldots, i_n]) \leq \frac{\xi}{(\text{diam } U)^{s^*}} (\text{diam } A)^{s^*},$$

and thus $H^{s^*}(K^*) \geq (\text{diam } U)^{s^*}/\xi > 0$, and consequently $\dim_H K^* \geq s^*$. $\square$

As an example we take a last look at the Sierpinski triangle Δ_∞. Here $U := \text{int } \Delta_0$ meets the conditions of Theorem 5.19 (see Figure 5.5). So we can sharpen our above observation by saying that the Hausdorff dimension of Δ_∞ equals $\frac{\log 3}{\log 2}$. This agrees with our intuitive feeling that Δ_∞ is "more than a line" but "less than a surface". Since its Hausdorff dimension exceeds the value of any reasonable topological (e.g. covering, inductive [23]) dimension, the Sierpinski triangle constitutes an example of a *fractal*. (See [23, 24] for a discussion of this latter notion coined by Mandelbrot.)

As a second example consider the two maps

$$T_1(x) = \frac{1}{3}x, \quad T_2(x) := \frac{1}{3}x + \frac{2}{3} \quad (5.16)$$

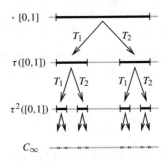

on $X := [0, 1]$ (cf. Figure 5.7). Here we may take $U :=]0, 1[$, and we find $\dim_H C_\infty = \frac{\log 2}{\log 3}$ for the attractor of the iterated function system $(T_i)_{i=1}^2$. Usually C_∞ is called the *Cantor middle-thirds set*; it is by now the classical prototype of all Cantor sets which we have seen in earlier chapters.

Figure 5.7. The construction of the classical Cantor middle-thirds set C_∞

So far we have only dealt with the geometry of $\text{supp } \mu^*$, where μ^* stands for the limit measure of an asymptotically stable Markov operator on measures induced by an iterated function system. But how about the measure μ^* itself? Since a thorough analysis of probability measures on metric spaces is rather demanding we shall merely discuss two simple examples here. Consider first the hyperbolic iterated function system (5.16) and the induced Markov operator on $\mathcal{M}_1(\mathbb{R})$,

$$P\mu := \frac{1}{2}T_1\mu + \frac{1}{2}T_2\mu. \quad (5.17)$$

We already know that this operator is weakly asymptotically stable, the unique invariant distribution μ^* being supported on the Cantor middle-thirds set C_∞. Furthermore $\dim_H \text{supp } \mu^* = s^* = \frac{\log 2}{\log 3}$. In order to proceed with our analysis we translate (5.17) to the language of distribution functions. Recall that we called $F_\mu(x) := \mu(]-\infty, x])$ the *distribution function* of the probability measure $\mu \in \mathcal{M}_1(\mathbb{R})$. From (5.17) we obtain

$$F_{P\mu}(x) = \frac{1}{2}F_\mu(3x) + \frac{1}{2}F_\mu(3x - 2). \quad (5.18)$$

Since $C_\infty \subseteq [0, 1]$ we rewrite and treat (5.18) on the subspace X of distribution functions in $C([0, 1])$, i.e. $X := \{f \in C([0, 1]) : f(0) = 0, f(1) = 1\}$ by means of the map

$$\Phi_P : \begin{cases} X & \to & X \\ f & \mapsto & \Phi_P(f) \end{cases}$$

with

$$\Phi_P(f)(x) := \frac{1}{2} f(\min\{3x, 1\}) + \frac{1}{2} f(\max\{3x - 2, 0\}) .$$

The relation (5.18) then reads $F_{P\mu} = \Phi_P(F_\mu)$. If we endow X with the standard metric

$$d(f, g) := \max_{x \in [0,1]} |f(x) - g(x)| \quad (f, g \in X)$$

then $d(\Phi_P(f), \Phi_P(g)) = \frac{1}{2}d(f, g)$ for all $f, g \in X$. Hence Φ_P is a contraction, its unique fixed point being F_{μ^*}. The geometrical meaning of iterating Φ_P can be grasped from Figure 5.8. The function F_{μ^*} is locally constant (hence differentiable)

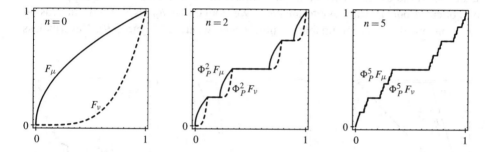

Figure 5.8. For any $\mu \in \mathcal{M}_1([0, 1])$ the sequence $(\Phi_P^n(F_\mu))_{n \in \mathbb{N}_0}$ uniformly converges to the devil's staircase F_{μ^*}.

on $[0, 1] \backslash C_\infty$ while strictly increasing on C_∞; it is called the *devil's staircase* and has found applications in various fields of mathematics and physics ([20]). Choosing $\mu_0 := \lambda^1|_{[0,1]}$ we have

$$P^n \mu_0 \big(T_{i_0} \circ \ldots \circ T_{i_m} ([0, 1]) \big) = 2^{-(m+1)} \quad \text{for all } i_0, \ldots, i_m \in \{1, 2\}, \ n \geq m + 1 .$$

Since F_{μ^*} is continuous $\mu^* \big(\partial T_{i_0} \circ \ldots \circ T_{i_m} ([0, 1]) \big) = 0$, and thus

$$\mu^* \big(T_{i_0} \circ \ldots \circ T_{i_m} ([0, 1]) \big) = 2^{-(m+1)} \quad \text{for all } i_0, \ldots, i_m \in \{1, 2\} .$$

Taking on trust the result $H^{s^*}(C_\infty) = 1$ (which is not difficult to achieve, see [24]) we see that μ^* and H^{s^*} agree on all sets of the form $T_{i_0} \circ \ldots \circ T_{i_m} ([0, 1]) \cap C_\infty$; by the Extension Theorem A.4 therefore $\mu^* = H^{s^*}$. Despite the complicated geometry of its support the limit measure μ^* thus is quite simple: it is just the uniform Hausdorff measure of the appropriate dimension.

As a second example let us consider the Markov operator

$$P\mu := p T_1 \mu + q T_2 \mu \tag{5.19}$$

on $\mathcal{M}_1(\mathbb{R})$ with $0 < p, q < 1$, $p + q = 1$ induced by the two similarities

$$T_1(x) := \frac{1}{2}x, \quad T_2(x) := \frac{1}{2}x + \frac{1}{2}.$$

Here $\operatorname{supp} \mu^* = [0, 1]$ – evidently a set of a rather tame geometric structure. Switching to distribution functions as before we must now iterate $\Psi_P : X \to X$ with

$$\Psi_P(f)(x) := pf(\min\{2x, 1\}) + qf(\max\{2x - 1, 0\}).$$

Things look much the same as before. If $p = q = \frac{1}{2}$ then $F_{\mu^*}(x) \equiv x$ which corresponds to the fact that $P\lambda^1|_{[0,1]} = \lambda^1|_{[0,1]}$ in this case. However, if $p \neq q$ we end up with a little surprise: the limiting measure μ^* is *much* more complicated than one would expect. This situation contrasts our first example where $\operatorname{supp} \mu^*$ was complicated but μ^* was easy to analyse. We notice that F_{μ^*} is a strictly increasing function (and thus is differentiable λ^1-almost everywhere, cf. Figure 5.9). Nevertheless

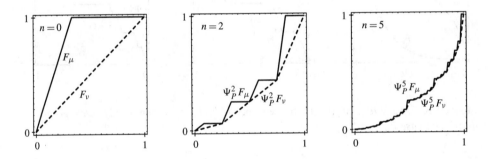

Figure 5.9. The fixed point of Ψ_P is the distribution function of a fractal measure μ^* with $\lambda^1 \perp \mu^*$ but $\operatorname{supp} \mu^* = [0, 1]$.

$\lambda^1 \perp \mu^*$, i.e., λ^1 and μ^* are mutually singular. In order to see this let us define a map $h : [0, 1] \to \Sigma_2$ assigning to each $x \in [0, 1]$ its binary expansion, hence $h(x) = (x_{n+1} + 1)_{n \in \mathbb{N}_0}$ for $x = \sum_{n=1}^{\infty} x_n 2^{-n}$. With the exception of a countable set this map is well defined and measurable. Consider two probability measures ν_1, ν_2 on $(\Sigma_2, \mathcal{B}(\Sigma_2))$ defined as

$$\nu_1 := \mu_{\pi_1, P_1} \quad \text{with} \quad \pi_1 := \left(\frac{1}{2}, \frac{1}{2}\right), \ P_1 := \begin{pmatrix} \frac{1}{2} & \frac{1}{2} \\ \frac{1}{2} & \frac{1}{2} \end{pmatrix},$$

$$\nu_2 := \mu_{\pi_2, P_2} \quad \text{with} \quad \pi_2 := (p, q), \ P_2 := \begin{pmatrix} p & q \\ p & q \end{pmatrix}.$$

For these measures we find $\nu_1 = h\lambda^1|_{[0,1]}$ and $\nu_2 = h\mu^*$. From Lemma 3.14 we know that ν_1 and ν_2 both are ergodic measures for the shift map. Therefore by Lemma 3.13

$\nu_1 \perp \nu_2$ which means that $\nu_1(A) = \nu_2(A^c) = 0$ for some set $A \in \mathcal{B}(\Sigma_2)$. But then $\lambda^1(h^{-1}(A)) = 0$ whereas $\mu^*(h^{-1}(A)) = 1$. Hence $\lambda^1|_{[0,1]}$ and μ^* are mutually singular. Notice however that supp $\mu^* = [0, 1] = \text{supp } \lambda^1|_{[0,1]}$. Despite its complicated structure the measure μ^* has found several applications in physics ([43]). It is usually referred to as an example of a *fractal measure* or a *multifractal*.

Our observations indicate that the dynamical evolution of measures gives rise to a great variety of phenomena. A rigorous analysis of the latter will open the door to many different fields of applied (and pure) mathematics. Nevertheless, recall that for us a major motivation for introducing the concepts of measure theory originated from our ambition to statistically describe chaotic dynamical systems. We shall therefore close this section by pointing out that the geometrically bizarre objects encountered here naturally constitute a place for chaos to occur. Given our earlier discussion (of the logistic family or the solenoid, say) this fact should not come as a complete surprise.

Suppose we are given a hyperbolic iterated function system $(T_i)_{i=1}^l$ with the contractions T_i being homeomorphisms. (All the examples studied so far meet this additional condition.) There then exist continuous inverse maps $S_i : T_i(K^*) \to K^*$ such that $T_i \circ S_i = \text{id}_{T_i(K^*)}$ and $S_i \circ T_i = \text{id}_{K^*}$ for all i. Define now a map $S : K^* \to K^*$ by $S(x) := S_{i(x)}(x)$ where the choice function $i : K^* \to \{1, \ldots, l\}$ satisfies $i(x) \in \{j : x \in T_j(K^*)\}$. If the sets $T_i(K^*)$ are disjoint then S is uniquely determined and continuous; otherwise we require that S be measurable. Recall that earlier we have defined a continuous map h from Σ_l to K_* as

$$h\left((x_n)_{n \in \mathbb{N}_0}\right) := \lim_{n \to \infty} T_{x_0} \circ \ldots \circ T_{x_n}(y),$$

where y denotes an arbitrary point in K_* (cf. the proof of Theorem 5.13).

As in Chapter Two the setting of $x_n := i(S^n(x))$ for $n \in \mathbb{N}_0$ may be interpreted as a coding procedure; and we observe that $h \circ \sigma^k\left((x_n)_{n \in \mathbb{N}_0}\right) = S^k(x)$ for all $k \in \mathbb{N}_0$. Therefore every orbit under S is the h-image of a shift orbit in Σ_l. If $T_i(K^*) \cap T_j(K^*)$ is empty whenever $i \neq j$ then the attractor K^* is totally disconnected, and h turns out to be a topological conjugacy between (Σ_l, σ) and (K^*, S). In this case chaotic dynamics occurs quite naturally on the attractor K^*.

If there is an overlap, i.e. $T_i(K^*) \cap T_j(K^*) \neq \emptyset$ for some $i \neq j$, one can likewise gain some insight by making the maps T_i act on a larger space. As a specific example, consider the hyperbolic iterated function system (5.15) where $K^* = [0, 1]$. Two admissible versions of S are depicted in Figure 5.10. Let $\tilde{X} := \mathbb{R}^2$ and

$$\tilde{T}_1(x_1, x_2) := \left(\frac{2}{3}x_1, \frac{1}{3}x_2\right), \quad \tilde{T}_2(x_1, x_2) := \left(\frac{2}{3}x_1 + \frac{1}{3}, \frac{1}{3}x_2 + \frac{2}{3}\right). \tag{5.20}$$

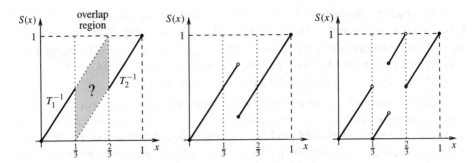

Figure 5.10. Two admissible versions of S (middle and right box)

Now the attractor $\tilde{K}^* \subseteq \mathbb{R}^2$ of $(\tilde{T}_i)_{i=1}^2$ is totally disconnected (cf. Figure 5.11), and $(\tilde{K}^*, \tilde{S})$ is topologically conjugate to the full shift (Σ_2, σ). Denoting by π the projection onto the first coordinate we have $[0, 1] = K^* = \pi(\tilde{K}^*)$; the orbits of S in $[0, 1]$ thus are projections of orbits of the chaotic system $(\tilde{K}^*, \tilde{S})$. We leave it as an exercise to generalize this construction to arbitrary hyperbolic iterated function systems (Exercise 5.12).

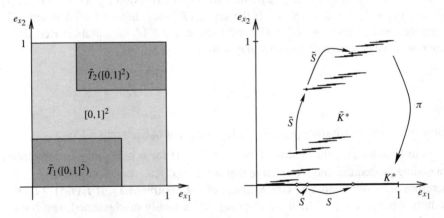

Figure 5.11. The totally disconnected attractor $\tilde{K}^*$ of (5.20)

Sometimes the map S as defined above may be extended to larger parts (or even all) of X. Then K^* is repelling under S and hence gives rise to a chaotic motion of transient type. Unfortunately, not every complicated invariant set in dynamics is thus easily analysed. As we have seen earlier, non-linear maps from applications may generate complicated sets that are attractors in the sense of Chapter Two (corresponding to non-transient chaos like for example on the solenoid). A rigorous analysis of these sets typically is *much* more demanding than the simple hyperbolic case explained here. In fact, only very few representatives from the large class of commonly known attractors have rigorously been proved chaotic so far ([20, 27]). However, even our elementary observations indicate that geometric complexity may – under certain circumstances – come along with complicated dynamics.

5.4 Three final examples

We close our treatment of the dynamics of measures by presenting three different examples loosely connected to this field. The presentation mainly intends to stimulate the interest of the curious and therefore is rather informal and superficial. However, we shall concludingly gain some additional insights concerning topics discussed earlier in this book.

5.4.1 Searching for non-normal numbers

Recall that we called a real number $x \in [0, 1]$ *normal* with respect to a base $b \in \mathbb{N} \setminus \{1\}$ if all possible digits (and all finite blocks of digits having the same length) occur with the same asymptotic frequency in the b-adic expansion of x. As a corollary of the Birkhoff Ergodic Theorem we found that a typical number (in the sense of Lebesgue measure) from the unit interval is *absolutely normal*, i.e. normal with respect to every base. Unfortunately, it is not easy to explicitly construct normal numbers. On the other hand non-normal numbers, though rare exceptions in the sense of Lebesgue measure, abound in $[0, 1]$. In the sequel we shall present a quantification of the latter assertion.

Let X again denote a complete separable metric space and write $B_r(x)$ for the closed ball of radius r centered at x, i.e. $B_r(x) := \{y : d(y, x) \leq r\}$. The following simple observation concerns coverings of bounded sets by means of such balls. Assume that the family $\left(B_{r_i}(x_i)\right)_{i \in I}$ is contained in a bounded subset of X. Setting $\rho_1 := \sup_{i \in I} r_i$ we may find $i_1 \in I$ such that $r_{i_1} > \frac{3}{4}\rho_1$. In an inductive manner let

$$\rho_n := \sup\left\{r_i : i \in I \setminus \{i_1, \ldots, i_{n-1}\}, \ B_{r_i}(x_i) \cap B_{r_{i_l}}(x_{i_l}) = \emptyset \text{ for all } 1 \leq l < n\right\}$$

and take $i_n \in I$ such that $r_{i_n} > \frac{3}{4}\rho_n$. This procedure may or may not terminate. In either case the members of the family $\left(B_{r_{i_n}}(x_{i_n})\right)_{n \in \mathbb{N}}$ are disjoint, and $\bigcup_i B_{r_i} \subseteq \bigcup_n B_{4r_{i_n}}$. We may thus select from $\left(B_{r_i}(x_i)\right)_{i \in I}$ a countable disjoint subfamily which – after an appropriate enlargement – covers the original family.

Lemma 5.20. *Let $A \in \mathcal{B}(X)$ and $\mu \in \mathcal{M}_1(X)$.*

(i) *If $\overline{\lim}_{r \to 0} \mu\left(B_r(x)\right)/r^s < c$ holds for some $c \in \mathbb{R}^+$ and all $x \in A$ then $H^s(A) \geq \mu(A)/c$.*

(ii) *If $\overline{\lim}_{r \to 0} \mu\left(B_r(x)\right)/r^s > c$ holds for some $c \in \mathbb{R}^+$ and all $x \in A$ then $H^s(A) \leq 8^s/c$.*

Proof. In order to prove (i) take $0 < \delta < 1$ and set

$$A_\delta := \{x \in A : \mu(B_r(x)) < cr^s \text{ for } 0 < r \le \delta\}.$$

If $(U_n)_{n \in \mathbb{N}}$ is a δ-cover of A then $B_{\text{diam } U_n}(x) \supseteq U_n$ for every $x \in A \cap U_n$. Consequently

$$\mu(A_\delta) \le \sum_{n:U_n \cap A_\delta \ne \emptyset} \mu(U_n) \le c \sum_n (\text{diam } U_n)^s$$

and therefore $\mu(A_\delta) \le cH_\delta^s(A)$. Since $A = \bigcup_{0<\delta<1} A_\delta$ and the sets A_δ increase with decreasing δ we have $\mu(A) \le cH^s(A)$, hence (i).

We tackle (ii) by first assuming A to be bounded. Fix $\delta > 0$ and consider the family of balls

$$\{B_r(x) : x \in A, 0 < r \le \delta, \mu(B_r(x)) > cr^s\}$$

which by assumption provides a 2δ-cover of A. By virtue of our above observation there exists a subfamily $(B_{r_n}(x_n))_{n \in \mathbb{N}}$ of disjoint sets such that $(B_{4r_n}(x_n))_{n \in \mathbb{N}}$ still covers A. Therefore

$$H_{8\delta}^s(A) \le \sum_n (8r_n)^s = 8^s \sum_n r_n^s < \frac{8^s}{c} \sum_n \mu(B_{r_n}(x_n)) \le \frac{8^s}{c}.$$

If A is not bounded, an approximation of A by bounded sets yields (ii). $\square$

As a corollary of the above lemma we note that for $\mu(A) > 0$ the relation $\dim_H A = \lim_{r \to 0} \log \mu(B_r(x))/\log r$ holds if only this limit exists for all $x \in A$. In application this corollary is sometimes used to estimate the dimension of complicated attractors: one tries to estimate $\mu(A \cap B_r(x))$ for a few small values of r and then plots the graph of $\log \mu(A \cap B_r(x))$ against $\log r$ (see [24] for details).

We are now in a position to analyse some special sets containing only non-normal numbers. To this end fix $b \in \mathbb{N} \setminus \{1\}$ as a base. Given $x \in [0, 1]$ let $h_i(x, n)$ denote the relative frequency of the digit $i \in \{0, 1, \ldots, b-1\}$ in the first n digits of the b-adic expansion of x, that is

$$h_i(x, n) := \frac{1}{n}\#\{1 \le k \le n : x_k = i\} \quad \text{where} \quad x = \sum_{k=1}^{\infty} x_k b^{-k}. \tag{5.21}$$

Observe that we have to remove a countable set from $[0, 1]$ in order to uniquely define the quantities h_i. Now choose a non-negative vector $(p_0, \ldots, p_{b-1}) \in \mathbb{R}^b$ such that $\sum_{i=0}^{b-1} p_i = 1$ and define

$$X_{(p_0,\ldots,p_{b-1})} := \{x \in [0, 1] : \lim_{n \to \infty} h_i(x, n) = p_i \text{ for } 0 \le i \le b - 1\}. \tag{5.22}$$

Clearly $X_{(p_0,\ldots,p_{b-1})} \cap X_{(q_0,\ldots,q_{b-1})} = \emptyset$ if $(p_0, \ldots, p_{b-1}) \ne (q_0, \ldots, q_{b-1})$; furthermore $\lambda^1(X_{(p_0,\ldots,p_{b-1})}) = 0$ whenever $(p_0, \ldots, p_{b-1}) \ne (b^{-1}, \ldots, b^{-1})$. On the other hand $X_{(p_0,\ldots,p_{b-1})}$ is easily seen to be dense in $[0, 1]$. In fact, we can give a much more accurate description of these sets.

Theorem 5.21. *With the notation of* (5.21) *and* (5.22)

$$\dim_H X_{(p_0,\ldots,p_{b-1})} = -\frac{1}{\log b} \sum_{i=0}^{b-1} p_i \log p_i .$$

Proof. Let us first define a probability measure μ on the unit interval according to

$$\mu(\{x : x_1 = i_1, \ldots, x_n = i_n\}) := p_{i_1} \ldots p_{i_n}$$

for all $i_1, \ldots, i_n \in \{0, \ldots, b-1\}$ and $n \in \mathbb{N}$. Furthermore let T_b denote the map $x \mapsto bx \pmod 1$ on $[0, 1]$. From our studies on Markov maps we know that $([0, 1], T_b)$ is isomorphic to (Σ_l, σ) if we endow the former space with the measure μ and the latter with $\mu_{\pi,P}$ where $\pi := (p_i)_{i=0}^{b-1}$ and $P := (p_j)_{i,j=0}^{b-1}$. By virtue of the Birkhoff Ergodic Theorem

$$h_i(x, n) = \frac{1}{n} \sum_{k=0}^{n-1} \mathbb{1}_{\left[\frac{i}{b}, \frac{i+1}{b}\right[} \circ T_b^k(x) \rightarrow p_i \quad [\mu]$$

and thus $\mu(X_{(p_0,\ldots,p_{b-1})}) = 1$. Fix $x \in X_{(p_0,\ldots,p_{b-1})}$ and define a sequence of sets by $A_n := \{y : x_1 = y_1, \ldots, x_n = y_n\}$ for every $n \in \mathbb{N}$. Then

$$\frac{1}{n} \log \frac{\mu(A_n)}{(\operatorname{diam} A_n)^s} = \sum_{i=0}^{b-1} h_i(x, n) \log p_i + s \log b \rightarrow \sum_{i=0}^{b-1} p_i \log p_i + s \log b$$

as $n \rightarrow \infty$. Thus if we set $s^* := -\frac{1}{\log b} \sum_{i=0}^{b-1} p_i \log p_i$ then

$$\lim_{n \rightarrow \infty} \frac{\mu(A_n)}{(\operatorname{diam} A_n)^s} = \begin{cases} 0 & \text{if } s < s^*, \\ +\infty & \text{if } s > s^*; \end{cases}$$

and a look at a slightly modified version of Lemma 5.20 (taking into account that x need not be the midpoint of A_n) completes the proof. $\square$

The unit interval thus contains within a set of Lebesgue measure zero an uncountable union of uncountable sets which exclusively comprise non-normal numbers. One should keep in mind results like this in order to not succumb to the temptation of regarding the unit interval as a small or even familiar set!

5.4.2 The fractal nature of Brownian paths

In the course of earlier chapters we repeatedly dealt with the family of non-deterministic Markov operators on $\mathcal{M}_1(\mathbb{R})$ given by

$$\frac{dP_t\mu}{d\lambda^1}(x) := \frac{1}{\sqrt{4\pi t}} \int_{\mathbb{R}} e^{-\frac{(x-y)^2}{4t}} d\mu(y) \quad (t > 0).$$

Our interpretation of this family originated from the observation that the density $u_\mu(x, t) := \frac{dP_t\mu}{d\lambda^1}(x)$ is a solution of the heat equation $\frac{\partial u}{\partial t} = \frac{\partial^2 u}{\partial x^2}$ for $t > 0$ and that furthermore $P_t \mu \overset{w}{\to} \mu$ as $t \to 0$. We therefore regarded $P_t \mu$ as a description of what had happened to the initial distribution μ of heat within t units of time. Especially, we considered

$$P_t \delta_0([a, b]) = \frac{1}{\sqrt{4\pi t}} \int_a^b e^{-\frac{x^2}{4t}} \, dx \quad (t > 0) \tag{5.23}$$

as the total amount of heat contained in $[a, b]$ at time t after one unit of heat had been released at the origin at $t = 0$. In the sequel we are going to adopt a different point of view: rather than studying such global objects as distributions we shall focus on tiny quantities of energy which we assume to move on the line according to a simple symmetric random walk. When put into the appropriate framework this microscopic point of view not only yields (5.23) in a natural way but also provides additional information about phenomena not visible to the macroscopic description via the Markov operators P_t.

Let ξ_n describe the n-th step of a simple symmetric random walk on $\mathbb{Z}$ starting at the origin, i.e. $\mathbb{P}(\xi_0 = i) = \delta_{0i}$ and

$$\mathbb{P}(\xi_{n+1} = i + 1 | \xi_n = i) = \mathbb{P}(\xi_{n+1} = i - 1 | \xi_n = i) = \frac{1}{2}.$$

In order to obtain a continuous curve for the history $(n, \xi_n)_{n \in \mathbb{N}_0}$ we rescale and linearly interpolate the latter (Figure 5.12).

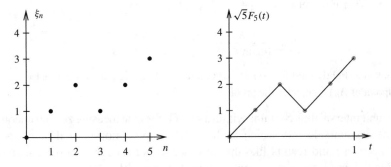

Figure 5.12. The piecewise linear interpolation of the random walk $(\xi_n)_{n \in \mathbb{N}_0}$

More formally, let us introduce for every $n \in \mathbb{N}$ and $(i_0, \ldots, i_n) \in \mathbb{Z}^{n+1}$ a continuous function $F_{(i_0, \ldots, i_n)}$ on the unit interval according to

$$F_{(i_0, \ldots, i_n)}(t) := \frac{1}{\sqrt{n}} \Big((1 - nt + \lfloor nt \rfloor) i_{\lfloor nt \rfloor} + (nt - \lfloor nt \rfloor) i_{\lfloor nt \rfloor + 1} \Big)$$

which linearly interpolates the points $(0, i_0/\sqrt{n}), (1/n, i_1/\sqrt{n}), \ldots, (1, i_n/\sqrt{n})$. Setting $F_n := F_{(\xi_0, \ldots, \xi_n)}$ thus defines a random variable in $C([0, 1])$ with 2^n equally likely outcomes. Consequently we may associate with F_n a probability measure μ_n on $C([0, 1])$ defined as

$$\mu_n(\{f\}) := \begin{cases} 2^{-n} & \text{if } f = F_{(0,i_1,\dots,i_n)} \text{ for some } (i_1, \dots, i_n) \in \mathbb{Z}^n, \\ 0 & \text{otherwise}. \end{cases}$$

For large n the measure μ_n may be thought of as the law governing the random motion of a particle on the line where a large number of tiny steps is performed within one unit of time. If there was a sensible limit for the sequence $(\mu_n)_{n \in \mathbb{N}}$, we would expect this limit to give a microscopic description of the diffusion process discussed above.

Theorem 5.22. *There exists a unique probability measure* $\mathbb{W}$ *on* $C([0, 1])$, *usually referred to as the Wiener measure, such that* $\mu_n \overset{w}{\to} \mathbb{W}$.

Although this assertion is easily written down, its proof turns out to be difficult and rather technical. We shall therefore omit the details (see [10]) and merely concentrate on a few aspects relevant to our earlier discussion. Since $F_n(0) = 0$ for all n, Wiener measure is supported on the closed subspace $\{f \in C([0, 1]) : f(0) = 0\}$. Moreover, $\mathbb{W}$ is non-atomic which means that $\mathbb{W}(\{f\}) = 0$ for any $f \in C([0, 1])$.

In the light of the physical interpretation of $\mathbb{W}$ we are most interested in the way this measure quantifies the oscillations of continuous functions. Given $0 \leq t_1 < t_2 \leq 1$ it may be shown that for all $a < b$

$$\mathbb{W}(\{f : a < f(t_2) - f(t_1) < b\}) = \lim_{n \to \infty} \mu_n(\{f : a < f(t_2) - f(t_1) < b\}). \quad (5.24)$$

In order to evaluate the right-hand side of (5.24) we just have to count the sample paths lying in the corresponding set of continuous functions. To be more specific assume $t_1 = k_1 2^{-l}$, $t_2 = k_2 2^{-l}$ with $0 \leq k_1 < k_2 \leq 2^l$ and some $l \in \mathbb{N}$. If $2n > l$ then

$$\mu_{2^{2n}}(\{f : a < f(t_2) - f(t_1) < b\}) =$$

$$2^{-2^{2n}} \sum_{i \in \mathbb{Z}} \#\{\text{sample paths } f \text{ with } f(t_1) = i2^{1-n}, f(t_2) - f(t_1) \in]a, b[\}.$$

It is a simple matter of counting to evaluate the right-hand side by virtue of the Markov property for the random walk $(\xi_n)_{n \in \mathbb{N}_0}$. We have

$$\mu_{2^{2n}}(\{f : a < f(t_2) - f(t_1) < b\}) = 2^{-2^{2n}} N_1 N_2 N_3$$

where the numbers N_1, N_2, N_3 are given by

$$N_1 = 2^{k_1 2^{2n-l}}, \quad N_2 = \sum_{j \in 2^{n-1}]a,b[\cap \mathbb{Z}} \binom{(k_2 - k_1)2^{2n-l}}{(k_2 - k_1)2^{2n-l-1} + j}, \quad N_3 = 2^{2^{2n} - k_2 2^{2n-l}},$$

respectively. Via the deMoivre–Laplace Theorem ([55]) which asserts that

$$\lim_{n \to \infty} \sum_{i : a < \frac{i-n}{\sqrt{n/2}} < b} \binom{2n}{i} 2^{-2n} = \frac{1}{\sqrt{2\pi}} \int_a^b e^{-\frac{x^2}{2}} dx$$

one finds by a straightforward calculation

$$\lim_{n \to \infty} \mu_{2^{2n}}(\{f : a < f(t_2) - f(t_1) < b\}) = \frac{1}{\sqrt{2\pi (t_2 - t_1)}} \int_a^b e^{-\frac{x^2}{2(t_2 - t_1)}} dx .$$

By an approximation argument this relation holds for all $0 \leq t_1 < t_2 \leq 1$. The measure $\mathbb{W}$ thus assigns a normal distribution with mean zero and variance $t_2 - t_1$ to the increment $f(t_2) - f(t_1)$. A completely analogous result holds for any finite number of increments: given $0 \leq t_1 < \ldots < t_{n+1} \leq 1$ and real intervals $I_1, \ldots, I_n$ the same procedure as above yields

$$\mathbb{W}\big(\{f : f(t_2) - f(t_1) \in I_1, \ldots, f(t_{n+1}) - f(t_n) \in I_n\}\big)$$

$$= \prod_{i=1}^n \frac{1}{\sqrt{2\pi (t_{i+1} - t_i)}} \int_{I_i} e^{-\frac{x^2}{2(t_{i+1} - t_i)}} dx .$$

We may therefore think of $\mathbb{W}$ as describing a stochastic process on the line which starts at the origin and has independent increments over disjoint intervals of time. Furthermore, the increment occurring within $[t, t']$ is normally distributed with mean zero and variance $t' - t$. This stochastic process is commonly called a one-dimensional *Brownian motion*. Taking a snapshot of the motion at time t by means of the continuous projection

$$\pi_t : \begin{cases} C([0, 1]) & \to & \mathbb{R} \\ f & \mapsto & f(t) \end{cases}$$

we obtain a probability measure $\nu_t := \pi_t \mathbb{W}$ on the line. Explicitly we find

$$\nu_t(]a, b[) = \mathbb{W}(\{f : a < f(t) < b\}) = \frac{1}{\sqrt{2\pi t}} \int_a^b e^{-\frac{x^2}{2t}} dx = P_{\frac{t}{2}} \delta_0(]a, b[)$$

and have thereby returned to our starting point (5.23): a family in $\mathcal{M}_1(\mathbb{R})$ describing the diffusion of energy initially concentrated at the origin. As announced earlier, the microscopic point of view thus naturally yields a macroscopic description of the whole process. (Unfortunately, most situations in physics do not permit such a straightforward passage from one point of view to the other.) But it clearly yields much more by enhancing our understanding of the things that happen on the microscopic scale. For example, our construction of the Wiener measure provides a method of simulating Brownian motion via drawing and appropriately rescaling the graphs of a simple symmetric random walk. As can be seen from Figure 5.13 these graphs become increasingly irregular. Taking into account the probabilistic structure of Brownian motion such an effect can hardly be a surprise. Quite on the contrary, we expect typical paths (in the sense of Wiener measure) to exhibit a wildly oscillatory behaviour even on the smallest time scale. In the sequel we are going to express this more precisely.

A preliminary observation tells us that most Brownian paths (by which notion we denominate elements of $C([0, 1])$ endowed with the Wiener measure) have infinite

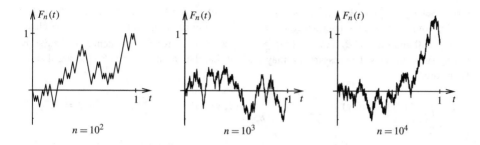

Figure 5.13. The random walk approximation of Brownian motion

variation. Indeed, defining

$$v_n(f) := \sum_{i=1}^{2^n} \left| f\left(\frac{i}{2^n}\right) - f\left(\frac{i-1}{2^n}\right) \right|$$

for all $n \in \mathbb{N}$ and $f \in C([0,1])$ we deduce from the Chebyshev inequality A.12 (vi) that for any $K > 0$

$$\mathbb{W}(\{f : V_{[0,1]}f \geq K\}) \geq \overline{\lim}_{n \to \infty} \mathbb{W}(\{f : v_n(f) \geq K\}) = 1 \, ;$$

here $V_{[0,1]}f$ denotes the (possibly infinite) *variation* of $f \in C([0,1])$, that is

$$V_{[0,1]}f := \sup_n \sum_{k=1}^{n} |f(t_k) - f(t_{k-1})|$$

with the supremum taken over all finite partitions $0 = t_0 < t_1 < \ldots < t_n = 1$. Consequently $\mathbb{W}(\{f : V_{[0,1]}f < \infty\}) = 0$, from which we get a first impression of the irregularity of Brownian paths. Subsequently we shall give a more quantitative description of the latter in terms of Hausdorff dimension. To this purpose a little preparatory work has to be done.

Fix again a probability measure μ on a complete separable metric space X and assign to that measure and $s \geq 0$ the (possibly infinite) *s-energy*

$$I_s(\mu) := \int_X \int_X d(x,y)^{-s} d\mu(x) \, d\mu(y) \, .$$

Although not evident at first sight this concept has some connections with Hausdorff dimension and therefore plays a prominent role in fractal geometry.

Lemma 5.23. *Let μ denote a probability measure supported on $A \subseteq X$. If $I_s(\mu) < \infty$ then $\dim_H A \geq s$.*

Proof. Set $A_1 := \{x \in A : \overline{\lim}_{r \to 0} \mu(B_r(x))/r^s > 0\} \subseteq A$. If $x \in A_1$ we have $\mu(B_{r_n}(x)) \geq \varepsilon r_n^s$ for some $\varepsilon > 0$ and a sequence of radii converging to zero. Since $I_s(\mu) < \infty$ clearly $\mu(\{x\}) = 0$ and thus $\mu(B_{r_n}(x) \backslash B_{r_n'}(x)) \geq \frac{1}{4}\varepsilon r_n^s$ with an appropriate sequence $0 < r_n' < r_n$. Taking subsequences if necessary we may assume that the annuli $B_{r_n}(x) \backslash B_{r_n'}(x)$ are disjoint. Therefore

$$\int_X d(x, y)^{-s} d\mu(y) \geq \sum_n \int_{B_{r_n}(x) \backslash B_{r_n'}(x)} d(x, y)^{-s} d\mu(y) \geq \sum_n 2^{-2} r_n^{-s} \varepsilon r_n^s = \infty$$

implying that $\mu(A_1) = 0$. On $A \backslash A_1$ by definition $\overline{\lim}_{r \to 0} \mu(B_r(x))/r^s = 0$ holds and thus by Lemma 5.20

$$H^s(A) \geq H^s(A \backslash A_1) \geq \frac{\mu(A \backslash A_1)}{c} = \frac{1}{c}$$

for every $c > 0$. From this we see that $\dim_H A \geq s$. $\square$

Let us now turn towards the analysis of Brownian paths. We already know that they typically exhibit infinite variation. However, there is also a certain rest of regularity.

Lemma 5.24. *Suppose $0 < \lambda < \frac{1}{2}$. Then*

$$\mathbb{W}(\{f : \overline{\lim}_{h \to 0} \frac{|f(t+h) - f(t)|}{h^\lambda} < C_f \text{ for all } t \in [0, 1[\}) = 1.$$

Proof. A straightforward calculation yields

$$\mathbb{W}(\{f : |f(t+h) - f(t)| > h^\lambda\}) = \sqrt{\frac{2}{\pi}} \int_{h^{\lambda-1/2}}^\infty e^{-\frac{x^2}{2}} dx \leq c_1 \int_{h^{\lambda-1/2}}^\infty e^{-x} dx \leq c_2 h^2$$

with c_1, c_2 only depending on λ. But then

$$\mathbb{W}(\{f : \left|f\left(\frac{k}{2^l}\right) - f\left(\frac{k-1}{2^l}\right)\right| > 2^{-\lambda l} \text{ for some } l \geq L \text{ and } 1 \leq k \leq 2^l\})$$

$$\leq c_2 \sum_{l=L}^\infty 2^{-2l} 2^l = c_2 2^{-L+1}$$

which implies that with probability one there exists an $L_0(f) \in \mathbb{N}$ such that

$$\left|f\left(\frac{k}{2^l}\right) - f\left(\frac{k-1}{2^l}\right)\right| \leq 2^{-\lambda l} \text{ for all } l \geq L_0(f) \text{ and } 1 \leq k \leq 2^l.$$

If $h < 2^{-L_0(f)}$ the interval $]t, t+h[$ may be expressed as a countable union of binary intervals with lengths $2^{-l} \leq h$ taking not more than two of each length. Denoting by k the least integer with $2^{-k} \leq h$ and also taking into account the continuity of f we have with probability one

$$|f(t+h) - f(t)| \leq 2 \sum_{l=k}^\infty 2^{-\lambda l} = \frac{2^{1-k\lambda}}{1 - 2^{-\lambda}} \leq \frac{2h^\lambda}{1 - 2^{-\lambda}},$$

and thus the lemma is proved. $\square$

The above result may be rephrased by saying that a Brownian path is – with probability one – Hölder continuous with every exponent less than $\frac{1}{2}$. Keeping this fact in mind we now are in a position to prove our main result concerning the geometric structure of Brownian paths. To this end we consider the graph of $f \in C([0,1])$ as a compact subset of the Euclidean plane by setting

$$\text{graph } f := \{(t, f(t)) : t \in [0,1]\} \subseteq \mathbb{R}^2.$$

Theorem 5.25. *A typical Brownian path has Hausdorff dimension 1.5, i.e.*

$$\mathbb{W}(\{f : \dim_H \text{graph } f = 1.5\}) = 1.$$

Proof. Let us first prove that $\dim_H \text{graph } f \leq 1.5$ with probability one. If $\lambda < \frac{1}{2}$ then we typically have $|f(t+h) - f(t)| \leq C_f h^\lambda$ for small h. But then for any $0 \leq t_1 < t_2 < 1$

$$\sup_{t_1 \leq u, v \leq t_2} |f(u) - f(v)| \leq 3C_f |t_2 - t_1|^\lambda$$

if $|t_2 - t_1|$ is sufficiently small. Take $k \in \mathbb{N}$ and denote by N_k the number of squares of the k^{-1}-mesh (generated by $k^{-1}\mathbb{Z}^2$) in the plane that intersect graph f. Carefully counting we find

$$N_k \leq \sum_{i=1}^{k} (2 + 3C_f k^{1-\lambda}) = 2k + 3C_f k^{2-\lambda} < Dk^{2-\lambda}$$

for sufficiently large k, where D does not depend on k. Consequently

$$H^s_{\sqrt{2}k^{-1}}(\text{graph } f) \leq N_k(\sqrt{2}k^{-1})^s < 2^{s/2}Dk^{2-\lambda-s}$$

from which we deduce that $\dim_H \text{graph } f \leq 2 - \lambda$. Since $\lambda < \frac{1}{2}$ was arbitrary $\dim_H \text{graph } f \leq 1.5$ with probability one.

In order to prove the reverse inequality let $1 < s < 2$ and observe that for all $t \in [0, 1[$ and $0 < h < 1 - t$

$$\sqrt{2\pi} \int_{C([0,1])} \left(|f(t+h) - f(t)|^2 + h^2 \right)^{-s/2} d\mathbb{W}(f) = \int_0^\infty (hu + h^2)^{-s/2} e^{-\frac{u}{2}} \frac{du}{\sqrt{u}}$$

$$\leq \int_0^h h^{-s} \frac{du}{\sqrt{u}} + \int_h^\infty h^{-s/2} u^{-\frac{s+1}{2}} du = \frac{2s}{s-1} h^{1/2-s}.$$

The continuous map $F : t \mapsto (t, f(t))$ induces a probability measure μ_f on the plane via $\mu_f := F\lambda^1|_{[0,1]}$ which evidently is supported on graph f. Calculating the s-energy of this measure yields

$$I_s(\mu_f) = \int_{\mathbb{R}^2} \int_{\mathbb{R}^2} \|x - y\|_2^{-s} d\mu_f(x) d\mu_f(y) = \int_0^1 \int_0^1 \left((u-v)^2 + (f(u) - f(v))^2 \right)^{-s/2} du \, dv.$$

By means of this expression we obtain

$$\int_{C([0,1])} I_s(\mu_f) d\mathbb{W}(f) \leq \sqrt{\frac{2}{\pi}} \frac{s}{s-1} \int_0^1 \int_0^1 |u - v|^{\frac{1}{2}-s} du \, dv < \infty$$

if only $1 < s < \frac{3}{2}$. This in turn implies that for such s

$$\mathbb{W}(\{f : \dim_H \text{ graph } f \geq s\}) = 1$$

by virtue of Lemma 5.23, and the proof is complete. $\square$

Theorem 5.25 gives quantitative support to our observation concerning the oscillatory behaviour of Brownian paths. As one might imagine, a lot of related properties of the latter have been studied extensively. With some care one can even drop the assumption that the increments be *independent*, and one thus enters the field of so-called *fractional Brownian motion* ([24, 43]). By appropriately choosing certain parameters it is possible to have typical paths exhibit any Hausdorff dimension between one and two. The corresponding stochastic processes have found a lot of applications in physics, biology and the mathematics of finance. After all, it have been real-world problems such as the motion of small particles suspended in a liquid or the time evolution of stock prices that caused the introduction and analysis of Brownian motion.

5.4.3   Patterns of congruence in the Pascal triangle

Our final example deals with the most basic (and fascinating) objects of mathematics: we shall investigate simple relations between certain natural numbers. Although not obvious at first sight this topic turns out to be intimately related to what has been discussed so far in the present chapter. First of all, recall the construction of the *Pascal triangle*, the k-th entry in the n-th row, $\binom{n}{k} = \frac{n!}{k!(n-k)!}$, being the sum of its two upper neighbours. We now fix a prime number p and symbolize by black dots those numbers in the Pascal triangle that are not multiples of p while we leave blank the other places (see Figure 5.14). To compare the resulting patterns of black dots we rescale them to an equilateral triangle. As can be seen from Figure 5.15 there is – at least in the cases $p = 2$ and $p = 3$ – a regular pattern emerging from that procedure. In the sequel we shall rigorously analyse our visual impression.

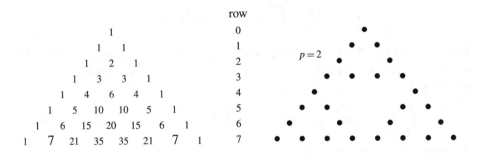

Figure 5.14. The top of the Pascal triangle (left) and its colouring modulo two

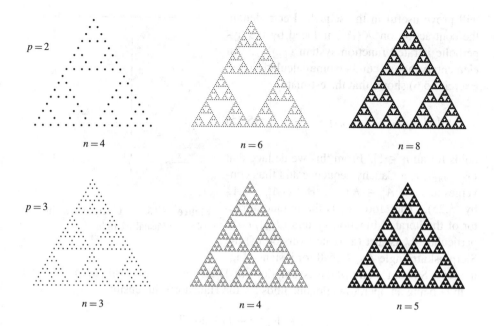

Figure 5.15. The rescaled coloured Pascal triangles $A_{p,n}$ seem to converge.

Lemma 5.26. *Given $n \in \mathbb{N}$ and a prime number p as well as $i, j, k, l \in \mathbb{N}_0$ with $0 \leq i, k \leq p^n - 1$ and $0 \leq l \leq j \leq p - 1$ the following equivalence holds:*

$$\binom{i + jp^n}{k + lp^n} \text{ is a multiple of } p \iff \binom{i}{k} \text{ is a multiple of } p.$$

We leave the elementary proof of this fact to the reader (Exercise 5.6 or [43]). In order to describe the patterns of congruence introduced above we formally define the coloured Pascal triangles $A_{p,n} \subseteq \mathbb{R}^2$ as

$$A_{p,n} := \left\{ \frac{1}{p^n - 1} \left(r, \frac{2s - r}{\sqrt{3}} \right) : r, s \in \mathbb{N}_0, \ 0 \leq s \leq r \leq p^n - 1, \right.$$

$$\left. \binom{r}{s} \text{ is not a multiple of } p \right\}.$$

Figure 5.16 illustrates the idea behind this definition. Simply speaking, $A_{p,n}$ is just an appropriately scaled version of the coloured Pascal triangle modulo p; obviously $A_{p,n} \in \mathcal{K}(\mathbb{R}^2)$. In addition, the family of $\frac{1}{2}p(p + 1)$ similarities $T_{j,l} : \mathbb{R}^2 \to \mathbb{R}^2$ with

$$T_{j,l}(x) := \frac{1}{p} \left(x + \left(j, \frac{2l - j}{\sqrt{3}} \right) \right) \quad 0 \leq l \leq j \leq p - 1$$

will prove useful in the sequel. Let τ denote the contraction on $\mathcal{K}(\mathbb{R}^2)$ induced by the hyperbolic iterated function system $(T_{j,l})_{j,l}$. An elementary though cumbersome calculation (see Exercise 5.6) shows that the estimate

$$d_H\big(A_{p,n+1}, \tau(A_{p,n})\big) \leq \frac{4}{p^n} \qquad (5.25)$$

holds for all $n \in \mathbb{N}$. From this we deduce that $(A_{p,n})_{n \in \mathbb{N}}$ is a Cauchy sequence and thus converges to some $A_p^* \in \mathcal{K}(\mathbb{R}^2)$. But $\tau(A_p^*) = A_p^*$ by (5.25). Therefore A_p^* is the unique attractor of the iterated function system $(T_{j,l})_{j,l}$. In particular, A_2^* is just (a rotated version of) the Sierpinski triangle Δ_∞ (recall for instance Fig-

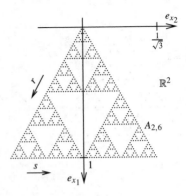

Figure 5.16.   Coordinates in the coloured Pascal triangle

ure 5.5). Since, for any p, the assumptions of Theorem 5.19 are easily seen to be met with U being an open triangle, one finds for the Hausdorff dimension of A_p^*

$$\dim_H A_p^* = 1 + \frac{\log(p+1) - \log 2}{\log p} \in \,]1, 2[\,.$$

If congruences of the Pascal triangle modulo p^r with $r \in \mathbb{N} \setminus \{1\}$ rather than modulo p are considered things will grow slightly more complicated. However, appropriately rescaled versions of the resulting patterns still converge to an attractor $\tilde{A}_p^*$. Surprisingly $\dim_H \tilde{A}_p^* = \dim_H A_p^*$; although the specific form of $\tilde{A}_p^*$ depends on r, its Hausdorff dimension does not ([43]). Finally, given $q = p_1^{r_1} \ldots p_m^{r_m}$ (the p_i being distinct primes) a natural number is divisible by q if and only if it is divisible by $p_i^{r_i}$ for all i. Therefore the pattern of congruence modulo q is the union of the patterns modulo $p_i^{r_i}$.

Not surprisingly, all the results above may also be obtained by other methods (such as coding or cellular automata [43]). However, recall that we observed chaotic dynamics to occur naturally on attractors of hyperbolic iterated function systems and that we just have constructed a series of such attractors from patterns in the Pascal triangle. Putting it more concisely (and, perhaps, more controversially), we have found the structure of chaos even in the undoubtedly simplest infinite set of mathematics, namely the set of natural numbers. It seems as if there was no way of coming around the old saying of Pythagoras: *Everything is number.* Within the whole history of science one can hardly find quotations still proving as accurate after more than two and a half thousand years!

Exercises

(5.1) Define a sequence $(\mu_n)_{n\in\mathbb{N}}$ of measures in $\mathcal{M}_1([0, 1])$ by $\frac{d\mu_n}{d\lambda}(x) := 2\sin^2 \pi nx$. Show that $(\mu_n)_{n\in\mathbb{N}}$ converges weakly and calculate the limit. Is there also a strong limit?

(5.2) Let the continuous map $T : \mathbb{R} \to \mathbb{R}$ satisfy $T(x) \neq x$ for all $x \in \mathbb{R}$. Prove that there does not exist any finite invariant measure for the associated Frobenius–Perron operator P_T. Does this assertion remain valid if *continuous* is replaced by *measurable*?

(5.3) Let X denote a compact metric space. For $\delta > 0$ and $A \subseteq X$ define $N_\delta(A)$ to be the smallest number of closed balls with radius δ that cover A. We call

$$\underline{\dim}_B A := \underline{\lim}_{\delta\to 0} \frac{\log N_\delta(A)}{-\log \delta} \quad \text{and} \quad \overline{\dim}_B A := \overline{\lim}_{\delta\to 0} \frac{\log N_\delta(A)}{-\log \delta}$$

the *lower* and *upper box-counting dimension* of A, respectively. If $\underline{\dim}_B A$ and $\overline{\dim}_B A$ agree then this common value is simply called the *box-counting dimension* of A. Prove the following basic properties:

(i) if $T : X \to X$ is a Hölder map of exponent α then $\underline{\dim}_B T(A) \leq \alpha^{-1}\underline{\dim}_B A$ and $\overline{\dim}_B T(A) \leq \alpha^{-1}\overline{\dim}_B A$; hence lower and upper box-counting dimensions are metric invariants;

(ii) $\dim_B [0, 1]^d = d$;

(iii) $\dim_H A \leq \underline{\dim}_B A \leq \overline{\dim}_B A$ for all $A \subseteq X$;

(iv) $\underline{\dim}_B A = \underline{\dim}_B \overline{A}$ and $\overline{\dim}_B A = \overline{\dim}_B \overline{A}$ where $\overline{A}$ denotes the closure of the set $A \subseteq X$.

In applied sciences box-counting dimension is often preferred to Hausdorff dimension because it is calculated more easily in many cases. Not too seldom $\dim_H$ and $\dim_B$ agree. However, some features of the box-counting dimension make it less attractive for mathematical purposes. For example, set $X_\alpha := \{0\} \cup \{n^{-\alpha} : n \in \mathbb{N}\}$ with $\alpha > 0$ and show that

$$\dim_B X_\alpha = \frac{1}{1+\alpha} > 0$$

whereas $\dim_H X_\alpha = 0$.

(5.4) Let $(T_i)_{i=1}^l$ be a hyperbolic iterated function system on X which additionally satisfies $d(T_i(x), T_i(y)) \geq \gamma d(x, y)$ for all $x, y \in X$, $i \in \{1, \ldots, l\}$ and some $\gamma > 0$. Denote by K^* the attractor of $(T_i)_{i=1}^l$. Assume that $T_i(K^*) \cap T_j(K^*) = \emptyset$ if $i \neq j$. As discussed in Section 5.3 there exists a unique continuous map $S : K^* \to K^*$ which satisfies $T_i \circ S = \mathrm{id}_{T_i(K^*)}$ and $S \circ T_i = \mathrm{id}_{K^*}$ for all i; furthermore S is chaotic on K^*. Prove the following version of a *shadowing theorem*: given $\varepsilon > 0$ there exists a $\delta > 0$ such that every δ-pseudo-orbit ε-shadows a true orbit under S.

(5.5) Verify that a typical Brownian path is not differentiable at any point, that is

$$\mathbb{W}(\{f : f'(t_0) \text{ exists for at least one } t_0 \in [0, 1]\}) = 0;$$

here $\mathbb{W}$ denotes the Wiener measure on $C([0, 1])$.

(5.6) Prove Lemma 5.26 and inequality (5.25). Deduce that $(A_{p,n})_{n \in \mathbb{N}}$ is in fact a Cauchy sequence with respect to the Hausdorff metric.

(5.7) Consider a sequence $(\mu_n)_{n \in \mathbb{N}}$ of probability measures on $\mathbb{N}_0$ (taken by a subspace of $\mathbb{R}$). Define $a_i(n) := \mu_n(\{i\}) \geq 0$ and assume that $\alpha_i := \lim_{n \to \infty} a_i(n)$ exists for every $i \in \mathbb{N}_0$. Prove the following elementary version of Scheffé's Theorem: if $\sum_{i \in \mathbb{N}_0} \alpha_i = 1$ then

$$\mu_n \to \mu := \sum_{i \in \mathbb{N}_0} \alpha_i \delta_i .$$

Show that this conclusion may fail if $\sum_{i \in \mathbb{N}_0} \alpha_i \neq 1$. As an application prove Poisson's Theorem which asserts that for

$$B_{n,p_n}(\{i\}) := \binom{n}{i} p_n^i (1 - p_n)^{n-i} \quad (0 \leq p_n \leq 1),$$

$$P_\lambda(\{i\}) := \frac{e^{-\lambda} \lambda^i}{i!},$$

one has $B_{n,p_n} \to P_\lambda$ provided that $\lim_{n \to \infty} np_n = \lambda$.

(5.8) Let μ_n, μ denote probability measures on the real line and consider the corresponding distribution functions. Demonstrate that the following statements are equivalent:

(i) $\mu_n \overset{w}{\to} \mu$;

(ii) $\lim_{n \to \infty} F_{\mu_n}(x) = F_\mu(x)$ if F_μ is continuous at x.

Give an example showing that $F_{\mu_n}(x) \to F_\mu(x)$ need not hold for *every* $x \in \mathbb{R}$. Prove however, that the convergence in (ii) is in fact *uniform* if F_μ is continuous.

(5.9) Once again study Newton's map (5.10) for calculating $\sqrt{a}$. Assume now that the perturbation is *multiplicative*, i.e., take $T(x, \xi) = N_p(x) \cdot \xi$ in (5.8) where ξ denotes a positive random quantity. Does there exist an invariant distribution for the corresponding Markov operator on measures as given by (5.9)? Is this operator asymptotically stable?

(5.10) Let μ_n, μ be probability measures on the metric space X. Prove that the following statements are equivalent (a fact that is sometimes referred to as the *Portmanteau Theorem*):

(i) $\mu_n \overset{w}{\to} \mu$;

(ii) $\overline{\lim}_{n \to \infty} \mu_n(F) \leq \mu(F)$ for every *closed* set $F \subseteq X$;

(iii) $\underline{\lim}_{n \to \infty} \mu_n(G) \geq \mu(G)$ for every *open* set $G \subseteq X$.

Deduce that $\lim_{n \to \infty} \mu_n(A) = \mu(A)$ if $\mu_n \overset{w}{\to} \mu$ and $\mu(\partial A) = 0$; here ∂A denotes the boundary of $A \in \mathcal{B}(X)$.

(5.11) A continuous curve $\gamma : [0, 1] \to X$ in the complete separable metric space X is called *rectifiable* if

$$l(\gamma) := \sup_n \sum_{i=1}^n d\big(\gamma(t_{i-1}), \gamma(t_i)\big) < \infty$$

with the supremum taken over all finite partitions $0 = t_0 < t_1 < \ldots < t_n = 1$. Prove that $\dim_H \gamma([0, 1]) = 1$ for every rectifiable curve.

As depicted in Figure 5.17 let $D_0 \subseteq \mathbb{R}^2$ denote an isosceles triangle with angle $\frac{2\pi}{3}$ and consider two similarities $T_{1,2}$ which rotate D_0 and scale it by a factor $1/\sqrt{3}$. Show that the attractor D_∞ of the corresponding hyperbolic iterated function system $(T_i)_{i=1}^2$ is a non-rectifiable continuous curve(!) without self-intersections; traditionally, D_∞ is termed the *von Koch curve*.

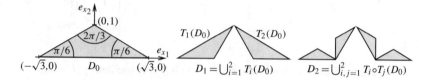

Figure 5.17. The first steps towards the *von Koch curve* D_∞

(5.12) Let $(T_i)_{i=1}^l$ denote a hyperbolic iterated function system on X and assume that each map T_i has a continuous inverse. Define l maps $\tilde{T}_i$ according to

$$\tilde{T}_i : \begin{cases} X \times \Sigma_l & \to \quad X \times \Sigma_l, \\ (x, (x_0, x_1, \ldots)) & \mapsto \quad (T_i(x), (i, x_0, x_1, \ldots)). \end{cases}$$

Show that $(\tilde{T}_i)_{i=1}^l$ yields a hyperbolic iterated function system on $X \times \Sigma_l$ if this space is endowed with the metric

$$d\big((x, (x_0, x_1, \ldots)), (y, (y_0, y_1, \ldots))\big) := d_X(x, y) + d_{\Sigma_l}\big((x_0, x_1, \ldots), (y_0, y_1, \ldots)\big).$$

The attractor $\tilde{K}^*$ of the system $(\tilde{T}_i)_{i=1}^l$ is totally disconnected because $\tilde{T}_i(X \times \Sigma_l) \cap \tilde{T}_j(X \times \Sigma_l) = \emptyset$ if $i \neq j$. As outlined in Section 5.3 the induced map $\tilde{S} : \tilde{K}^* \to \tilde{K}^*$ is continuous and chaotic. Verify that however $S : K^* \to K^*$ is defined in the overlap regions, any orbit under S is the projection onto the first coordinate of an orbit under $\tilde{S}$.

(5.13) Exercise 4.12 dealt with eigenfunctions of the Frobenius–Perron operator $P_{T_{(k)}}$ associated with the statistically stable maps $T_{(k)} : x \mapsto kx$ on $\mathbb{T}^1$ where $k \in \mathbb{N} \setminus \{1\}$. Specifically, fix $s \in]0, 1[$ and consider

$$g_s(x) := \sum_{n=0}^{\infty} k^{-sn} \cos(2\pi k^n x).$$

In the terminology of Exercise 4.12 $g_s = \Re f_{k^{-s},1}$ and hence g_s is a real-valued eigen-function of $P_{T_{(k)}}$ corresponding to the eigenvalue k^{-s}. Though a continuous function, g_s is highly oscillatory. As a quantification of this statement, prove that for large k the box-counting dimension (cf. Exercise 5.3) of graph g_s is given by

$$\dim_B \text{ graph } g_s = 2 - s > 1;$$

the plotting of sufficiently accurate partial sums of g_s may be helpful for visualizing the oscillatory behaviour of this function (Figure 5.18). At present it is not known whether $\dim_H \text{ graph } g_s = \dim_B \text{ graph } g_s$ holds for all sensible values of k and s ([24]).

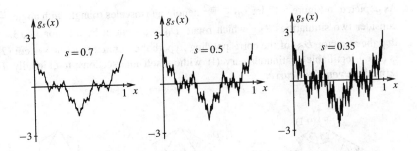

Figure 5.18. For small s the graph of g_s is highly oscillatory ($k = 3$).

(5.14) Find the Hausdorff dimension of the *solenoid* $\Gamma \subseteq \mathbb{R}^3$ as defined in Section 2.6. (You may wish to consult [24] for a *lower* bound on $\dim_H \Gamma$.)

(5.15) Let X denote a metric space and $T : X \rightarrow X$ a continuous map which preserves $\mu_n \in \mathcal{M}_1(X)$ for all $n \in \mathbb{N}$. Assume that $(\mu_n)_{n \in \mathbb{N}}$ converges weakly, to μ say. Is μ also preserved by T? And if so, does it follow that

$$h_{\mu_n}(T) \rightarrow h_\mu(T) \quad \text{as } n \rightarrow \infty ?$$

Appendix A
The toolbox

Throughout this text we use abstract measure and integration theory as a tool – just as, for example, calculus is used within the theory of differential equations. As developed from Chapter Three on, our statistical analysis of dynamical systems cannot satisfyingly be grasped without a basic knowledge of measure theory. A short review of some elementary definitions and results might therefore be desirable. The exposition in Sections A.2 to A.4 is by no means exhaustive, nor does it intend to replace a serious course on the topics covered. Its aim is merely to outline the desirable prerequisites, and also to indicate where to start from when exploring the literature. For any details, we refer to the excellent textbooks [6] and [47].

A.1 A survey of notations

For the convenience of the reader this short preliminary section gathers the notations used throughout the book, although most of them are standard in mathematical text-books.

We denote by $\mathbb{N}$, $\mathbb{Z}$, $\mathbb{Q}$, $\mathbb{R}$ and $\mathbb{C}$ the sets of natural, integer, rational, real and complex numbers, respectively; the set of non-negative integers is symbolized by $\mathbb{N}_0$, whereas $\mathbb{R}^+$ denominates the positive real numbers. Given $x \in \mathbb{R}$ the quantity $\lfloor x \rfloor$ stands for the largest integer less than or equal to x; analogously $\lceil x \rceil = -\lfloor -x \rfloor$ designates the smallest integer greater than or equal to x. For real numbers $x < y$ the closed interval having these endpoints is symbolized by $[x, y]$ whereas $]x, y[$ denotes the corresponding open interval; expressions like $]x, y]$ are to be read analogously. The sign of x, in symbols sign x, equals $+1$, 0 or -1 depending on whether $x > 0$, $x = 0$ or $x < 0$, respectively. Logarithms are understood as natural logarithms unless indicated otherwise by means of subscripts, e.g. $\log_2$, $\log_{10}$. Inverse trigonometric functions should be read as principal values: for example, $|\arctan x| \leq \frac{\pi}{2}$ for $x \in \mathbb{R}$, and $0 \leq \arccos x \leq \pi$ for $x \in [-1, 1]$. At rare occasions we also use the (real) Euler Gamma function $\Gamma :]0, +\infty[\to \mathbb{R}$ which interpolates the factorials in the sense that $\Gamma(n + 1) = n!$ for all $n \in \mathbb{N}$. Given a sequence $(x_n)_{n \in \mathbb{N}}$ of real numbers we denote by $\underline{\lim}_{n \to \infty} x_n$ the (possibly infinite) smallest of its accumulation points; correspondingly, the largest accumulation point of the sequence is $\overline{\lim}_{n \to \infty} x_n = -\underline{\lim}_{n \to \infty}(-x_n)$.

For any $z \in \mathbb{C}$ the real numbers $\Re z$, $\Im z$ and $|z| := \sqrt{(\Re z)^2 + (\Im z)^2}$ denote the real part, the imaginary part and the absolute value of $z = \Re z + i \Im z$, respectively, whereas $\bar{z} = \Re z - i \Im z$ stands for its complex conjugate. For $z \neq 0$ there is a unique angle $\arg z \in [0, 2\pi[$ such that $z = |z| e^{i \arg z}$; we refer to $\arg z$ as the argument of z.

By default $\mathbb{C}$ is endowed with the standard metric $d(z_1, z_2) := |z_1 - z_2|$. The unit circle $S^1 := \{z \in \mathbb{C} : |z| = 1\}$ then is a compact metric space and – with complex multiplication – also an abelian group. Points in the d-dimensional space $\mathbb{C}^d$ are typically written as row-vectors. Unless stated otherwise $\mathbb{C}^d$ is endowed with the hermitian inner product

$$\langle x, y \rangle := \sum_{j=1}^{d} x_j \overline{y}_j \quad \text{for all } x = (x_j)_{j=1}^{d}, y = (y_j)_{j=1}^{d} \in \mathbb{C}^d$$

and the corresponding norm $\|x\|_2 := \sqrt{\langle x, x \rangle}$. For $A \subseteq \mathbb{C}^d$ and $\alpha \in \mathbb{C}$ we write $\alpha A := \{\alpha x : x \in A\}$ for the sake of brevity.

Matrices $M = (m_{jk})_{j,k=1}^{d} \in \mathbb{C}^{d \times d}$ are always symbolized by capital Latin letters. By $(M^n)_{jk}$ we denote the entry at position (j, k) of the n-th power of M; $\det M$ and $\operatorname{tr} M$ stand for the determinant and the trace of M, respectively. The spectrum of M, symbolically $\sigma(M)$, is meant to designate the family of eigenvalues of M where each eigenvalue appears according to its multiplicity. Therefore $\sigma(M)$ always comprises d not necessarily distinct complex numbers; whenever possible we nevertheless treat $\sigma(M)$ as if it were a set.

Generally, $\emptyset$ symbolizes the empty set. For any non-empty set X the power set, i.e. the set of all subsets of X including $\emptyset$ and X itself, is denoted by $\mathcal{P}(X)$. If A is any subset of X then $\mathbf{1}_A : X \to \{0, 1\}$, the indicator function of A, is defined as

$$\mathbf{1}_A(x) := \begin{cases} 1 & x \in A, \\ 0 & x \notin A. \end{cases}$$

For us the cardinality of a set A is either a non-negative integer or infinity; it is symbolized by $\#A$. We write $A \backslash B := \{x : x \in A, x \notin B\}$ and $A \triangle B := A \backslash B \cup B \backslash A = (A \cup B) \backslash (A \cap B)$ for the difference and symmetric difference of any two sets $A, B \subseteq X$, respectively. Given two maps $T : X \to Y, S : Y \to Z$ on arbitrary sets we denote their composition by $S \circ T$, which means that $S \circ T(x) = S(T(x))$ for all $x \in X$. For $A \subseteq X$ and $B \subseteq Y$ we furthermore denominate by $T(A) := \{T(x) : x \in A\} \subseteq Y$ and $T^{-1}(B) := \{x : T(x) \in B\} \subseteq X$ the image of A and the pre-image of B under T, respectively. We call $T : X \to Y$ one-to-one (onto, respectively) if for each $y \in Y$ the set $T^{-1}(\{y\})$ contains at most (least) one point. Equivalently, T is one-to-one precisely if $T(x_1) = T(x_2)$ necessarily implies $x_1 = x_2$, and it is onto if and only if $T(X) = Y$. To each one-to-one and onto map $T : X \to Y$ we can uniquely assign an inverse map $T^{-1} : Y \to X$ by $T^{-1}(y) := x$ for $y = T(x)$. If $T : X \to X$ and $T(A) \subseteq A$ for some $A \subseteq X$ then $T|_A : A \to A$ stands for the restriction of T to A. Additionally, T^n denotes the n-fold composition of T with itself; by definition T^0 means the identity map of X which we write as id_X.

If X is a metric space and $A \subseteq X$ then $\overline{A}$ and ∂A denote the closure and the boundary of A, respectively. Furthermore $C(X)$ stands for the space of continuous real-valued functions on X. If X is compact then $C(X)$ is itself a complete metric space with the metric

$$d(f, g) := \max_{x \in X} |f(x) - g(x)| \quad (f, g \in C(X)).$$

A continuous, one-to-one and onto map between metric spaces, whose inverse also is continuous, is called a homeomorphism.

A.2 Basic facts from measure theory

Let X denote a set. One of the basic aims of abstract measure theory may be expressed as follows: assign a non-negative, possibly infinite *size* $\mu(A)$ to as many subsets $A \subseteq X$ as possible (or, at least, to all members of a reasonable family of subsets). We require that the rule μ satisfy $\mu(\emptyset) = 0$ as well as

$$\mu\left(\bigcup_{n \in \mathbb{N}} A_n\right) = \sum_{n \in \mathbb{N}} \mu(A_n)$$

for any sequence of sets with $A_n \cap A_m = \emptyset$ for $n \neq m$. This latter property, usually referred to as σ-*additivity*, generalizes one's intuitive feeling that a quantity called size, mass, etc. should (at least) be finitely additive.

Definiton A.1 A non-empty family $\mathcal{A} \subseteq \mathcal{P}(X)$ is called a σ-*algebra* (in X) if

(i) $A^c \in \mathcal{A}$ whenever $A \in \mathcal{A}$;

(ii) $\bigcup_{n \in \mathbb{N}} A_n \in \mathcal{A}$ for any sequence of sets with $A_n \in \mathcal{A}$ for all n.

The pair $(X, \mathcal{A})$ is called a *measurable space*.

Any σ-algebra $\mathcal{A}$ is easily seen to be closed under all countable combinations of elementary set-theoretical operations ($\cup, \cap, \triangle, {}^c$ etc.); evidently one has the inclusions $\{\emptyset, X\} \subseteq \mathcal{A} \subseteq \mathcal{P}(X)$. Furthermore, for any $Y \subseteq X$ the family $Y \cap \mathcal{A} := \{Y \cap A : A \in \mathcal{A}\}$ constitutes a σ-algebra in Y. Since the intersection of σ-algebras again is a σ-algebra, for any family $\mathcal{E} \subseteq \mathcal{P}(X)$ there is a unique smallest σ-algebra $A_\sigma(\mathcal{E})$ containing $\mathcal{E}$, called the σ-algebra *generated* by $\mathcal{E}$. In general there will be many generators for $A_\sigma(\mathcal{E})$, some being more useful than others.

Definiton A.2 Let $\mathcal{A}$ denote a σ-algebra in X. A map $\mu : \mathcal{A} \to [0, +\infty]$ which is σ-additive and satisfies $\mu(\emptyset) = 0$ is called a *measure* on $(X, \mathcal{A})$. The triple $(X, \mathcal{A}, \mu)$ is called a *measure space*.

Many rules of assigning size (mass, etc.) to certain sets may be subsumed under the concept of a measure. Some elementary observations are contained in the following lemma.

Lemma A.3. *Let* $(X, \mathcal{A}, \mu)$ *be a measure space and* $A, B, A_n \in \mathcal{A}$.

(i) *If $A \subseteq B$ then $\mu(A) \leq \mu(B)$;*

(ii) $\mu\left(\bigcup_{n\in\mathbb{N}} A_n\right) \leq \sum_{n\in\mathbb{N}} \mu(A_n)$;

(iii) *If $(A_n)_{n\in\mathbb{N}}$ is monotonically increasing, which means that $A_1 \subseteq A_2 \subseteq \ldots$, then*
$\mu\left(\bigcup_{n\in\mathbb{N}} A_n\right) = \lim_{n\to\infty} \mu(A_n)$;

(iv) *If $(A_n)_{n\in\mathbb{N}}$ is monotonically decreasing, i.e. $A_1 \supseteq A_2 \supseteq \ldots$, and if additionally $\mu(A_{n_0}) < \infty$ for some n_0 then $\mu(\bigcap_{n\in\mathbb{N}} A_n) = \lim_{n\to\infty} \mu(A_n)$.*

A measure μ is said to be finite if $\mu(X) < \infty$; it is called a *probability* (or *normed*) *measure* if $\mu(X) = 1$. A weaker notion is that of σ-*finiteness*, in which case $X = \bigcup_{n\in\mathbb{N}} A_n$ for some sequence of sets $A_n \in \mathcal{A}$ with $\mu(A_n) < \infty$ for all $n \in \mathbb{N}$. If $Y \in \mathcal{A}$ then $\mu|_Y$ (rather than the cumbersome $\mu|_{Y\cap\mathcal{A}}$) is used to designate the restriction of μ to the measurable space $(Y, Y \cap \mathcal{A})$.

Since in many cases σ-algebras tend to be rather large, the explicit definition of $\mu(A)$ for all $A \in \mathcal{A}$ often is an impossible task. This naturally raises the question of extension: given a family of sets $\mathcal{E}$ and a measure-like rule ν on $\mathcal{E}$, does there exist a (hopefully unique) measure μ on $\mathcal{A}_\sigma(\mathcal{E})$ such that $\mu(E) = \nu(E)$ for all $E \in \mathcal{E}$? Although in full generality the answer to this question is definitely negative, some mild additional preconditions may help.

Theorem A.4 (Extension Theorem). *Let $\mathcal{E} \subseteq \mathcal{P}(X)$ denote a non-empty family of subsets in X and ν be σ-additive on $\mathcal{E}$. If*

(i) $\mathcal{E}$ *is $\cap$-stable, i.e. $E_1 \cap E_2 \in \mathcal{E}$ for any $E_1, E_2 \in \mathcal{E}$, and*

(ii) $X = \bigcup_{n\in\mathbb{N}} E_n$ *with $E_n \in \mathcal{E}$ and $\nu(E_n) < \infty$*

then there is a uniquely determined measure μ on $\mathcal{A}_\sigma(\mathcal{E})$ such that $\mu(E) = \nu(E)$ for every $E \in \mathcal{E}$.

A few examples may give an idea of the vast variety of measures.

- Fix a point $x_0 \in X$ and an arbitrary σ-algebra $\mathcal{A}$ in X. Then $\delta_{x_0}(A) := \mathbf{1}_A(x_0)$ defines a probability measure on $(X, \mathcal{A})$ called the *Dirac measure* concentrated at x_0.

- Again let $(X, \mathcal{A})$ be arbitrary. By

$$\#A := \begin{cases} n & \text{if } A \text{ contains exactly } n \geq 0 \text{ elements,} \\ \infty & \text{if } A \text{ is infinite,} \end{cases}$$

the *counting measure* on $(X, \mathcal{A})$ is obtained.

- Assume X to be at most countable, i.e. finite or countable. Then any measure on $(X, \mathcal{P}(X))$ is easily seen to be of the form $\mu = \sum_{x\in X} p_x \delta_x$ with $0 \leq p_x := \mu(\{x\}) \leq \infty$; obviously, μ is a probability measure if and only if $\sum_{x\in X} p_x = 1$.

- Let $\mathcal{E} \subseteq \mathcal{P}(\mathbb{R})$ denote the family of finite half-open intervals, that is $\mathcal{E} :=$ $\{\,]a,b]\,:\,-\infty < a \leq b < \infty\}$, and define $\nu(]a,b]) := b - a$. There exists a unique extension λ^1 of ν to $\mathcal{B}^1 := A_\sigma(\mathcal{E})$, the family of one-dimensional Borel sets. (The non-trivial part in applying the Extension Theorem here concerns the σ-additivity of ν.) The measure λ^1 therefore extends the elementary notion of length to the very large family $\mathcal{B}^1$; it is called the one-dimensional *Lebesgue measure*. An analogous construction in $\mathbb{R}^d$ yields the d-dimensional Lebesgue measure λ^d on $\mathcal{B}^d := A_\sigma\left(\prod_{i=1}^d \mathcal{E}\right)$, which generalizes the elementary d-dimensional volume. The same procedure applies mutatis mutandis also to the unit circle S^1; the corresponding probability measure λ_{S^1} may consequently be thought of as the normalized arc-length extended to $\mathcal{B}(S^1)$ which by definition is the smallest σ-algebra containing all arcs.

The last example also provides a motivation for entering the technicalities of σ-algebras. Intuitively, measuring *all* subsets of $\mathbb{R}^d$ could be considered most desirable. Clearly, this would boil down to defining a measure μ^d on all of $\mathcal{P}(\mathbb{R}^d)$. Since μ^d is assumed to behave like elementary volume we certainly would require that this measure be invariant under isometries; furthermore $\mu^d([0,1]^d) = 1$ should hold. Unfortunately, no such measure exists (not even if the assumptions are weakened [6]). Reducing the family of measurable sets by dealing with σ-algebras other than $\mathcal{P}(\mathbb{R}^d)$ therefore turns out to be inevitable.

Instead of being a mere set the spaces under consideration frequently bear some additional structure which may interact with measures defined on them in a more or less friendly way. Specifically, many interesting measure spaces at the same time are also topological spaces. In this case the *Borel σ-algebra* $\mathcal{B}(X)$ in X, which by definition is the σ-algebra generated by the family of open sets, provides a natural place for measures to live in. For example, with the usual topology $\mathcal{B}(\mathbb{R}^d) = \mathcal{B}^d$ as defined above. Questions concerning the specific behaviour of such *Borel measures* on special (e.g. open, compact) sets have been studied extensively ([47]). One notion which we frequently use from this context concerns the support of a measure. For our purposes it is sufficient to introduce this term for the case where X is a metric space and μ is a locally finite measure on $\mathcal{B}(X)$. In this situation the *support* of μ, defined as

$$\operatorname{supp}\mu := \{x \in X : \mu(U) > 0 \text{ for every open } U \text{ containing } x\},$$

may be characterized as the smallest closed set of full measure. In other words, if $A \subseteq X$ is a closed set and $\mu(X \setminus A) = 0$ then $A \supseteq \operatorname{supp}\mu$.

The following method of constructing new measures from given ones is most important in measure and probability theory. Let $(X_i, \mathcal{A}_i, \mu_i)_{i=1,2}$ denote two σ-finite measure spaces. The family of sets $A_1 \times A_2$ with $A_i \in \mathcal{A}_i$ generates a σ-algebra in $X_1 \times X_2$ which is called the *product σ-algebra* $\mathcal{A}_1 \otimes \mathcal{A}_2$. With the definition

$$\mu_1 \otimes \mu_2(A_1 \times A_2) := \mu_1(A_1) \cdot \mu_2(A_2) \quad \text{for all } A_1 \in \mathcal{A}_1, A_2 \in \mathcal{A}_2$$

the Extension Theorem yields a σ-finite measure on $(X_1 \times X_2, \mathcal{A}_1 \otimes \mathcal{A}_2)$ referred to as the *product measure* $\mu_1 \otimes \mu_2$ of μ_1 and μ_2; the measure space $(X_1 \times X_2, \mathcal{A}_1 \otimes \mathcal{A}_2,$

$\mu_1 \otimes \mu_2)$ is symbolized by $\bigotimes_{i=1}^{2}(X_i, \mathcal{A}_i, \mu_i)$. If μ_1, μ_2 are probability measures then so is $\mu_1 \otimes \mu_2$. If the μ_i are considered the governing mechanism of random experiments ξ_i with outcomes in X_i then $\mu_1 \otimes \mu_2$ naturally may be looked at as the corresponding mechanism for the compound (ξ_1, ξ_2) where the coordinates are *independent*.

In the case of probability spaces the outlined construction generalizes to arbitrary families $(X_i, \mathcal{A}_i, \mu_i)_{i \in I}$. If I is countable, say $I = \mathbb{N}$, the product $\prod_{i \in \mathbb{N}} X_i$ may be regarded the space of sequences $(x_i)_{i \in \mathbb{N}}$ with $x_i \in X_i$ for all i. Furthermore, $\bigotimes_{i \in \mathbb{N}} \mathcal{A}_i$ is understood as the σ-algebra generated by the family of *finite cylinder sets*

$$[A_1, A_2, \ldots, A_n] := A_1 \times A_2 \times \ldots \times A_n \times X_{n+1} \times X_{n+2} \times \ldots$$
$$= \{(x_i)_{i \in \mathbb{N}} : x_i \in A_i \text{ for } i = 1, 2, \ldots, n\}.$$

Requiring that for these sets

$$\bigotimes_{i \in \mathbb{N}} \mu_i([A_1, A_2, \ldots, A_n]) := \mu_1(A_1) \cdot \mu_2(A_2) \cdot \ldots \cdot \mu_n(A_n)$$

uniquely yields a probability measure $\bigotimes_{i \in \mathbb{N}} \mu_i$ on $\left(\prod_{i \in \mathbb{N}} X_i, \bigotimes_{i \in \mathbb{N}} \mathcal{A}_i \right)$. The resulting probability space $\bigotimes_{i \in \mathbb{N}}(X_i, \mathcal{A}_i, \mu_i)$ may be considered the formal joint description of a *sequence* of independent experiments.

A.3   Integration theory

The structure-preserving morphisms within the measurable category are assigned their own name (just as *continuous* and *linear* maps on topological and linear spaces, respectively).

Definiton A.5 Let $(X', \mathcal{A}')$, $(X'', \mathcal{A}'')$ denote measurable spaces. A map $T : X' \to X''$ is $\mathcal{A}'$-$\mathcal{A}''$-*measurable* if $T^{-1}(A'') \in \mathcal{A}'$ for all $A'' \in \mathcal{A}''$. If $(X'', \mathcal{A}'')$ equals $(\mathbb{R}, \mathcal{B}^1)$ or $(\mathbb{C}, \mathcal{B}(\mathbb{C}))$ then the σ-algebras $\mathcal{B}^1$ and $\mathcal{B}(\mathbb{C})$ are commonly suppressed in the notation of measurability.

A map is easily seen to be measurable if and only if $T^{-1}(\mathcal{E}'') \subseteq \mathcal{A}'$ for some generator $\mathcal{E}''$ of $\mathcal{A}''$. Furthermore, any measure μ' on $(X', \mathcal{A}')$ naturally induces a measure $T\mu'$ on $(X'', \mathcal{A}'')$ via $T\mu'(A'') := \mu'(T^{-1}(A''))$ for all $A'' \in \mathcal{A}''$.

Simply speaking, abstract integration theory tries to assign a reasonable *mean* $\int f \, d\mu$ to as many $\mathcal{A}$-measurable real- or complex-valued functions f on $(X, \mathcal{A}, \mu)$ as possible. It turns out that many elementary notions of counting, summing and integrating are in fact covered by the same abstract concept.

Theorem A.6. *Let* $(X, \mathcal{A}, \mu)$ *denote a measure space and write* $\mathcal{E}^*(X, \mathcal{A})$ *for the set of all non-negative* $\mathcal{A}$-*measurable functions on* X. *There exists a unique rule of assigning a number* $\int_X f \, d\mu \in [0, +\infty]$ *to each* $f \in \mathcal{E}^*(X, \mathcal{A})$ *such that*

(i) $\int_X (\alpha f + \beta g) \, d\mu = \alpha \int_X f \, d\mu + \beta \int_X g \, d\mu$ *for all* $\alpha, \beta \geq 0$ *and all functions* $f, g \in \mathcal{E}^*(X, \mathcal{A})$;

(ii) $\int_X f\,d\mu \le \int_X g\,d\mu$ if $f \le g$;

(iii) $\int_X \mathbf{1}_A d\mu = \mu(A)$ for all $A \in \mathcal{A}$;

(iv) $\lim_{n\to\infty} \int_X f_n d\mu = \int_X f\,d\mu$ if $f_n, f \in \mathcal{E}^*(X, \mathcal{A})$ and $\big(f_n(x)\big)_{n\in\mathbb{N}}$ is monotonically increasing to $f(x)$ everywhere on X.

The quantity $\int_X f\,d\mu$ is called the integral of f with respect to μ.

In order to define an integral for complex-valued $\mathcal{A}$-measurable functions it is convenient to first introduce the positive and negative part of *real*-valued measurable functions according to

$$f^+ := \max\{0, f\} \quad\text{and}\quad f^- := \max\{0, -f\},$$

respectively; clearly f^+ and f^- are both measurable and non-negative. Taking a real number $p \ge 1$ and setting

$$\mathcal{L}^p(X, \mathcal{A}, \mu) := \{f \mid f : X \to \mathbb{C} \text{ is } \mathcal{A}\text{-measurable and } \int_X |f|^p d\mu < \infty\},$$

one can define for all $f \in \mathcal{L}^1(X, \mathcal{A}, \mu)$

$$\int_X f\,d\mu := \int_X (\Re f)^+ d\mu - \int_X (\Re f)^- d\mu$$
$$+ i \int_X (\Im f)^+ d\mu - i \int_X (\Im f)^- d\mu \in \mathbb{C}.$$

If there is no danger of confusion, $\mathcal{L}^p(X, \mathcal{A}, \mu)$ is abbreviated as $\mathcal{L}^p(\mu)$. Properties of functions in $\mathcal{L}^p(\mu)$, e.g. non-negativity or continuity, are said to hold μ-*almost everywhere* (symbolically $[\mu]$) if they hold everywhere outside a set of μ-measure zero. Since $\int_X |f|^p d\mu = 0$ if and only if $f = 0$ $[\mu]$, it turns out to be convenient to identify functions in $\mathcal{L}^p(\mu)$ if they agree μ-almost everywhere, more formally

$$f \sim g :\Longleftrightarrow f = g\ [\mu].$$

The set of equivalence classes $L^p(\mu) := \mathcal{L}^p(\mu)/_\sim = \{[f]_\sim : f \in \mathcal{L}^p(\mu)\}$ can be given a linear structure by $[f]_\sim + [g]_\sim := [f + g]_\sim$ and $\alpha[f]_\sim := [\alpha f]_\sim$. The map $\|\cdot\|_p : L^p(\mu) \to [0, +\infty[$ defined according to

$$\|f\|_p := \Big(\int_X |f|^p d\mu\Big)^{\frac{1}{p}} \quad\text{for } f \in L^p(\mu)$$

is then easily seen to be a norm on $L^p(\mu)$. The same procedure also works for

$$\mathcal{L}^\infty(\mu) := \{f \mid f : X \to \mathbb{C} \text{ is } \mathcal{A}\text{-measurable and } |f| \le \beta_f\ [\mu] \text{ for some } \beta_f \in \mathbb{R}\}$$

together with $\|f\|_\infty := \inf\{\alpha : \mu(|f| > \alpha) = 0\}$. Here and in the sequel we shall follow the usual practice of using the same symbols for individual functions and their equivalence classes unless this may cause confusion. It is common sense to refer to the elements of $L^p(\mu)$ as *functions* as long as the choice of a particular member of $[f]_\sim$ is irrelevant. Throughout the text we furthermore write $\|f\|$ instead of $\|f\|_1$ for the sake of brevity.

A notion of convergence naturally arises from the definition of the norm $\|\cdot\|_p$. We say that a sequence $(f_n)_{n\in\mathbb{N}}$ of $L^p(\mu)$-functions converges *strongly* to $f \in L^p(\mu)$ if $\lim_{n\to\infty} \|f_n - f\|_p = 0$. Additionally, $(f_n)_{n\in\mathbb{N}_0}$ is said to converge *weakly* if $f_n g \to fg$ in $L^1(\mu)$ for all $g \in L^{p'}(\mu)$ where $p' := p/(p-1)$ for $p > 1$ and $p' := \infty$ for $p = 1$. The spaces $L^p(\mu)$ provide a large class of examples extensively studied in functional analysis. For example, $L^{p'}(\mu)$ may be proved to be the dual of $L^p(\mu)$ so that the notion of weak convergence used here naturally fits into the framework of functional analysis. It is not only in this context that the following two general theorems are most important; after all, they demonstrate the great flexibility of the abstract integral with respect to passages to a limit.

Theorem A.7 (Riesz, Fischer). *For any $p \in [1, +\infty]$ the normed space $(L^p(\mu), \|\cdot\|_p)$ is complete, hence a Banach space.*

Theorem A.8 (Dominated Convergence Theorem). *Let $1 \le p < \infty$ and suppose that the sequence $(f_n)_{n\in\mathbb{N}}$ of $L^p(\mu)$-functions converges μ-almost everywhere in X to some function f. If there exists $g \in L^p(\mu)$ such that $|f_n| \le g$ $[\mu]$ for all n then the limit function f is in $L^p(\mu)$ and $\|f_n - f\|_p \to 0$ as $n \to \infty$; especially*

$$\lim_{n\to\infty} \int_X f_n \, d\mu = \int_X f \, d\mu.$$

In practice the problem arises how to evaluate integrals. The following examples sketch partial answers to this question.

- Let X be at most countable and consider the measure $\mu := \sum_{x\in X} p_x \delta_x$ on $\mathcal{P}(X)$ with positive finite weights $0 < p_x < \infty$. Then $f \in L^p(\mu)$ if and only if $\sum_{x\in X} |f(x)|^p p_x < \infty$, and $\int_X f \, d\mu = \sum_{x\in X} f(x) p_x$. If one specifically sets $X = \mathbb{N}$ and $f_n := \frac{1}{n} \sum_{i=1}^n p_i^{-1} \mathbf{1}_{\{i\}} \in L^1(\mu)$ then $f_n(x) \to 0$ for all $x \in X$ but

$$\lim_{n\to\infty} \int_X f_n \, d\mu = 1 \ne 0 = \int_X 0 \, d\mu;$$

  the existence of an *integrable* dominating function in Theorem A.8 must therefore not be omitted.

- If $(X, \mathcal{A}, \mu) = ([a, b], [a, b] \cap \mathcal{B}^1, \lambda^1|_{[a,b]})$ then the integral $\int_X f \, d\mu$ can be calculated as an ordinary Riemann integral in many cases. More precisely, if $f : [a, b] \to \mathbb{R}$ is $[a, b] \cap \mathcal{B}^1$-measurable and the (proper) Riemann integral $\int_a^b f(x) \, dx$ exists then both integrals coincide. On the other hand, many functions turn out to be integrable with respect to $\lambda^1|_{[a,b]}$ although their Riemann integral does not exist (take $f := \mathbf{1}_{\mathbb{Q}\cap[a,b]}$ as an example). The abstract integral may thus be considered a generalization of the integral of classical calculus.

- Though not necessarily advantageous for explicit calculations there is a general way of representing abstract integrals as integrals on the real line. More precisely, if $f : X \to [0, \infty[$ denotes a non-negative measurable function on the probability space $(X, \mathcal{A}, \mu)$ then

$$\int_X f \, d\mu = \int_0^{+\infty} \mu(\{x : f(x) > t\}) \, dt \, .$$

Given a non-negative $\mathcal{A}$-measurable function f on $(X, \mathcal{A}, \mu)$ one can easily see that

$$\nu(A) := \int_A f \, d\mu := \int_X f 1_A d\mu \quad (A \in \mathcal{A})$$

defines another measure on $(X, \mathcal{A})$. Usually f is called the *density* of ν with respect to μ. Since functions are dealt with more comfortably than measures it is important to know whether a given measure may be expressed by means of a density with respect to another measure.

Theorem A.9 (Radon, Nikodym). *Let μ, ν denote two σ-finite measures on $(X, \mathcal{A})$. The following statements are equivalent:*

(i) *ν has a density with respect to μ;*

(ii) *ν is absolutely continuous with respect to μ, symbolically $\nu \ll \mu$, which means that $\nu(A) = 0$ whenever $\mu(A) = 0$.*

The density according to (i) turns out to be unique almost everywhere; it is denoted by $\frac{d\nu}{d\mu}$ and referred to as the *Radon–Nikodym derivative* of ν. By means of the latter, integrals with respect to ν can be evaluated as μ-integrals because $\int g \, d\nu = \int g \frac{d\nu}{d\mu} d\mu$ whenever theses integrals exist. This fact may be highly advantageous if ν is rather complicated but μ is well understood. Measures that are absolutely continuous with respect to a well-behaved measure are therefore quite easily dealt with.

Two measures μ, ν are *equivalent* if $\nu \ll \mu$ as well as $\mu \ll \nu$. Equivalent measures assign zero to precisely the same sets, and – in the σ-finite case – are characterized by $0 < \frac{d\nu}{d\mu} < \infty \, [\mu]$.

The opposite notion of absolute continuity is that of mutual singularity: the measures μ, ν on $(X, \mathcal{A})$ are said to be *mutually singular*, symbolically $\mu \perp \nu$, if $\mu(A) = \nu(A^c) = 0$ for some $A \in \mathcal{A}$. Intuitively speaking, mutually singular measures live on different parts of the space though this interpretation should not be taken too literally: for example, every countable dense set $X := \{x_n : n \in \mathbb{N}\} \subseteq \mathbb{R}$ yields the probability measure $\mu := \sum_{n \in \mathbb{N}} 2^{-n} \delta_{x_n}$ which is singular to λ^1 for $\mu(\mathbb{R} \backslash X) = \lambda^1(X) = 0$; nevertheless $\operatorname{supp} \mu = \mathbb{R} = \operatorname{supp} \lambda^1$.

Theorem A.10 (Lebesgue Decomposition Theorem). *Let μ, ν denote two σ-finite measures on $(X, \mathcal{A})$. There exist two uniquely determined measures ν_{ac}, ν_s on $(X, \mathcal{A})$ such that*

$$\nu = \nu_{ac} + \nu_s \quad \text{with} \quad \nu_{ac} \ll \mu \quad \text{and} \quad \nu_s \perp \mu \, .$$

The measures v_{ac} and v_s are, respectively, called the absolutely continuous and the singular part of v with respect to μ.

A.4   Conditional expectations

Conditional expectations provide a useful and far-reaching generalization of conditional probability. In order to get some impression of this fact let $(X, \mathcal{A}, \mu)$ be a probability space and suppose that there exists a countable partition $\mathcal{B} = \{B_n : n \in \mathbb{N}\}$ of X with $\mu(B_n) > 0$ for all n. For any point $x \in X$ let $\mathcal{B}(x)$ denote the unique set B_n with $x \in B_n$. Given $C \in \mathcal{A}$ the conditional probability

$$\mu\big(C|\mathcal{B}(x)\big) := \frac{\mu\big(C \cap \mathcal{B}(x)\big)}{\mu\big(\mathcal{B}(x)\big)} = \sum_{n \in \mathbb{N}} \frac{\mu(C \cap B_n)}{\mu(B_n)} \mathbf{1}_{B_n}(x)$$

describes the probability of the event C relative to the measurable partition $\mathcal{B}$ which might be thought of as the available information about the actual position of x. Observe that

$$\int_{B_n} \mu\big(C|\mathcal{B}(x)\big) \, d\mu = \mu(C \cap B_n) = \int_{B_n} \mathbf{1}_C d\mu \quad \text{for all } B_n \in \mathcal{B},$$

and $\mu\big(C|\mathcal{B}(x)\big)$ is the only $A_\sigma(\mathcal{B})$-measurable function with this property.

Definiton A.11 Let $(X, \mathcal{A}, \mu)$ denote a probability space and $\mathcal{B} \subseteq \mathcal{A}$ a sub-σ-algebra. Given $f \in L^1(X, \mathcal{A}, \mu)$ there exists a uniquely determined element $\mathbb{E}(f|\mathcal{B})$ of $L^1(X, \mathcal{B}, \mu)$ which satisfies

$$\int_B \mathbb{E}(f|\mathcal{B}) \, d\mu = \int_B f \, d\mu \quad \text{for all } B \in \mathcal{B}.$$

The $\mathcal{B}$-measurable function $\mathbb{E}(f|\mathcal{B})$ is called the *conditional expectation* of f under $\mathcal{B}$.

According to the above observation $\mathbb{E}(f|\mathcal{B})$ may be considered a version of f which is smoothened with respect to the available information $\mathcal{B}$. Clearly $\mathbb{E}(f|\{\emptyset, X\}) = \int_X f \, d\mu$ and $\mathbb{E}(f|\mathcal{A}) = f$ in $L^1(\mu)$. The following properties are fairly obvious.

Theorem A.12. *For any $f \in L^1(X, \mathcal{A}, \mu)$ and sub-σ-algebras $\mathcal{C} \subseteq \mathcal{B} \subseteq \mathcal{A}$ the following relations hold:*

(i) $\mathbb{E}(\cdot|\mathcal{B})$ *is a positive linear operator;*

(ii) $\mathbb{E}(gf|\mathcal{B}) = g\mathbb{E}(f|\mathcal{B})$ *for all $g \in L^\infty(X, \mathcal{B}, \mu)$;*

(iii) $\mathbb{E}(\mathbb{E}(f|\mathcal{B})|\mathcal{C}) = \mathbb{E}(f|\mathcal{C})$;

(iv) *If $f \in L^2(X, \mathcal{A}, \mu)$ then $\mathbb{E}(f|\mathcal{B})$ equals the orthogonal projection of f onto $L^2(X, \mathcal{B}, \mu)$;*

(v) $|\mathbb{E}(f|\mathcal{B})| \le \mathbb{E}(|f| \,|\mathcal{B})$ *and thus* $\|\mathbb{E}(f|\mathcal{B})\| \le \|f\|$;

(vi) *Setting* $A_\alpha := \{x \in X : \mathbb{E}(f|\mathcal{B}) > \alpha\}$ *with* $\alpha > 0$ *yields* $\mu(A_\alpha) \le \frac{1}{\alpha} \int_X |f| \, d\mu$;

(vii) *If the map* $T : X \to X$ *preserves* μ, *then* $\mathbb{E}(f|\mathcal{B}) \circ T = \mathbb{E}(f \circ T | T^{-1}\mathcal{B})$.

By (iv) the conditional expectation $\mathbb{E}(f|\mathcal{B})$ may be considered the least-squares approximation of f in $L^2(X, \mathcal{B}, \mu)$. Property (vi) is commonly referred to as *Doob's* (if $\mathcal{B} \subsetneq \mathcal{A}$) or *Chebyshev's inequality*. It turns out to be the key ingredient for proving the following important result.

Theorem A.13. *Let* $f \in L^1(X, \mathcal{A}, \mu)$ *and* $(\mathcal{A}_n)_{n \in \mathbb{N}}$ *a sequence of sub-σ-algebras in* $\mathcal{A}$.

(i) *If* $\mathcal{A}_1 \supseteq \mathcal{A}_2 \supseteq \ldots$ *then* $\lim_{n \to \infty} \mathbb{E}(f|\mathcal{A}_n) = \mathbb{E}(f | \bigcap_{n \in \mathbb{N}} \mathcal{A}_n)$ *in* $L^1(\mu)$ *and almost everywhere ("Decreasing Martingale Theorem");*

(ii) *If* $\mathcal{A}_1 \subseteq \mathcal{A}_2 \subseteq \ldots$ *then* $\lim_{n \to \infty} \mathbb{E}(f|\mathcal{A}_n) = \mathbb{E}\big(f | \mathcal{A}_\sigma(\mathcal{A}_n | n \in \mathbb{N})\big)$ *in* $L^1(\mu)$ *and almost everywhere ("Increasing Martingale Theorem").*

Appendix B
A student's guide to the literature

The number of books that deal with (deterministic or stochastic) aspects of dynamical systems, topological dynamics and ergodic theory in one way or another is overwhelming. In fact, these books could fill a library of their own. However, often it is anything but easy to choose from this wealth the literature most appropriate to one's own needs and level of knowledge. The following short compilation intends to guide a student's choice by listing a number of books that may accompany, supplement and extend this text. Basically, the list contains books from which the author received help and inspiration. The comments thereon certainly reflect the author's personal view and thus are subject to discussion. For more extensive and authoritative references on dynamical systems, both of historical and bibliographical nature, the reader may wish to look up the notes in [31].

Is chaos theory science fiction?

[48] Ruelle, D.: *Chance and Chaos*, Princeton University Press, 1991.

> Pointed and easy-to-read essay, written with a sort of dry humor, showing the surprisingly multifarious appearance of *chance* in modern science; it is not by chance that the title of the text in hand resembles that of Ruelle's book.

[57] Stewart, I.: *Does God Play Dice?*, Penguin, 1990.

> More technical and going into more detail than Ruelle's book, rather emphasizing deterministic aspects; historical notes on Newton, Poincaré and all the other heroes of dynamics round off the book.

[52] Sigmund, K.: *Games of life*, Oxford University Press, 1993.

> What help is mathematics for the understanding of (biological) life? Witty and with lots of details the author comes to the insight that nowhere – and least in real life – the *stimulating* effects of chaos and chance can be excluded.

Discrete dynamical systems (general aspects)

[20] Devaney, R. L.: *An Introduction to Chaotic Dynamical Systems*, Addison-Wesley, 1989.

A friendly introductory book, on which parts of the present text are based; also contains a slightly more advanced chapter on dynamics in the complex plane; necessitates a certain sophistication on the part of the reader but provides a sound basic knowledge concerning discrete dynamical systems.

[46] **Robinson, C.:** *Dynamical Systems; Stability, Symbolic Dynamics, and Chaos*, CRC Press, 1995.

A careful and extensive graduate text which focuses on stability, bifurcations, chaos and its measurement, as well as many other deterministic aspects of dynamics; also addresses the global theory of hyperbolic systems. The reader will possibly consider this book an ideal continuation of Devaney's, and also of Chapter Two of the present text.

[31] **Katok, A.; Hasselblatt, B.:** *Introduction to the Modern Theory of Dynamical Systems*, Cambridge University Press, 1995.

Definitely a standard reference book on the subject, impressive by both style and extent; deals with a great many topics from dynamics in breadth *and* depth; provides nevertheless – not least by virtue of many elegant proofs – also delightful reading.

Concerning more specialized questions one should also consult the fine monographs **[16, 39]** (one-dimensional maps) and **[8, 40]** (conformal dynamics). Differential equations upon which we touched occasionally as a source of inspiring examples, may equally exhibit chaotic behaviour; this multifaceted theme is thoroughly discussed in **[27, 61]**.

Ergodic theory

[44] **Pollicott, M.; Yuri, M.:** *Dynamical Systems and Ergodic Theory*, Cambridge University Press, 1998.

Deals with a diversified list of topics from ergodic theory and topological dynamics, well accessible not least due to a division into relatively short chapters; unfortunately, the presentation suffers from quite a number of misprints – in case of doubt one should consult the respective cited literature or the list of errata on the authors' webpage.

[62] **Walters, P.:** *An Introduction to Ergodic Theory*, Springer, 1982.

A highly readable standard reference text of ergodic theory; requires a certain maturity on the part of the reader but provides a solid overview of ergodic theory and topological dynamics as well as interrelations thereof with other branches of mathematics.

[17] Cornfeld, I. P.; Fomin, S. V.; Sinai, Ya. G.: *Ergodic Theory*, Springer, 1982.

> A standard reference book of ergodic theory, too; treats in some detail classical topics as spectral theory and dynamical systems from algebra and number theory; also contains a separate chapter on billiards.

[3] Arnold, V. I.; Avez, A.: *Problèmes Ergodiques de la Mechanique Classique*, Gauthier-Villars, 1967.

> Discusses statistical aspects of classical mechanics, thereby focussing on Hamiltonian systems; a lot of examples and details are contained in the 34(!) appendices.

[13] Boyarsky, A.; Góra, P.: *Laws of Chaos*, Birkhäuser, 1997.

> Contrary to what the title might suggest this book nearly exclusively deals with the statistical aspects of one-dimensional maps; problems of regularity and approximation are discussed at some length. The applications in the last chapter may not be highly spectacular, but there are a lot of interesting problems, most of them with hints or solutions.

[35] Lasota, A.; Mackey, M. C.: *Chaos, Fractals and Noise*, Springer, 1994.

> Focusses on stochastic aspects of deterministic systems under a broader perspective than [13]; the text thoroughly develops the mathematical tools before exhibiting a number of interesting examples, mostly from physics and chemistry; also deals with more general stochastic processes.

To the reader who wishes to see more examples and applications or to get closer to more recent topics of ergodic theory we recommend **[42, 50, 63]**.

Probability theory

[7] Bauer, H.: *Probability Theory*, de Gruyter, 1996.

> One of those rare books that were originally written in German and later attained international appreciation; provides an elegant treatment of the core fields of modern probability theory and also deals with some aspects of stochastic processes. A reader having at his/her disposal a sound knowledge of measure theory will certainly enjoy this book.

[19] Denker, M.; Woyczyński, W.: *Introductory Statistics and Random Phenomena*, Birkhäuser, 1998.

> Well-motivated introduction to the mathematical modelling and treatment of uncertainty and complexity; a large number of computer experiments also permits down-to-earth statistical experience by means of data provided via the Internet. The formal level will possibly not satisfy the sophisticated reader.

[41] Norris, J. R.: *Markov Chains*, Cambridge University Press, 1997.

> Concise and easy-to-read introduction to the theory of countable Markov processes requiring no previous knowledge; contains many examples as well as interesting outlooks on present-day applications (networks, mathematics of finance, Monte Carlo methods).

A definitive treatment of the foundations and applications of measure theory may be looked for in the excellent textbooks **[6, 47]**.

Fractal geometry

[23] Edgar, G. A.: *Measure, Topology, and Fractal Geometry*, Springer, 1990.

> An introductory book which requires nearly no previous knowledge; rigorously treats some aspects of measure theory, topology and Hausdorff dimension and thus provides a sound basis for further reading; also contains a wealth of (partly challenging) exercises.

[24] Falconer, K. J.: *Fractal geometry*, Wiley, 1990.

> A standard reference text on fractal geometry; concisely deals with quite a lot of different topics from both pure and applied mathematics; necessitates a portion of sophistication on the part of the reader but is certainly helpful for obtaining a general view of the subject.

Bibliography

[1] Abramowitz, M.; Stegun, I. A.: *Handbook of Mathematical Functions*, Dover, New York, 1965.

[2] Arnold, V. I.: *Mathematical Methods of Classical Mechanics*, Springer, New York–Berlin–Heidelberg, 1989.

[3] Arnold, V. I.; Avez, A.: *Problèmes Ergodiques de la Méchanique Classique*, Gauthier-Villars, Paris, 1967.

[4] Barnsley, M.: *Fractals Everywhere*, Academic Press, Boston, 1988.

[5] Barnsley, M.; Demko, S. (eds.): *Chaotic Dynamics and Fractals*, Academic Press, San Diego, 1986.

[6] Bauer, H.: *Measure and Integration Theory*, Walter de Gruyter, Berlin, 2001.

[7] Bauer, H.: *Probability Theory*, Walter de Gruyter, Berlin, 1996.

[8] Beardon, A. F.: *Iterations of Rational Functions*, Springer, New York–Berlin–Heidelberg, 1991.

[9] Berger, A.: Regular and chaotic motion of a kicked pendulum: A Markovian approach, *Z. Angew. Math. Mech.* **81** (2001), to appear.

[10] Billingsley, P.: *Convergence of Probability Measures*, Wiley, New York, 1968.

[11] Block, L. S.; Coppel, W. A.: *Dynamics in One Dimension*, Springer, Berlin–Heidelberg–New York, 1992.

[12] Boldrighini, C.; Keane, M.; Marchetti, F.: Billiards in Polygons, *Ann. Probab.* **6** (1978), 532–540.

[13] Boyarsky, A.; Góra, P.: *Laws of Chaos*, Birkhäuser, Boston–Basel–Berlin, 1997.

[14] Brémaud, P.: *Markov chains*, Springer, New York–Berlin–Heidelberg, 1999.

[15] Bunimovich, L. A.: On the Ergodic Properties of Nowhere Dispersing Billiards, *Comm. Math. Phys.* **65** (1979), 295–312.

[16] Collet, P.; Eckmann, J.-P.: *Iterated maps on the unit interval as dynamical systems*, Birkhäuser, Boston–Basel–Berlin, 1980.

[17] Cornfeld, I. P; Fomin, S. V.; Sinai, Ya. G.: *Ergodic Theory*, Springer, Berlin–Heidelberg–New York, 1982.

[18] Denker, M.; Grillenberger, C.; Sigmund, K.: *Ergodic Theory on Compact Spaces*, Springer, Berlin–Heidelberg–New York, 1976.

[19] Denker, M.; Woyczyński, W.: *Introductory Statistics and Random Phenomena*, Birkhäuser, Boston–Basel–Berlin, 1998.

[20] Devaney, R. L.: *An Introduction to Chaotic Dynamical Systems*, Addison-Wesley, Redwood City, 1989.

[21] Dorfman, J. R.: *An Introduction to Chaos in Nonequilibrium Statistical Mechanics*, Cambridge University Press, Cambridge, 1999.

[22] Drmota, M.; Tichy, R. F.: *Sequences, Discrepancies and Applications*, Springer, Berlin–Heidelberg–New York, 1997.

[23] Edgar, G. A.: *Measure, Topology, and Fractal Geometry*, Springer, New York–Berlin–Heidelberg, 1990.

[24] Falconer, K. J.: *Fractal geometry*, Wiley, Chichester, 1990.

[25] Fu, X.-C., et al.: Chaotic Properties of Subshifts Generated by a Nonperiodic Recurrent Orbit, *Internat. J. Bifur. Chaos Appl. Sci. Engrg.* **10** (2000), 1067–1073.

[26] Gantmacher, F. R.: *The theory of matrices*, Chelsea, New York, 1989.

[27] Guckenheimer, J.; Holmes, P.: *Nonlinear Oscillations, Dynamical Systems and Bifurcations of Vector Fields*, Springer, New York–Berlin–Heidelberg, 1983.

[28] Heuser, H.: *Funktionalanalysis*, Teubner, Stuttgart, 1986.

[29] Holmgren, R. A.: *A first Course in Discrete Dynamical Systems*, Springer, New York–Berlin–Heidelberg, 1994.

[30] Irwin, M. C.: *Smooth Dynamical Systems*, Academic Press, London, 1980.

[31] Katok, A.; Hasselblatt, B.: *Introduction to the Modern Theory of Dynamical Systems*, Cambridge University Press, Cambridge, 1995.

[32] Keller, G.: *Equilibrium States in Ergodic Theory*, Cambridge University Press, Cambridge, 1997.

[33] Kemeny, J. G.; Snell, J. L.: *Finite Markov Chains*, Springer, New York–Berlin–Heidelberg, 1976.

[34] Kołodziej, R.; The rotation number of some transformation related to billiards in an ellipse, *Studia Math* **81**(1985), 293–302.

[35] Lasota, A.; Mackey, M.: *Chaos, Fractals and Noise*, Springer, New York–Berlin–Heidelberg, 1993.

[36] Lind, D.; Marcus, B.: *An Introduction to Symbolic Dynamics and Coding*, Cambridge University Press, Cambridge, 1995.

[37] MacKay, R. S.; Meiss, J. D.: *Hamiltonian Dynamical Systems*, Adam Hilger, Bristol, 1987.

[38] Mañé, R.: *Ergodic Theory and Differentiable Dynamics*, Springer, New York–Berlin–Heidelberg, 1987.

[39] de Melo, W.; van Strien, S.: *One-Dimensional Dynamics*, Springer, Berlin–Heidelberg–New York, 1993.

[40] Milnor, J. W.: *Dynamics in One Complex Variable*, Vieweg, Wiesbaden, 1999.

[41] Norris, J. R.: *Markov Chains*, Cambridge University Press, Cambridge, 1997.

[42] Parry, W.: *Topics in ergodic theory*, Cambridge University Press, Cambridge, 1981.

[43] Peitgen, H.-O.; Jürgens, H.; Saupe, D.: *Fractals for the classroom I,II*, Springer, New York–Berlin–Heidelberg, 1992.

[44] Pollicott, M.; Yuri, M.: *Dynamical Systems and Ergodic Theory*, Cambridge University Press, Cambridge, 1998.

[45] Renardy, M.; Rogers, R. C.: *An Introduction to Partial Differential Equations*, Springer, New York–Berlin–Heidelberg, 1993.

[46] Robinson, C.: *Dynamical systems; Stability, Symbolic Dynamics, and Chaos*, CRC Press, Boca Raton, 1995.

[47] Rudin, W.: *Real and Complex Analysis*, McGraw-Hill, New York, 1987.

[48] Ruelle, D.: *Chance and Chaos*, Princeton University Press, Princeton, 1991.

[49] Ruelle, D.: *Chaotic Evolution and Strange Attractors*, Cambridge University Press, Cambridge, 1989.

[50] Schmidt, K.: *Dynamical Systems of Algebraic Origin*, Birkhäuser, Boston–Basel–Berlin, 1995.

[51] Schmidt, W. M.: Über die Normalität von Zahlen zu verschiedenen Basen, *Acta Arith.* **7** (1962), 299–309.

[52] Sigmund, K.: *Games of life*, Oxford University Press, Oxford, 1993.

[53] Sinai, Ya. G.: *Introduction to ergodic theory*, Princeton University Press, Princeton, 1977.

[54] Sinai, Ya. G. (ed.): *Dynamical Systems II (Encyclopaedia of Mathematical Sciences)*, Springer, Berlin–Heidelberg–New York, 1989.

[55] Sinai, Ya. G.: *Probability Theory*, Springer, Berlin–Heidelberg–New York, 1992.

[56] Smale, S.: Differentiable Dynamical Systems, *Bull. Amer. Math. Soc.* **73** (1967), 747–817.

[57] Stewart, I.: *Does God Play Dice?*, Penguin, Harmondsworth, 1990.

[58] Strogatz, S. H.: *Nonlinear Dynamics and Chaos*, Addison-Wesley, Redwood City, 1994.

[59] Szász, D.: Boltzmann's ergodic hypothesis, a conjecture for centuries?, *Studia Sci. Math. Hungar.* **31** (1996), 299–322.

[60] Troger, H.; Steindl, A.: *Nonlinear Stability and Bifurcation Theory*, Springer, Wien–New York, 1991.

[61] Verhulst, F.: *Nonlinear Differential Equations and Dynamical Systems*, Springer, Berlin–Heidelberg–New York, 1990.

[62] Walters, P.: *An Introduction to Ergodic Theory*, Springer, Berlin–Heidelberg–New York, 1982.

[63] Ward, T.: *Lecture Notes on Entropy*, preprint based on a course given in 1994 (notes at www.mth.uea.ac.uk/~h720/lecture_notes/lecture_notes.html).

[64] Weyl, H.: Mean Motion, *Amer. J. Math.* **60** (1938), 889–896.

Index